Resource Provision of the Sustainable Development under Global Shocks

Resource Provision of the Sustainable Development under Global Shocks

Editors

Pavel Tcvetkov
Nikolay Didenko

MDPI • Basel • Beijing • Wuhan • Barcelona • Belgrade • Manchester • Tokyo • Cluj • Tianjin

Editors
Pavel Tcvetkov
Department of Economics,
Organization and Management,
Saint Petersburg Mining
University,
21st Line of the VI, 2, Saint
Petersburg, Russia

Nikolay Didenko
Peter the Great St. Petersburg
Polytechnic University,
Saint Petersburg, Russia

Editorial Office
MDPI
St. Alban-Anlage 66
4052 Basel, Switzerland

This is a reprint of articles from the Special Issue published online in the open access journal *Resources* (ISSN 2079-9276) (available at: https://www.mdpi.com/journal/resources/special_issues/Resource_Provision).

For citation purposes, cite each article independently as indicated on the article page online and as indicated below:

LastName, A.A.; LastName, B.B.; LastName, C.C. Article Title. *Journal Name* **Year**, *Volume Number*, Page Range.

ISBN 978-3-0365-4617-9 (Hbk)
ISBN 978-3-0365-4618-6 (PDF)

Contents

Preface to "Resource Provision of the Sustainable Development under Global Shocks"

The fundamental basis of sustainable development is the efficient and rational use of mineral and fuel resources. However, ensuring the necessary levels of efficiency requires stable logistics chains, as well as natural resource recovery, which is an extremely acute industrial problem. In an attempt to identify possible solutions to these problems, this book is focused on the development of the resource potential of the Arctic, as one of the most promising territories in terms of provision of sustainable development with mineral and energy resources.

This book, which consists of 11 articles written by research experts in their topic of interest, reports the most recent research on the development of raw materials markets and logistic chains. Several novel and fascinating methods related to the sustainable development of Arctic region and companies are introduced.

Pavel Tcvetkov and Nikolay Didenko

Editors

1773

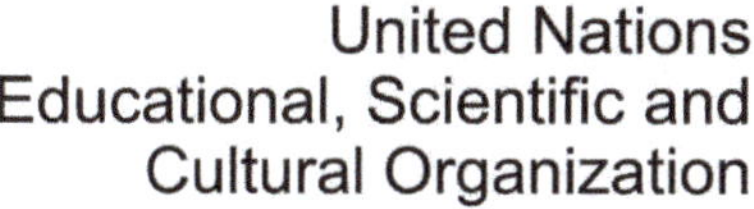

United Nations
Educational, Scientific and
Cultural Organization

International Competence Centre
for Mining-Engineering Education
under the auspices of UNESCO

 *resources*

Review

The Role of Hydrocarbons in the Global Energy Agenda: The Focus on Liquefied Natural Gas

Vladimir Litvinenko

Doctor of Engineering, Academic Supervisor of Research Center "The Arctic", Saint Petersburg Mining University, Saint Petersburg 199106, Russia; river4020@yandex.ru

Received: 17 April 2020; Accepted: 19 May 2020; Published: 21 May 2020

Abstract: Presently, there is a paradoxical situation in the global energy market related to a gap between the image of hydrocarbon resources (HCR) and their real value for the economy. On the one hand, we face an increase in expected HCR production and consumption volumes, both in the short and long term. On the other hand, we see the formation of the image of HCR and associated technologies as an unacceptable option, without enough attention to the differences in fuels and the ways of their usage. Due to this, it seems necessary to take a step back to review the vitality of such a political line. This article highlights an alternative point of view with regard to energy development prospects. The purpose of this article is to analyse the consistency of criticism towards HCR based on exploration of scientific literature, analytical documents of international corporations and energy companies as well as critical assessment of technologies offered for the HCR substitution. The analysis showed that: (1) it is impossible to substitute the majority of HCR with alternative power resources in the near term, (2) it is essential that the criticism of energy companies with regard to their responsibility for climate change should lead not to destruction of the industry but to the search of sustainable means for its development, (3) the strategic benchmarks of oil and coal industries should shift towards chemical production, but their significance should not be downgraded for the energy sector, (4) liquified natural gas (LNG) is an independent industry with the highest expansion potential in global markets in the coming years as compared to alternative energy options, and (5) Russia possesses a huge potential for the development of the gas industry, and particularly LNG, that will be unlocked if timely measures on higher efficiency of the state regulation system are implemented.

Keywords: energy sector; hydrocarbon resources; liquified natural gas; LNG; alternative energy technologies; state regulation; market conditions; energy balance; sustainable development

---

1. Introduction

Recent decades have demonstrated an intensive development of theories, concepts and approaches related to various aspects of sustainable development such as circular economy [1], waste management [2], cleaner production [3], environmental economics [4], etc. Within these approaches, a special place is occupied by the fight against global warming, which is caused by human activities, namely by higher emissions of man-made greenhouse gases, mainly CO_2 [5].

As a result of multiple environmental studies carried out in relation to the above concepts, it has been found that the energy sector is one of the main sources of greenhouse gas emissions among all types of human activity [6]. This fact has become one of the main reasons for combatting the already existing energy system [7], which is founded on hydrocarbon resources (HCR).

Examples of this combatting are seen in the political arena, where it is openly declared that the era of hydrocarbons is over, and we need to switch to alternative energy sources as quickly as possible. This assertion can be found in the declarations of the G7, the World Bank, the European Investment Bank and many other international corporations that restrict access of production and geological survey projects to the investment capital. An example of large-scale "anti-hydrocarbon" policy based on environmental taxes (Figure 1), can be seen in Europe, which has been continually criticized by the industry companies due to worsening investment climate and lower production profitability [8].

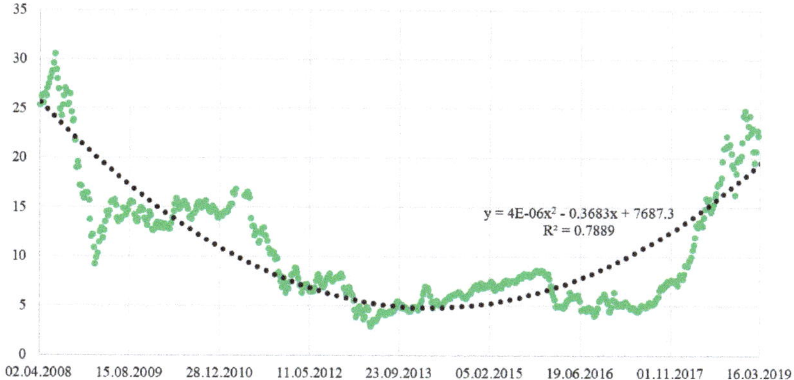

Figure 1. Average carbon tax rate in Europe, Euro/t. CO_2 [9].

It is remarkable that the real activities of these international corporations significantly differ from their declared policy, as shown, for instance in [10–12]. For example, despite declarations, they continue to provide financial support for hydrocarbon energy projects, although indirectly.

An increase of oil production volumes in the United States of America (USA) by 8410 thous. bbl/day for the period of 2015–2018, which is twice as high as total global increase in production over the same period, is a clear example of such two-track policy (Figure 2). At the same time, a major part of USA production is shale oil, which is much more cost-intensive than conventional oil. Additionally, according to the data of the State Energy Department [13], the USA has been implementing an extensive program in support of gas hydrates studies, which is related to prospective HCR.

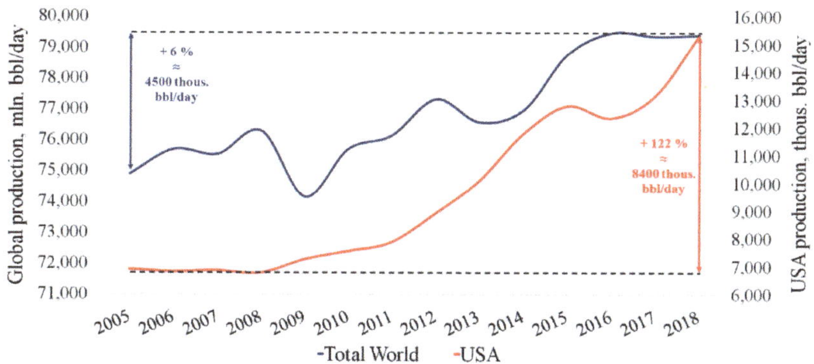

Figure 2. Global and USA oil production dynamics. Based on BP Energy Overviews.

Additionally, the collapse of oil prices in March 2020 has triggered a proactive purchase of cheap Arab Light oil by the USA and Urals oil by China to refill strategic reserves. These examples clearly show that there is no intention to decrease our activity in the hydrocarbon energy sector.

The real state and trends of the energy sector can be seen when considering forecasts of leading global companies, for example British Petroleum (BP), over different years (Table 1).

Table 1. The comparison of BP Energy Overviews (EO) for the period of 2014–2019.

Indicator	EO14	EO15	EO16	EO17	EO18	EO19
Proportion of HCR as primary energy sources in 2025, incl.:						
oil, %	30	30	30	30	31	31
gas, %	24	25	24	24	25	24
coal, %	30	29	27	27	25	24
Proportion of HCR as primary energy sources in 2035, incl.:						
oil, %	27	28	29	29	29	29
gas, %	26	27	26	25	27	26
coal, %	27	27	25	24	22	21
Aggregate consumption of energy resources in 2035, bln toe	17.6	17.5	17.3	17.2	17.1	17

The Table shows that the forecast proportion of HCR in the global energy balance varies slightly despite milestones with respect to their substitution for alternative energy sources. The highest decrease is shown by coal (−6%), while the proportion of oil went up by 1–2%. The share of natural gas demonstrates minor fluctuations, contributing as little as 1–2%. Taken as a whole, and based on the latest BP report, the proportion of HCR (oil, coal and natural gas) in the global energy balance will amount to nearly 76% in 2035, which is equal to 12.9 bln toe. This means a guaranteed demand for HCR, as well as tougher competition and intensification of the struggle for access to them, resulting in HCR earning a status as a geopolitical resource that can be seen even today.

A somewhat different forecast based on plotting "Hubbert curves" is provided in [14], whereby the production peak of oil will occur in 2009–2021, of gas in 2024-2046, and of coal in 2042–2062. It should be noted that these results should be treated with some care, since Hubbert curves may be modified when affected by economic and technology factors. Nevertheless, similar results have been obtained by other authors [15,16]. Despite some variations within the ranges in years, all of the above forecasts came to the same conclusion: the production peak of the majority of HCR is still ahead of us.

It should be understood that all of the above scenarios expect that, within the next few years, oil and gas production volumes will increase due to fields with complex subsurface conditions or located in adverse climatic conditions. Pursuant to [17], the cost for development of such reserves can be 3–4 times greater than the conventional reserves of fields in the Middle East (production cost below 10 $/bbl). Taking into account current prices for oil (around 30 $/bbl), the development of such reserves may become economical in the case of state support or where there is intensification of technical progress rates in the production of hydrocarbons, incl. due to the digitalization of operating activities [18].

At this point, the main question is whether global regulators will be able to create the conditions required for reaching and passing the HCR production peak with regard to sustainable development principles [19], or whether industries will survive amid discriminatory financing of potentially attractive but presently non-competitive technologies.

This study presents the analysis of alternatives offered today for substitution of HCR and an assessment of their viability. The further contents of this article are organized as follows: Section 2 describes the main critical comments for HCR; Section 3 reviews the main alternative energy technologies and their strengths and weaknesses; Section 4 gives a consolidated description of the current position of HCR and a detailed analysis of liquified natural gas (LNG) development prospects as one of the most prospective HCRs; Section 5 contains conclusions and reasoning of the author.

2. Climate Change and Public Image of HCR as a Source of CO_2 Emission

Climate change is definitely a significant issue on a global scale. There are two opposing opinions about its nature. The first opinion is that the main reason for Global Warming is the increase in the emission of technogenic greenhouse gases, mainly CO_2. This is the most widespread point of view, which is partly explained by possibility of studying this phenomenon. To date, there have been many compelling studies [20], proving that mankind is provoking climate change on the planet through its activities. This has been reflected in strategic intergovernmental agreements aimed at decreasing the rate of temperature growth, such as the Paris agreement, which, however, may not be enough [21].

The main concern is that human activity may lead to the passing of a tipping point, after which climate stabilization will become impossible, even if the emissions of technogenic greenhouse gases were to be reduced [22]. Moreover, it has been shown [23] that there are many tipping elements that could lead to irreversible climate consequences. This creates confusion in the process of defining the strategic aims of global environmental activities and the ways to achieve them [24].

According to the International Energy Agency (IEA), volumes of CO_2 emissions in 2019 amounted to 33 Gt, which is the peak value of all time. In addition, a significant part of these emissions is formed as a result of HCR use [25]. Multiple studies have been dedicated to the assessment of indirect losses [26] from the use of various HCR, for example: the use of land for the construction of power plants, accidents at production enterprises, servicing of nuclear waste sites, the higher probability of military conflicts due to rights of access to raw material resources, and many others. In accordance with several obsolete assessments, such factors may increase the cost of using energy resources by 0.29 $ (wind energy sector)–14.87 $ (coal) per 1 kWh [27]. However, to be relevant and useful, such assessments should be conducted on a regular basis and for different regions, with respect to the specifics of local legislation.

The second position on Global Warming suggests that the main drivers of climate change are not related to human activity, but are connected with natural reasons. For example, NASA believes that global warming is mainly caused by changes in the Earth's orbit related to the sun. This opinion is based on the Milankovitch theory [28], which was initially ignored by the scientific community, but after the release of the study of Hays et al. [29], proving its validity in many aspects, much more attention has been given to it. However, it is extremely difficult to push this idea forward, due to objective technical and technological reasons.

Another example is related to the influence of natural disasters. In accordance with [30], the annual volumes of CO_2 emissions caused by volcanic activity may amount to almost 0.05–0.3 Gt per year. Another CO_2 source is fires. For example, in 2010, forest fires in Russia resulted in CO_2 emissions of more than 0.25 Gt [31]. Following the results of [32], the annual volume of CO_2 emissions caused by forest fires in Russia amounted to more than 0.12 Gt for the period of 1998–2010. Forest fires in Indonesia in 1997 produced from 0.81 to 2.57 Gt of CO_2 [33]. CO_2 emissions as a result of burning peatlands in South-Eastern Asia are estimated at 0.637–2.255 Gt per year [34]. These are some examples of natural factors influencing the higher volume of greenhouse gas emissions, while forest fires occur regularly, especially on peatlands. In addition, according to [35], the entire volume of carbon accumulated in the atmosphere is less than the basis point of total carbon on the planet, while 99% is underground. Emissions of other greenhouse gases (CH4, NOx, etc.) have been studied to a very limited extent [36–39], making it quite complicated to assess their volumes in a substantiated way.

The main critical argument towards the volumes is that they make up less than 10% of total man-made CO_2 emissions. However, it should be taken into account that volcanic activity, as well as fires, has taken place for thousands of years, while mankind only started to emit such CO_2 volumes in recent decades. At the same time, it is obvious that additional technogenic emission is not a way to balance the global carbon cycle [40], and the contribution of mankind has become more visible in recent centuries [41].

Based on this, society is fed the idea that we should immediately reject usage of HCR without details about the dependence of our economic stability on these resources. Such an approach is questionable, because, in fact, we have to find a consensus on the need to improve the environmental safety of the processes in this industry and to support its development.

Therefore, if the significance of HCR in the economy is indisputable, the political rhetoric in this field and the attempts to create a strongly negative perception of all HCR among the public, without attention given to the differences between fuels and between the technologies of their production and use, is a point of debate. Hence, it is a very important question: "Should we continue a discrimination policy towards the HCR due to their contribution to a global carbon emission, or should conditions for the sustainable development of this industry be created?", since forecasted HCR production shows that they will play a significant role in the global economy for several decades in the future.

3. Alternative Energy Technologies. Problems and Prospects

Technological transformations, especially within such a key area as the energy sector, go hand in hand with the transition to a new technological order, which is a very long-term process that may take up to several decades [42]. The creation of conditions for the implementation of these transformations and transitions with minimum losses will be of high importance, both for the environment and the economy, i.e., in view of the sustainable development principles [43]. The variety of terms defining sustainable development in modern publications does not change its common nature, offered in 1987 [44], while the principles and methods of sustainable development mainly depend on a regulator, i.e., on politics and policy [45], and may be implemented on national, regional, and local and lower levels [46,47].

The entire list of factors influencing the implementation of transformations at every level can be roughly divided into the technology-readiness level, the regulatory-readiness level, and the market-readiness level [48]. However, despite the fact that the focus of key policy provisions can be made on the basis of climatic (Kyoto Protocol) and technological (Paris Agreement) considerations, the objective fact is that there are no reliable scientific results showing how exactly technological factors influence policy development [49]. In other words, the extent of the influence of technology readiness on the formation of policy and, as a result, on the readiness of the regulatory framework, is unknown.

Elimination of this gap in knowledge requires an interdisciplinary approach [50] with studies executed by international scientific groups. This is related to a need for deep modernization of the already established global infrastructure, and requires some strategic decisions on the support of certain technological niches (technological groups) to be supported by scientifically substantiated and realistic recommendations on the determination of overall vector of activities [51] and specific measures. For further analysis, the most promising technological niches were selected from among those available today [52–54], and are reviewed below.

3.1. Nuclear Energy Sector

Nuclear industry, accounting for 4.5% in the global energy balance, has already made a considerable contribution to energy sector development (Figure 3), due to the relatively low cost of electricity generation [55]. The development dynamics of the nuclear energy sector differ greatly in various countries, because of varying available resources and access to alternative sources of energy, as well as project experience.

Prospects for its further development are questionable. On the one hand, the low production cost of energy generation, as well as the higher efficiency of the plants (the majority having a capacity use rate of more than 80%) [56], could mitigate the technological transformation of the energy sector. On the other hand, the diagram shows that sharp declines in nuclear energy generation are caused by accidents at plants that directly influence not only the energy generation process, but also produce fears on the expediency of the further use of nuclear energy. Here, for example, a final decision about the immediate decrease in the share of the nuclear energy in the country balance was taken by Germany in 2011 after the accident at Fukushima and the first-step measures included the shutdown of power units commissioned before the 1980s.

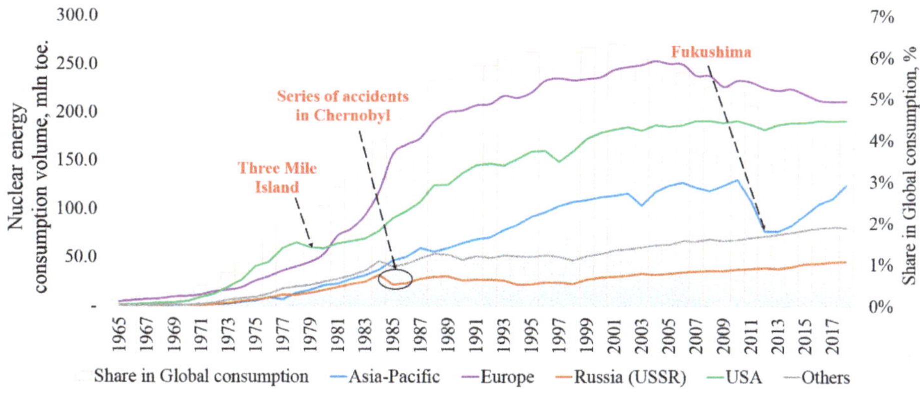

Figure 3. Dynamics of nuclear energy generation volumes. Based on BP Energy Overviews.

The problem of existing physical depreciation of energy generating facilities is topical for the nuclear energy sector even today. As of the end of 2018, the number of nuclear reactors in the world amounted to 450 [57], 281 of which (62.4%) were more than 30 years old, 26% were aged 10 to 30 years, and only 11.6% were younger than 10 years old. In accordance with IEA data [58], the highest average ages of nuclear plants are seen in the USA (39 years old), Europe (35 years old), Russia (31 years old) and Japan (29 years old). The highest shares of nuclear plants under 10 years old are observed in China (80%), India (40%), Russia (25%) and Korea (24%).

Strengthening security measures taken during the construction of the third-generation reactors enables the mitigation of risk of accidents, but increases the unitary cost of the generated energy (more than doubling the cost, compared to fully depreciated facilities) [59]. In general, with the increasing wear and tear of the reactor, the risk of unforeseen incidents increases, and therefore, activities in the field of nuclear energy should be accompanied by the availability of an appropriate license confirming the safety of power generating facilities.

It can be said that nuclear energy has some advantages compared to other energy technologies. First of all, a comparatively low energy generation cost with a minimal carbon footprint. Secondly, technology readiness with high technical efficiency. Thirdly, the technological scalability of both small- and large-scale capacities, which may ensure stable and uninterrupted energy generation.

Despite this, there are three key barriers that do not allow a significant expansion of nuclear plant usage. Firstly, the development of nuclear energy generation is associated with the development of nuclear weapons, which is a very negative and destabilizing factor [60].

Secondly, notwithstanding their higher reliability, global history knows examples of man-made catastrophes, which has produced a very adverse effect upon acceptance of the technology. To restore its image, great pains must be taken in terms of marketing and notifying society about the safety measures taken at modern energy generating facilities [61].

Thirdly, one of the most complicated problems is the handling of nuclear waste [62], requiring significant financial costs on the one hand and availability of hi-tech facilities preventing radioactive contamination of the environment on the other.

3.2. Renewable Energy Sector

Renewable energy sources (RES) are promising in the long-term development of the energy sector. As such, the question of their high use efficiency is one of the most discussed issues of recent decades. RES have several key features that ensure their high demand:

1. RES are a group of primary energy resources that are considered inexhaustible for foreseeable future consumption. Moreover, they have an extensive geographical range to allow the creation

and scaling of decentralized energy supply systems [63] and, as a result, avoid the problems related to the creation of centralized infrastructures.

2. Theoretically, RES technologies are carbon-free, with some exceptions. Nevertheless, when considering them in view of the entire lifecycle, including the production and use of worn equipment, environmental issues become more controversial [64].

Despite differences in nature, the majority of RES have similar problems that do not allow them to be considered a complete substitution for HCR at the current stage of development.

1. Similar to HCR, some RES have non-uniform geographical distribution. Significant differentiation can be seen even in one country [65]. Resources of wind and solar energy show the most wide-spread distribution in the world, though they do have certain restrictions.

2. Another issue is dependence on favourable weather conditions, prediction of which is a complex scientific task. The most predictable sources are biomass, geothermal energy and hydro-energy due to their low dependence (or independence) on changeable weather conditions. This issue also determines the complexity of energy transformation process control and maintenance of required levels of effectiveness [66].

3. The third issue is related to the immaturity of most RES-based technologies, which affects energy generation cost. The most cost-intensive options are solar and oceanic energy plants, with the electricity cost being an order of magnitude higher than that of HCR. The most competitive are large-scale hydro and wind plants, the energy cost of which is only a few times higher than that of HCR. Additionally, a comparatively low production cost of energy can be obtained during the processing of biomass, including peat, due to the high rate of its reserve base replenishment [67]. Low competitiveness in terms of price necessitates seeking ways of defining the project's potential based on theoretical prospects, rather than on the completeness of technologies and potential profitability [68].

4. The fourth, and probably main, issue with RES is the accumulation and storage of energy. Today, many technologies have been developed in this area, but there are objective problems with their effectiveness and possible storage period [69]. In accordance with market needs, RES development is impossible without these key stages of energy generation [70].

Coping with these four barriers is the main task of RES development proponents. Due to this, it is one of the most invested-in sectors today [71–73]. However, it should be taken into account that while seeking higher volumes of investment, one may forget that the efficiency of studies and innovation progress may grow disproportionately with the investment volume [74].

There are no studies of investment volume influencing their efficiency, due to objective problems of scientific and innovation activity assessment. Similar to market mechanisms, unlimited financing leads to a loss of competition, and this would result in lower quality and efficiency. The same can be applied in the global scientific sector; however, additional studies will be required to prove this.

A step towards the short-term global transition to RES is an attractive perspective, but hard to access due to some objective reasons [75]. Gradual transition is required, which can be objectively traced against a gradual decrease in the share of hydrocarbons in the global energy balance. Additionally, this transition must occur under the influence of objective competitive factors other than political influence and creation of artificial conditions for the displacement of existing hydrocarbon technologies.

3.3. Hydrogen Technologies

Hydrogen fuel is a potential solution for several energy generation issues typical of both RES and HCR. On the one hand, the production of hydrogen will enable the conversion, accumulation and storage of energy generated from any primary source in the long run. This will enable the use of hydrogen as a mobile energy carrier [76]. On the other hand, hydrogen is recognized as an environmentally clean secondary energy resource due to the lack of any pollutant emissions during its use [77]. Additionally, its energy intensity in relation to weight units is much higher than of any HCR.

Despite this, the hydrogen energy sector is in its primary stage of development and is not ready for large-scale implementation in the global energy system. There are several main factors hindering this process.

1. Unlike HCR, hydrogen is a secondary energy resource since its production from the natural environment is hard to implement. Many concepts and technologies have been developed for generation of hydrogen [78], which can be provisionally divided into two groups: (1) based on RES, and (2) based on HCR. The first group uses thermochemical or biological processes. The second group uses HCR reforming and pyrolysis processes. The problem is that the majority of accessible technologies are not sufficiently tested for their industrial implementation.

2. The cost of hydrogen production is quite high due to the initial stage of technology development and it does not allow competition with traditional HCR-based technologies. The study [79] includes a comparative analysis of 19 technologies for the production of hydrogen fuel. Based on this analysis, the authors drew several important conclusions in terms of the potential of the development of hydrocarbon resources. First of all, the reforming of the hydrocarbon feed has the highest energy efficiency among all of the options considered. Secondly, the exergy efficiency of hydrocarbon reforming is one of the highest (45–50%), with only biomass gasification being ahead of it (60%). Thirdly, it was shown that the cheapest hydrogen could also be obtained from the hydrocarbon feed, with a price of nearly 0.75 $/kg H2. The use of technologies such as water electrolysis will enable generation of hydrogen with a cost 1.5 times higher and more.

Price-specific advantages of hydrogen generation based on hydrocarbons can be seen in [80], as well, pursuant to which the generation based on natural gas varies from 1.34 $/kg (without CO_2 sequestration) to 2.27 $/kg (with sequestration); and with coal: 1.34 $/kg to 1.64 $/kg; while the majority of other generation methods are 1.5–6 times more expensive. The generation of hydrogen based on methane pyrolysis is the most economically efficient option, enabling a price of 1.22 $/kg. The authors did not point out a single potentially leading production technology, but stated that the preference should be given to hybrid methods.

Equivalent results have also been obtained on the basis of a fuzzy logic method [81], proving that the highest technical and economic efficiency, as well as level of readiness and level of reliability, was typical for technical chains based on hydrocarbon resources.

3. Transportation and storage processes are the foils of the hydrocarbon energy sector [82]. An increase in the efficiency of the processes is related to the solution of two key issues: the transformation of hydrogen into a form with higher density (for example, liquefaction), and an increase in the safety of tanks and delivery systems. In addition, while the first problem already has some practical solutions, the issues of safe hydrogen handling have not yet been studied. However, nearly 70 mln t are produced, presently, and as a rule is used during the processing of metals.

Therefore, despite the prospects for hydrogen in the long run, the current level of global economic readiness for the development of hydrogen infrastructure is less than that of the renewable energy sector, due to: a complete absence of market mechanisms; the lack of technologies and infrastructure enabling the efficient generation, distribution and storage of hydrogen; huge problems with the safety of its use; and other [83] issues typical of technologies at the initial stages of development.

4. Hydrocarbon Resources. Focus on LNG

4.1. The Role of Hydrocarbons in the Energy Sector

HCR have played an important role in the global economy for centuries. They are real drivers for the development of both industry and society as a whole. Currently, the most influential HCRs for the economy are coal, oil and natural gas. Due to external reasons related to the "greening" of global economies (not only the reduction of greenhouse gases emission), multiple technological transformations of industries will take place in the near future, mainly in the oil and coal industries, for the purpose of enhancing the raw material conversion ratio due to stronger vertical integration links with enterprises in the chemical industry. These transformations will lead to a change in the role of hydrocarbons in the global economy (Figure 4) and, most probably, to a partial substitution of some of them with alternative energy technologies.

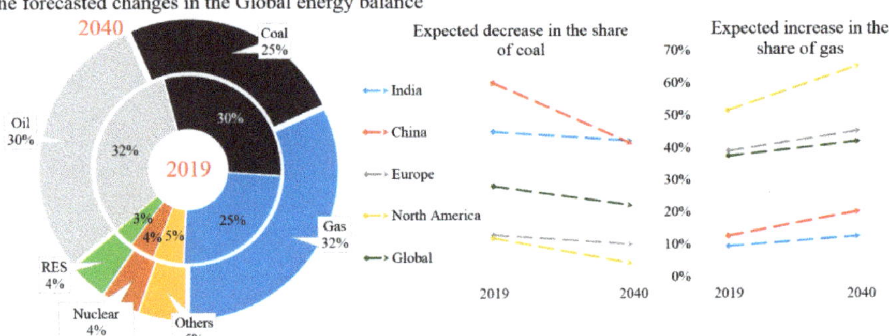

Figure 4. Forecasted change in the global energy balance. Based on [84].

HCR released due to such substitutions may be redirected to chemical facilities, but these processes require the development of international norms and standards of coal use, in order to improve its purity as an energy and chemical resource.

The economical prerequisites of such processes can be clearly seen in the oil industry, as they are associated with some negative factors that hinder its further development:

1. There is no system for the international regulation of processes connected with the development of the traditional oil and gas sector, whether by increasing the effectiveness of internal processes or by integration with associated industries.
2. Rapid depletion of the raw materials base of easy-to-recover reserves and, as a result, higher operating and investment costs. By some estimates, the refilling of reserve volumes trails behind the volumes of produced oil and gas resources by 20%–30% [85], which will lead to industrial stagnation and destabilization of the raw-material and associated markets in the long term.

To some extent, these negative effects can be compensated by manufacturing of scientific-intensive products with high added value. Even today, oil and gas giants such as ExxonMobil (7th place in the C&EN 2018 rating of chemical companies [86]) implement large-scale investment programs in the area of chemical enterprises development (Figure 5).

3. The comparatively low level of oil recovery at multiple fields. The average value of the global recovery ratio is around 20%–40% [87]. In some cases, such rates are connected with the possibility of reorientation to new "effortless" reservoirs; however, such policies may become impracticable soon.
4. Lack of investments in the development and implementation of innovative technologies [89] dealing with shale oil and procedures concerning the development of arctic and deep-water fields [90], i.e., dealing with all sources of hydrocarbons that can be categorized as "alternative".
5. The lack of commonly accepted indicators for assessment of oil and gas companies' performance based on sustainable development principles [91]. For example, it is evident that using nuclear energy resources must be regulated and controlled at an international level. Due to this, it is not quite clear why hi-tech oil and gas industries bearing huge technological and environmental risks are completely self-regulated.

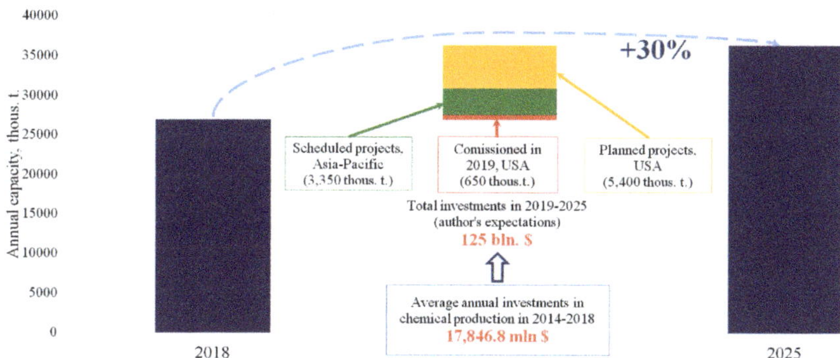

Figure 5. Increase in production of chemicals by ExxonMobil. Based on [88].

Unlike the oil and coal industries, the gas sector is on the rise and is demonstrating intensive expansion. This is due to both the cost performance of natural gas, which is slightly more expensive than oil, and environmental features. According to [92], CO_2 intensity during the generation of 1 kWh of electricity from natural gas is nearly 450 g, which is almost a mean value between hydro power plants (10 g/kWh) and coal (1000 g/kWh). Therefore, natural gas is a somewhat "win–win" solution as it makes it possible to balance the interests of the "green" energy sector and hydrocarbon energy sector supporters.

The gas industry is one of the most rapidly developing parts of the energy sector, which is mainly related to the intensive growth of capacities for the production of liquified natural gas (LNG), which have doubled every 10 years since 1998 [93]. Existing forecasts of LNG industry development provide for the increase in both demand and supply, though there is a certain probability of product surplus amounting up to 100 mln t per year [94], provided that all planned production facilities are commissioned and demand remains constant. If only the most probable projects are implemented, then by 2025 LNG shortage will amount to nearly 30 mln t. Figure 6 demonstrates these trends. Data were taken from various reports and publications of VYGON Consulting Co. [94], International Gas Union, International Group of Liquefied Natural Gas Importers (GIIGNL), and other agencies.

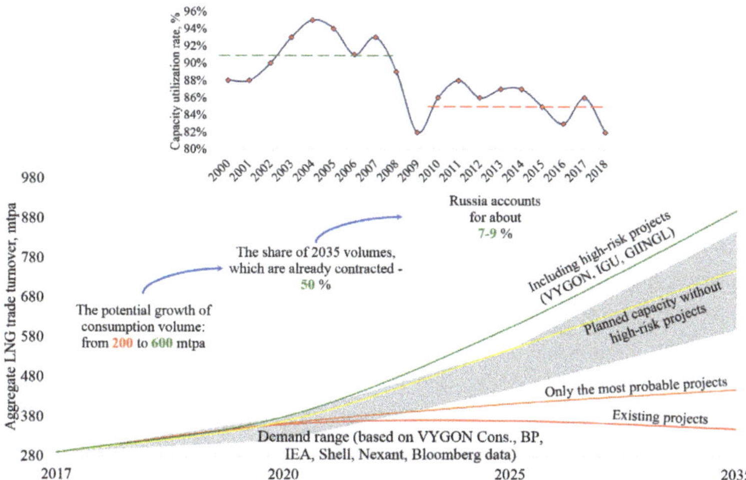

Figure 6. Expected production and consumption of LNG (bottom) and capacity use of LNG-plants in 2000–2018 (top).

When planning production volumes, certain problems with capacity use should also be considered. Before the economic crisis of 2008, the average capacity use of plants amounted to nearly 91%. After 2008, increased interest in LNG industry development resulted in the redistribution of investments for the development of new projects. This caused the reduction of capacity use in existing plants. This trend may continue in the medium term, with fiercer competition and an ambition to displace the active players by creating new hi-tech production facilities.

In accordance with Shell forecasts, LNG consumption may reach 800–900 bln m^3 by 2040 (Figure 7); however, the main drivers of LNG demand growth are the necessity to compensate decreased domestic gas recovery and the need to diversify supplies. The reliability of the forecasts is confirmed by the fact that almost half of the prospective production capacity of LNG plants in 2035 has already been contracted (7%–9% of the contracts in Russia).

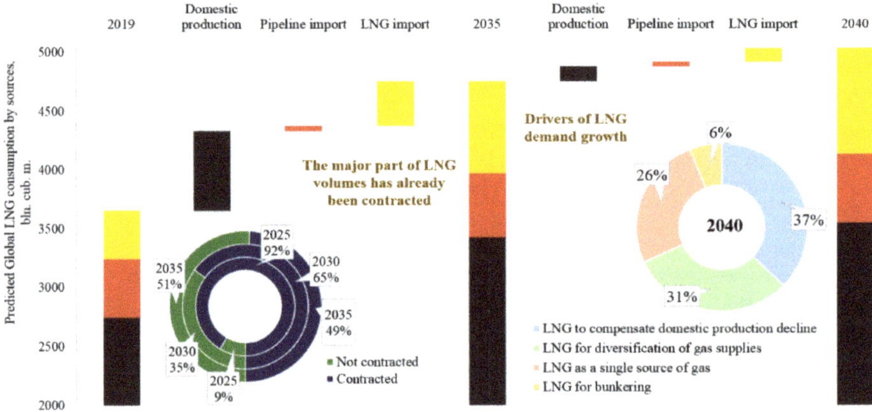

Figure 7. Expected growth in the consumption of natural gas by sources. Based on [95,96].

However, there is already a steady trend towards expanding the geography of LNG consumption (Figure 8), mainly due to Asia-Pacific Region (APR) countries (+220.6 bln m^3), especially China (+73 bln m^3).

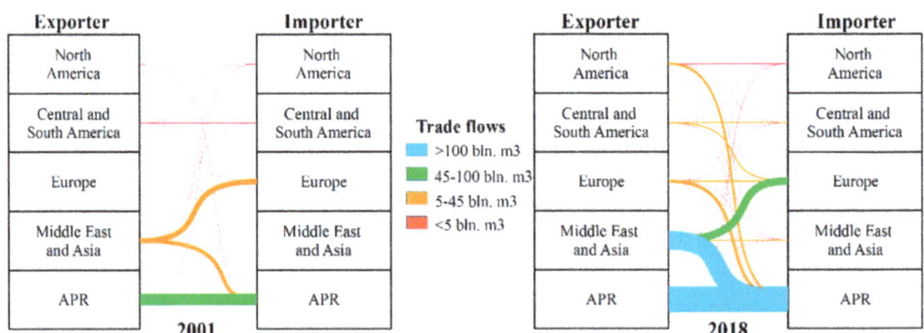

Figure 8. Change in geography of LNG flows. Based on BP Energy Overviews.

Higher demand for LNG import in Asian countries (Figure 9) has mainly been driven by a rapid increase in energy consumption, but if the growth in China can be partially covered by domestic production (more than 160 bln m^3) and pipeline supplies (nearly 110 bln m^3), the countries of Southern and Southeastern Asia will mainly depend on external supplies (about 50% of expected consumption) in the near future.

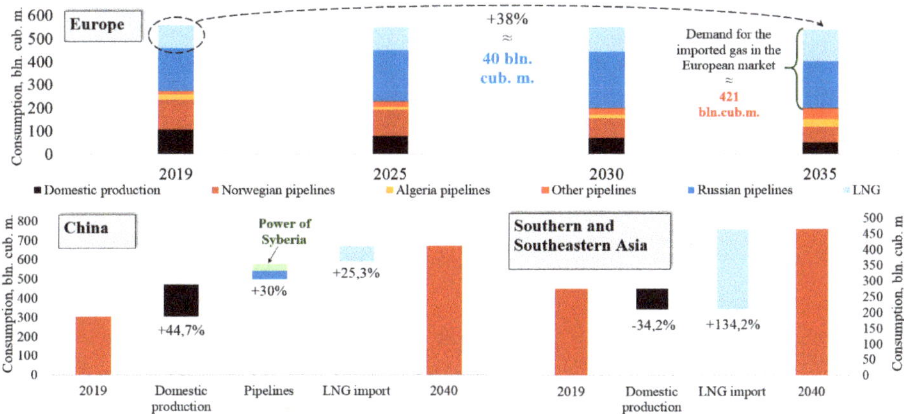

Figure 9. Expected gas consumption in various regions. Based on Shell LNG Outlooks.

In addition, the growing interest of the European Union in increasing LNG import is evident. The main reasons for increased interest in LNG in Europe are, firstly, the increasing energy needs of its countries. Secondly, a rapid decrease in the volume of domestic gas production at the Groningen field (the Netherlands) due to complex seismic conditions. While 53.87 bln m^3 was produced in 2013, only 18.83 bln m^3 was produced in 2018. The Dutch Government is still under pressure from the local community to completely stop natural gas production, which could happen as early as in 2022 [97]. Thirdly, a less rapid but stable decline in gas production in the United Kingdom and Norway, in the short term.

To compensate for the fall in domestic production and diversification in supplies, Europe is implementing an extensive program for the development of LNG facilities. Almost every European country with access to the sea has its own regasification terminals (15 countries), which imported 69 mln t of LNG in 2018 (15%–20% of the global market), with an average capacity use below 60% [98]. Total capacity of regasification terminals in Europe including pre-FEED projects may exceed 200 mln t per year in the nearest future [99].

4.2. The Role of Russia in the Development of the LNG Industry

The situation in the European market creates favourable conditions for increasing the share of Russian gas in the region. However, following the implementation of the "Nord Stream-2" (+55 bln m^3 per year) and "Turkish stream" (+31.5 bln m^3 per year) projects, a further growth in Russian pipeline gas supply seems unlikely, thus opening extensive prospects for the development of LNG. This suggests that Russian pipeline gas and LNG are not competitors on European markets.

At the same time, it is a mistake to believe that Russian gas supplies to Europe are completely protected from competition with the USA. In March 2020, the price of natural gas in the USA (Henry Hub) was 1.79 $/MBTU, and in the European market about 2.72 $/MBTU. When a spread of 2 or more $/MBTU is reached, LNG supplies from the USA could be profitable, and this situation has already been observed in 2017–2019. Given that gas consumption in Europe will grow at an accelerated pace in the coming years, it is likely that prices will rise, and competition risks will increase. This indicates the need to take urgent measures to increase the competitiveness of Russian natural gas export.

It is remarkable that even today a significant share of the increased LNG import to Europe is accounted for by Russia (with a more than 18% increase in 2018–2019), which has exclusively been regarded as pipeline gas supplier. The current situation demonstrates the inadequacy of this preconception, despite Russia's share in the global export of LNG, which is still relatively low (Figure 10).

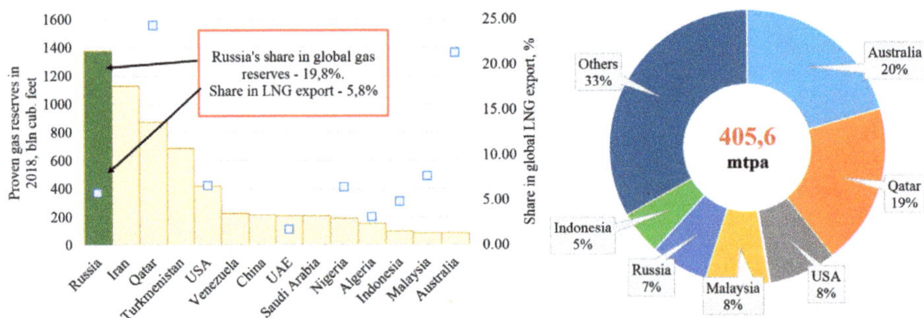

Figure 10. Russia's share in the global LNG market (on the left) and total LNG capacity of global plants in 2019 (on the right). Based on [100].

Similar prospects are also seen in Asian markets, the entry to which for a long time was only associated with the "Power of Siberia" (38 bln m^3 for export) project development. However, as shown in Figure 9, the share of the project is just a small part of the potential growth of China's gas consumption. Additionally, the markets of Southern and Southeastern Asia will be 50% supplied by imported LNG, which may be a favourable factor for Russia, since the efficiency of LNG production at Russian plants is among the highest in the world. This can be seen when comparing the cost of final products with delivery to APR (Figure 11).

As a whole, Russia is implementing an extensive program to develop production capacities and to realize the potential of the LNG industry on global markets. By 2025, the total annual capacity of plants will amount to 61 mln t, excluding low-tonnage facilities, and by 2035, about 140 mln t [101–105]. In total, there are three key centres of the growing gas industry in Russia (Figure 12): the Baltic region (with a focus on Europe), the Far East (with a focus on Asian market), and Arctic regions (with a focus on both Europe and Asia).

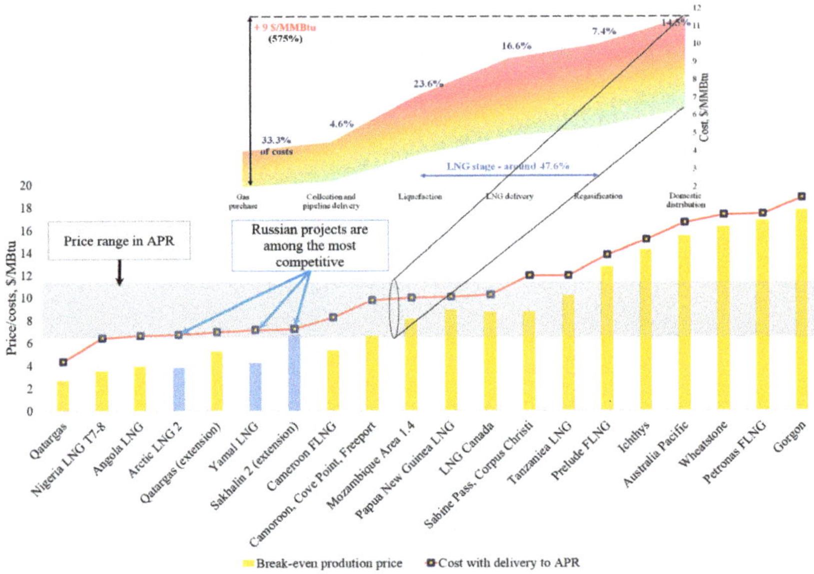

Figure 11. Comparison of LNG production costs with the delivery to APR. Based on [106,107].

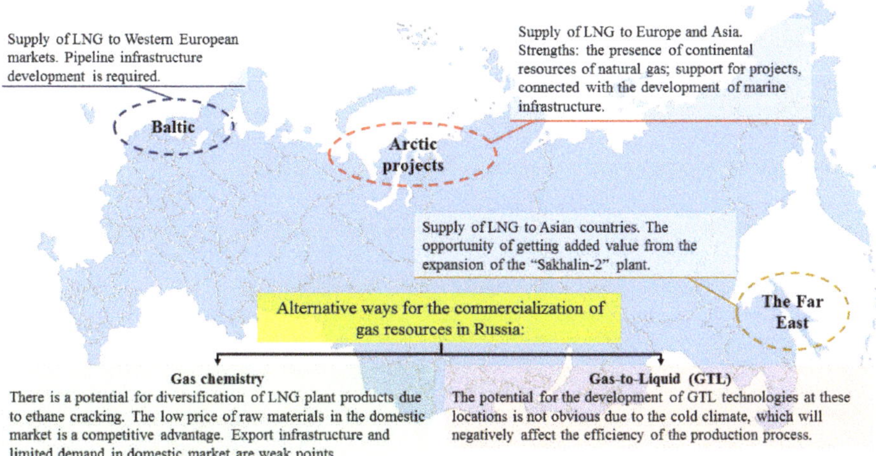

Figure 12. Prospective centres of the growing gas industry in Russia and options for gas resource commercialization.

A basic factor for the development of these growth centres is to define efficient methods for monetization of gas industry products (Figure 13). Based on the current level of technological development, it can be concluded that LNG production is very promising and can be diversified due to gas chemical enterprises, which currently have a capacity use level below 65% [108]. This is due to a lack of export infrastructure and limited demand in the domestic market [109]. On the other hand, the forecast for the development of the methanol production industry alone is indicative of potential market growth by 50% before 2025 [110].

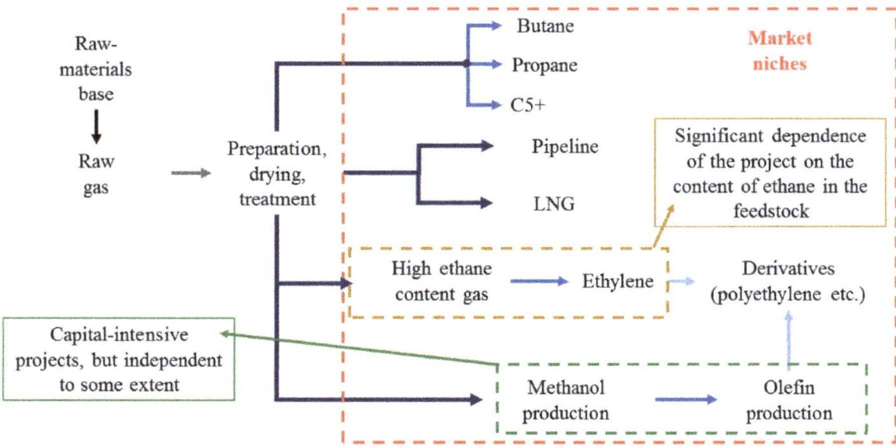

Figure 13. Conceptual scheme for commercialization of gas industry products.

To ensure the stability of gas monetization schemes, it is necessary to ensure the stability of the raw materials base for such projects for at least 25–30 years, both through their implementation at the most promising existing fields and through the search for new sources of raw materials, which is normally associated with the Arctic regions being quite an ambiguous factor.

On the one hand, the potential resources of natural gas in the Arctic amount to nearly 30% of the global volume [111]. However, a major portion of the resources is concentrated in the Western part of the Russian Arctic zone [112], which will enable the supply of plants with volumes of raw materials sufficient to produce around 150 mln t LNG per year. This will theoretically ensure nearly 30% market growth. The location is also favourable, as it will allow entry into both European and Asian markets, as well as contribute to the development of the Northern Sea Route, which is important for the social and economic development of all Arctic regions.

On the other hand, implementation of projects in the Arctic is a very labour- and capital-intensive process due to the influence of natural, climatic and infrastructural factors, the mitigation of which requires state support. Moreover, the unit costs for Arctic hydrocarbon production projects may be 2–10 times higher than that in Southern regions, especially at offshore deposits [113]. The development of practical solutions to reduce the cost for the development of such fields [114] will be one of the main drivers for the development of the Arctic hydrocarbons.

5. Conclusions and Discussion

5.1. Global Energy Markets

The influence of the fuel and energy sector on the global economy is impossible to overestimate, because energy is a basis for the development of enterprises, industry and society as a whole. Today's transformation processes in the energy sector need thorough control and attention from international regulators, as they are knowledge- and capital-intensive.

The allocation of significant volumes of financial, human and material resources for the development of alternative energy technologies is an integral part of the technological process [115], as confirmed by several studies in the field of building strategies and policies for the development of technological niches [116]. However, it should be understood that balanced political strategies must pay attention not only to projects with prospects in future decades, but to current operating and tactical needs.

In other words, instead of implementing a unilateral and comparatively simple policy of supporting one group of technologies, today's infrastructure and dynamics of markets should be taken into account. Scientifically substantiated portfolio strategies should be developed [117] to include not only potentially promising technologies, but also to take current economic needs into consideration. The strategies should also determine methods and procedures for a smooth transition from the traditional use of HCR to its advanced conversion and gradual transition to alternative energy technologies, as soon as they are ready and competitive.

Creation of favourable conditions for the rational use of HCR may ensure sustainable development of the fuel and energy sector and, more importantly, may create conditions for the development of alternative technologies. This thesis is confirmed by the existing investments of the largest oil and gas companies [118], while the reverse process is impossible under current conditions due to the low competitiveness of alternative energy options.

This study tried to expand the discussion field formed around energy sector development and balance it with critical analysis of existing image of HCR. The following conclusions were made on the basis of this analysis:

1. Today we face a paradoxical situation. International corporations and politicians calling for the need for an immediate transition from HCR to alternative energy technologies. However, the real activities of leading global economies show that they are not planning to decrease their footprint in the hydrocarbon energy sector and, moreover, they are expanding scientific and technical potential in geological surveying and the production of HCR.

2. Environmental protection is one of the main barriers to sustainable development in the energy sector, which is unlikely to be eliminated within the next few decades due to our current level of technological development. The risk of climatic catastrophe is extremely high. At the same

time, the lack of knowledge of the secondary theories of Global Warming mechanisms makes this issue even more complicated. On the one hand, we could have a chance to receive extra time to improve the environmental efficiency of our technologies. On the other hand, there is a risk that it is too late, and we will see the forecasted consequences much earlier than expected.

3. The most promising alternative energy options include RES, nuclear energy, and hydrogen. Theoretically, each of these resources is able to solve the HCR environmental issue, but there are some concurrent critical disadvantages that prevent them from being a complete alternative in the observed future:

    - the majority of RES-based technologies will be non-competitive within the next ten to twenty years due to insufficient efficiency of energy conversion and storage processes;
    - the generation of nuclear energy is associated with significant risks of a mainly technical (environmental) nature, mitigation of which is impossible at present;
    - hydrogen itself is not an energy resource, as much as a method for storage and transportation of primary energy and its broad application requires a solution to safe-storage and transportation issues, a higher efficiency of production, the creation of new infrastructure and the development of market-interaction mechanisms.

4. The global resource potential of the oil and gas industry as a whole is able to ensure the sustainable development of the world economy under the conditions of a better investment climate and the formation of an international regulation system for the sector. Here, special attention must be paid to ensuring good conditions for the development of hard-to-recover reserves in terms of geological and environmental factors.

5. The vector for the development of oil and coal sector has shifted towards integration with the chemical industry to enable the diversification of sectoral enterprises activities, but their role in the energy sector will not be reduced within the next few decades.

6. LNG production is the independent sector of the gas industry with the highest expansion potential in the long run. First of all, this can be attributed to the environmental and economical properties of LNG, which enable its consideration as a "win–win" solution for energy supply to satisfy even the strict requirements of European Union. Moreover, the flexibility of LNG logistic chains compared to pipelines is a very important factor.

Therefore, a comprehensive assessment of the potential for the implementation of new energy technologies shall be made in view of the needed preservation and development of the foundation of energy that is HCR. It should be noted that there has been no situation in the history of the energy sector in which new technology has completely displaced and substituted an older technology within a short period of time [119]. Gradual substitution is a conventional process that can be implemented at various rates, but on a long-term basis. Due to this, the immediate transition from HCR to alternative energy sources is not possible. The development of studies based on the agenda proposed in this article would create more realistic, viable and scientifically substantiated recommendations on the formation of climatic and energy policies, which are lacking today [120].

5.2. The Role of Russia in the Development of the LNG Industry

Russia has significant resource potential to develop LNG plants with a prospect to reach 140–160 mln t production capacity by 2030 (20% of the global market). Such intensive growth can be driven by limited prospects in the increase of Russian pipeline gas supplies to European countries as a result of the export's monopolistic nature.

Export monopoly is not only a barrier to entering foreign markets, but also prevents the creation of a competitive environment within the industry. Due to this, its rate of development is much slower than the worldwide average values, which is indicative of the stagnation of the sector.

Today, the export monopolization factor does not yet negatively affect the LNG industry. However, promising markets mentioned herein could be closed for Russian companies, if the situation with state regulation of the sector remains unchanged.

In this context, the speed of decision-making is crucial, and governs what share of markets Russian companies will be able to occupy. Taking into account the fact that nearly 50% of the designed capacities of LNG plants has already been contracted until 2035, it is quite realistic that after 2025, the second part of demand will be covered for the next decade, and it is unlikely that there will be potential for new companies to enter the market.

A transition to market methods for regulating industry activities seems necessary and timely, since direct involvement of the state in the production activities of companies has failed. This will improve the performance of companies and, as a result, additional advantages will be gained both for the country's budget and for the development of associated sectors, as well as social-economic infrastructure of the regions of the Russian Federation. Initially, switching to mechanisms for market regulation can be supported with tools such as golden shares, enabling the state to reserve the right to control only a few company operating aspects, but not to shape its future generally.

Funding: This research received no external funding.

Conflicts of Interest: The author declares no conflict of interest.

References

1. Lieder, M.; Rashid, A. Towards circular economy implementation: A comprehensive review in context of manufacturing industry. *J. Clean. Prod.* **2016**, *115*, 36–51. [CrossRef]
2. Kan, A. General characteristics of waste management: A review. *Energy Educ. Sci. Technol. Part A-Energy Sci. Res.* **2009**, *23*, 55–69.
3. Riaz, A.; Zahedi, G.; Klemeš, J.J. A review of cleaner production methods for the manufacture of methanol. *J. Clean. Prod.* **2013**, *57*, 19–37. [CrossRef]
4. Loiseau, E.; Saikku, L.; Antikainen, R.; Droste, N.; Hansjürgens, B.; Pitkänen, K.; Leskinen, P.; Kuikman, P.; Thomsen, M. Green economy and related concepts: An overview. *J. Clean. Prod.* **2016**, *139*, 361–371. [CrossRef]
5. Van der Ploeg, F.; Withagen, C. Global Warming and the Green Paradox: A Review of Adverse Effects of Climate Policies. *Rev. Environ. Econ. Policy* **2015**, *9*, 285–303. [CrossRef]
6. Abas, N.; Kalair, A.; Khan, N. Review of fossil fuels and future energy technologies. *Futures* **2015**, *69*, 31–49. [CrossRef]
7. Chu, S.; Majumdar, A. Opportunities and challenges for a sustainable energy future. *Nature* **2012**, *488*, 294–303. [CrossRef]
8. INEOS Group website. Open Letter to the European Commission President Jean-Claude Juncker. Available online: https://www.ineos.com/news/ineos-group/letter-to-the-european-commission-president-jean-claude-juncker/ (accessed on 12 February 2019).
9. Sandbag Climate Campaign CIC. Available online: https://ember-climate.org/carbon-price-viewer/ (accessed on 12 February 2019).
10. Asmelash, H. The G7's Pledge to End Fossil Fuel Subsidies by 2025: Mere Rhetoric or a Sign of Post-Paris Momentum? *Eur. Soc. Int. Law (Esil) Reflect.* **2016**, *5*, 1–7. [CrossRef]
11. Whitley, S.; Chen, H.; Doukas, A.; Gençsü, I.; Gerasimchuk, I.; Touchette, Y.; Worrall, L. *G7 Fossil Fuel Subsidy Scorecard*; International Institute for Sustainable Development: Winnipeg, MB, Canada, 2018. Available online: https://prod-edxapp.edx-cdn.org/assets/courseware/v1/0333e0d130fe95a336116caa1de2c98d/asset-v1:SDGAcademyX+CA001+1T2019+type@asset+block/6.R_Whitley__S__Chen__H__Doukas__A__et_al.__2018__G7_fossil_fuel_subsidy_scorecard._ODI_Report._London-_Overseas_Development_Institute.pdf (accessed on 12 February 2019).
12. Burrows, L.; Kotani, I.; Zorlu, P.; Popp, R.; Patuleia, A.; Littlecott, C. G7 Coal Scorecard–Fifth Edition Coal Finance Heads for the Exit. 2019. Available online: https://www.jstor.org/stable/pdf/resrep21850.pdf (accessed on 12 February 2019).

13. Department of Energy website. DOE Announces $3.8 Million Investment in New Methane Gas Hydrate Research. Available online: https://www.energy.gov/fe/articles/doe-announces-38-million-investment-new-methane-gas-hydrate-research (accessed on 12 September 2016).
14. Maggio, G.; Cacciola, G. When will oil, natural gas, and coal peak? *Fuel* **2012**, *98*, 111–123. [CrossRef]
15. Mohr, S.H.; Wang, J.; Ellem, G.; Ward, J.; Giurco, D. Projection of world fossil fuels by country. *Fuel* **2015**, *141*, 120–135. [CrossRef]
16. Wang, J.; Feng, L.; Tang, X.; Bentley, Y.; Höök, M. The implications of fossil fuel supply constraints on climate change projections: A supply-side analysis. *Futures* **2017**, *86*, 58–72. [CrossRef]
17. Dale, S. New economics of oil. *ONE J.* **2015**, *1*, 365.
18. Makhovikov, A.B.; Katuntsov, E.V.; Kosarev, O.V.; Tsvetkov, P.S. Digital transformation in oil and gas extraction. In Proceedings of the Innovation-Based Development of the Mineral Resources Sector: Challenges and Prospects-11th conference of the Russian-German Raw Materials, Potsdam, Germany, 7–8 November 2018; pp. 531–538.
19. Hoffert, M.I.; Caldeira, K.; Benford, G.; Criswell, D.R.; Green, C.; Herzog, H.; Jain, A.K.; Kheshgi, H.S.; Lackner, K.S.; Lewis, J.S.; et al. Advanced Technology Paths to Global Climate Stability: Energy for a Greenhouse Planet. *Science* **2002**, *298*, 981–987. [CrossRef] [PubMed]
20. Masson-Delmotte, V.; Zhai, P.; Pörtner, H.O.; Roberts, D.; Skea, J.; Shukla, P.R.; Pirani, A.; Moufouma-Okia, W.; Péan, C.; Pidcock, R.; et al. Global warming of 1.5 C. In *An IPCC Special Report on the Impacts of Global Warming of. 2018 October 8*; IPCC: Geneva, Switzerland, 2018.
21. Rogelj, J.; Den Elzen, M.; Höhne, N.; Fransen, T.; Fekete, H.; Winkler, H.; Schaeffer, R.; Sha, F.; Riahi, K.; Meinshausen, M. Paris Agreement climate proposals need a boost to keep warming well below 2 C. *Nature* **2016**, *534*, 631–639. [CrossRef]
22. Steffen, W.; Rockström, J.; Richardson, K.; Lenton, T.M.; Folke, C.; Liverman, D.; Summerhayes, C.P.; Barnosky, A.D.; Cornell, S.E.; Crucifix, M.; et al. Trajectories of the Earth System in the Anthropocene. *Proc. Natl. Acad. Sci. USA* **2018**, *115*, 8252–8259. [CrossRef]
23. Lenton, T.M.; Held, H.; Kriegler, E.; Hall, J.W.; Lucht, W.; Rahmstorf, S.; Schellnhuber, H.J. Tipping elements in the Earth's climate system. *Proc. Natl. Acad. Sci. USA* **2008**, *105*, 1786–1793. [CrossRef]
24. Schellnhuber, H.J.; Rahmstorf, S.; Winkelmann, R. Why the right climate target was agreed in Paris. *Nat. Clim. Chang.* **2016**, *6*, 649. [CrossRef]
25. Jiang, X.; Guan, D. Determinants of global CO_2 emissions growth. *Appl. Energy* **2016**, *184*, 1132–1141. [CrossRef]
26. Sovacool, B.K. What are we doing here? Analyzing fifteen years of energy scholarship and proposing a social science research agenda. *Energy Res. Soc. Sci.* **2014**, *1*, 1–29. [CrossRef]
27. Sundqvist, T. What causes the disparity of electricity externality estimates? *Energy Policy* **2004**, *32*, 1753–1766. [CrossRef]
28. Berger, A. Milankovitch theory and climate. *Rev. Geophys.* **1988**, *26*, 624–657. [CrossRef]
29. Hays, J.D.; Imbrie, J.; Shackleton, N.J. Variations in the Earth's orbit: Pacemaker of the ice ages. *Science* **1976**, *194*, 1121–1132. [CrossRef] [PubMed]
30. Fischer, T.P.; Arellano, S.; Carn, S.; Aiuppa, A.; Galle, B.; Allard, P.; Lopez, T.; Shinohara, H.; Kelly, P.; Werner, C.; et al. The emissions of CO_2 and other volatiles from the world's subaerial volcanoes. *Sci. Rep.* **2019**, *9*, 1–11. [CrossRef] [PubMed]
31. Guo, M.; Li, J.; Xu, J.; Wang, X.; He, H.; Wu, L. CO_2 emissions from the 2010 Russian wildfires using GOSAT data. *Environ. Pollut.* **2017**, *226*, 60–68. [CrossRef] [PubMed]
32. Shvidenko, A.Z.; Shchepashchenko, D.G.; Vaganov, E.A.; Sukhinin, A.I.; Maksyutov, S.S.; McCallum, I.; Lakyda, I.P. Impact of wildfire in Russia between 1998–2010 on ecosystems and the global carbon budget. *Dokl. Earth Sci.* **2011**, *441*, 1678–1682. [CrossRef]
33. Page, S.E.; Siegert, F.; Rieley, J.O.; Boehm, H.D.V.; Jaya, A.; Limin, S. The amount of carbon released from peat and forest fires in Indonesia during 1997. *Nature* **2002**, *420*, 61–65. [CrossRef]
34. Hooijer, A.; Page, S.; Canadell, J.G.; Silvius, M.; Kwadijk, J.; Wosten, H.; Jauhiainen, J. Current and future CO_2 emissions from drained peatlands in Southeast Asia. *Biogeosciences* **2010**, *7*, 1505–1514. [CrossRef]
35. Lee, C.T.A.; Jiang, H.; Dasgupta, R.; Torres, M. A Framework for Understanding Whole-Earth Carbon Cycling. In *Deep Carbon: Past to Present*; Cambridge University Press: Cambridge, UK, 2019; pp. 313–357.

36. Petrescu, A.M.R.; Van Beek, L.P.H.; Van Huissteden, J.; Prigent, C.; Sachs, T.; Corradi, C.A.R.; Dolman, A.J. Modeling regional to global CH4 emissions of boreal and arctic wetlands. *Global Biogeochemical Cycles* **2010**, *24*. [CrossRef]

37. Zhang, B.; Zhao, X.; Wu, X.; Han, M.; Guan, C.H.; Song, S. Consumption-Based Accounting of Global Anthropogenic CH4 Emissions. *Earth's Future* **2018**, *6*, 1349–1363. [CrossRef]

38. Tian, W.; Wu, X.; Zhao, X.; Ma, R.; Zhang, B. Quantifying global CH_4 and N_2O footprints. *J. Environ. Manag.* **2019**, *251*, 109566. [CrossRef]

39. Bloom, A.A.; Bowman, K.W.; Lee, M.; Turner, A.J.; Schroeder, R.; Worden, J.R.; Weidner, R.; McDonald, K.C.; Jacob, D.J. A global wetland methane emissions and uncertainty dataset for atmospheric chemical transport models (WetCHARTs version 1.0). *Geosci. Model Dev.* **2017**, *10*. [CrossRef]

40. Canadell, J.G.; Le Quéré, C.; Raupach, M.R.; Field, C.B.; Buitenhuis, E.T.; Ciais, P.; Conway, T.J.; Gillett, N.P.; Houghton, R.A.; Marland, G. Contributions to accelerating atmospheric CO_2 growth from economic activity, carbon intensity, and efficiency of natural sinks. *Proc. Natl. Acad. Sci. USA* **2007**, *104*, 18866–18870. [CrossRef]

41. Steffen, W.; Broadgate, W.; Deutsch, L.; Gaffney, O.; Ludwig, C. The trajectory of the Anthropocene: The great acceleration. *Anthr. Rev.* **2015**, *2*, 81–98. [CrossRef]

42. Fouquet, R. Path dependence in energy systems and economic development. *Nat. Energy* **2016**, *1*, 1–5.

43. Capros, P.; Tasios, N.; De Vita, A.; Mantzos, L.; Paroussos, L. Transformations of the energy system in the context of the decarbonisation of the EU economy in the time horizon to 2050. *Energy Strategy Rev.* **2012**, *1*, 85–96. [CrossRef]

44. Brundtland, G.; Khalid, M.; Agnelli, S.; Al-Athel, S.A.; Chidzero, B.; Fadika, L.M.; Hauff, V.; Lang, I.; Ma, S.; Marino de Botero, M.; et al. *Our Common Future: The World Commission on Environment and Development*; Oxford University Press: London, UK, 1987.

45. Glavič, P.; Lukman, R. Review of sustainability terms and their definitions. *J. Clean. Prod.* **2007**, *15*, 1875–1885. [CrossRef]

46. Vijayaraghavan, A.; Dornfeld, D. Automated energy monitoring of machine tools. *Cirp Ann.* **2010**, *59*, 21–24. [CrossRef]

47. Yoon, H.S.; Kim, E.S.; Kim, M.S.; Lee, J.Y.; Lee, G.B.; Ahn, S.H. Towards greener machine tools–A review on energy saving strategies and technologies. *Renew. Sustain. Energy Rev.* **2015**, *48*, 870–891. [CrossRef]

48. Kobos, P.H.; Malczynski, L.A.; La Tonya, N.W.; Borns, D.J.; Klise, G.T. Timing is everything: A technology transition framework for regulatory and market readiness levels. *Technol. Forecast. Soc. Chang.* **2018**, *137*, 211–225. [CrossRef]

49. Schmidt, T.S.; Sewerin, S. Technology as a driver of climate and energy politics. *Nat. Energy* **2017**, *2*, 17084. [CrossRef]

50. Rogge, K.S.; Reichardt, K. Policy mixes for sustainability transitions: An extended concept and framework for analysis. *Res. Policy* **2016**, *45*, 1620–1635. [CrossRef]

51. Meadowcroft, J. What about the politics? Sustainable development, transition management, and long term energy transitions. *Policy Sci.* **2009**, *42*, 323. [CrossRef]

52. Abergel, T.; Brown, A.; Cazzola, P.; Dockweiler, S.; Dulac, J.; Fernandez Pales, A.; West, K. *Energy Technology Perspectives 2017: Catalysing Energy Technology Transformations*; OECD: Paris, France, 2017.

53. Abergel, T.; Brown, A.; Cazzola, P.; Dockweiler, S.; Dulac, J.; Pales, A.F.; Gorner, M.; Malischek, R.; Masanet, E.R.; McCulloch, S.; et al. Energy technology perspectives 2017: Catalysing energy technology transformations. 2017. Available online: https://www.cleanenergyministerial.org/sites/default/files/2018-07/English-ETP-2017.pdf (accessed on 12 February 2019).

54. IRENA. *Global Energy Transformation: A Roadmap to 2050*; IRENA: Abu Dhabi, UAE, 2019.

55. Ram, M.; Child, M.; Aghahosseini, A.; Bogdanov, D.; Lohrmann, A.; Breyer, C. A comparative analysis of electricity generation costs from renewable, fossil fuel and nuclear sources in G20 countries for the period 2015-2030. *J. Clean. Prod.* **2018**, *199*, 687–704. [CrossRef]

56. Saito, S. Role of nuclear energy to a future society of shortage of e nergy resources and global warming. *J. Nucl. Mater.* **2010**, *398*, 1–9. [CrossRef]

57. International Atomic Energy Agency. *Nuclear Power Reactors in the World*; Reference Data Series No. 2; IAEA: Vienna, Austria, 2019.

58. IEA. *Nuclear Power in a Clean Energy System*; IEA: Paris, France, 2019. Available online: https://www.iea.org/reports/nuclear-power-in-a-clean-energy-system (accessed on 12 February 2019).

59. Lovering, J.R.; Yip, A.; Nordhaus, T. Historical construction costs of global nuclear power reactors. *Energy Policy* **2016**, *91*, 371–382. [CrossRef]

60. Dittmar, M. Nuclear energy: Status and future limitations. *Energy* **2012**, *37*, 35–40. [CrossRef]

61. Hossain, M.K.; Taher, M.A.; Das, M.K. Understanding Accelerator Driven System (ADS) Based Green Nuclear Energy: A Review. *World J. Nucl. Sci. Technol.* **2015**, *5*, 287. [CrossRef]

62. Ojovan, M.I.; Lee, W.E.; Kalmykov, S.N. *An Introduction to Nuclear Waste Immobilization*; Elsevier: Amsterdam, The Netherland, 2019.

63. Yaqoot, M.; Diwan, P.; Kandpal, T.C. Review of barriers to the dissemination of decentralized renewable energy systems. *Renew. Sustain. Energy Rev.* **2016**, *58*, 477–490. [CrossRef]

64. Asdrubali, F.; Baldinelli, G.; D'Alessandro, F.; Scrucca, F. Life cycle assessment of electricity production from renewable energies: Review and results harmonization. *Renew. Sustain. Energy Rev.* **2015**, *42*, 1113–1122. [CrossRef]

65. Cherepovitsyn, A.; Tcvetkov, P. Overview of the prospects for developing a renewable energy in Russia. In *2017 International Conference on Green Energy and Applications (ICGEA)*; IEEE: Piscataway, NJ, USA, 2017; pp. 113–117.

66. Dincer, I.; Acar, C. A review on clean energy solutions for better sustainability. *Int. J. Energy Res.* **2015**, *39*, 585–606. [CrossRef]

67. Tcvetkov, P.S. The history, present status and future prospects of the Russian fuel peat industry. *Mires Peat* **2017**, *19*, 1–12. [CrossRef]

68. Tan, R.R.; Aviso, K.B.; Ng, D.K.S. Optimization models for financing innovations in green energy technologies. *Renew. Sustain. Energy Rev.* **2019**, *113*, 109258. [CrossRef]

69. Sabihuddin, S.; Kiprakis, A.E.; Mueller, M. A numerical and graphical review of energy storage technologies. *Energies* **2015**, *8*, 172–216. [CrossRef]

70. Argyrou, M.C.; Christodoulides, P.; Kalogirou, S.A. Energy storage for electricity generation and related processes: Technologies appraisal and grid scale applications. *Renew. Sustain. Energy Rev.* **2018**, *94*, 804–821. [CrossRef]

71. Dutton, J.; Pilsner, L. Delivering Climate Neutrality: Accelerating Eu Decarbonisation with Research and Innovation Funding. 2019. Available online: https://www.jstor.org/stable/pdf/resrep21732.pdf (accessed on 12 February 2019).

72. Cabré, M.M.; Gallagher, K.P.; Li, Z. Renewable Energy: The Trillion Dollar Opportunity for Chinese Overseas Investment. *China World Econ.* **2018**, *26*, 27–49. [CrossRef]

73. Murdock, H.E.; Gibb, D.; André, T.; Appavou, F.; Brown, A.; Epp, B.; Kondev, B.; McCrone, A.; Musolino, E.; Ranalder, L.; et al. Renewables 2019 Global Status Report. Available online: https://www.ren21.net/wp-content/uploads/2019/05/gsr_2019_perspectives_en.pdf (accessed on 12 February 2019).

74. Kemeny, T. Does foreign direct investment drive technological upgrading? *World Dev.* **2010**, *38*, 1543–1554. [CrossRef]

75. Gielen, D.; Boshell, F.; Saygin, D.; Bazilian, M.D.; Wagner, N.; Gorini, R. The role of renewable energy in the global energy transformation. *Energy Strategy Rev.* **2019**, *24*, 38–50. [CrossRef]

76. Abdalla, A.M.; Hossain, S.; Nisfindy, O.B.; Azad, A.T.; Dawood, M.; Azad, A.K. Hydrogen production, storage, transportation and key challenges with applications: A review. *Energy Convers. Manag.* **2018**, *165*, 602–627. [CrossRef]

77. Hosseini, S.E.; Wahid, M.A. Hydrogen from solar energy, a clean energy carrier from a sustainable source of energy. *Int. J. Energy Res.* **2020**, *4*, 4110–4131. [CrossRef]

78. Cetinkaya, E.; Dincer, I.; Naterer, G.F. Life cycle assessment of various hydrogen production methods. *Int. J. Hydrog. Energy* **2012**, *37*, 2071–2080. [CrossRef]

79. Dincer, I.; Acar, C. Review and evaluation of hydrogen production methods for better sustainability. *Int. J. Hydrog. Energy* **2015**, *1109*, 4–11111. [CrossRef]

80. Nikolaidis, P.; Poullikkas, A. A comparative overview of hydrogen production processes. *Renew. Sustain. Energy Rev.* **2017**, *67*, 597–611. [CrossRef]

81. Acar, C.; Beskese, A.; Temur, G.T. Sustainability analysis of different hydrogen production options using hesitant fuzzy AHP. *Int. J. Hydrog. Energy* **2018**, *43*, 18059–18076. [CrossRef]

82. MA, J.; LIU, S.; ZHOU, W.; PAN, X. Comparison of Hydrogen Transportation Methods for Hydrogen Refueling Station. *J. Tongji Univ. (Nat. Sci.)* **2008**, *5*, 615–619.

83. Acar, C.; Dincer, I. Review and evaluation of hydrogen production options for better environment. *J. Clean. Prod.* **2019**, *218*, 835–849. [CrossRef]

84. Royal Dutch Shell plc. Shell LNG Outlook 2020. 2020. Available online: https://www.shell.com/energy-and-innovation/natural-gas/liquefied-natural-gas-lng/lng-outlook-2020.html#iframe=L3dlYmFwcHMvTE5HX291dGxvb2sv
(accessed on 12 February 2019).

85. Litvinenko, V.S.; Kozlov, A.V.; Stepanov, V.A. Hydrocarbon potential of the Ural–African transcontinental oil and gas belt. *J. Pet. Explor. Prod. Technol.* **2017**, *7*, 1–9. [CrossRef]

86. Tullo, A.H. C&EN's Global Top 50 chemical companies of 2018. *Chem. Eng. News* **2019**, 30–35. Available online: https://cen.acs.org/content/dam/cen/97/30/WEB/globaltop50-2018.pdf (accessed on 12 February 2019).

87. Muggeridge, A.; Cockin, A.; Webb, K.; Frampton, H.; Collins, I.; Moulds, T.; Salino, P. Recovery rates, enhanced oil recovery and technological limits. *Philos. Trans. R. Soc. A Math. Phys. Eng. Sci.* **2014**, *372*, 20120320. [CrossRef]

88. Exxon Mobil Financial Operating Review. 2018. Available online: https://corporate.exxonmobil.com/-/media/Global/Files/annual-report/2018-Financial-and-Operating-Review.pdf (accessed on 12 February 2019).

89. Litvinenko, V.S.; Sergeev, I.B. Innovations as a Factor in the Development of the Natural Resources Sector. *Stud. Russ. Econ. Dev.* **2019**, *30*, 637–645. [CrossRef]

90. Cherepovitsyn, A.; Lipina, S.; Evseeva, O. Innovative approach to the development of mineral raw materials of the arctic zone of the Russian federation. *Zap. Gorn. Inst. /J. Min. Inst.* **2018**, *232*, 438. [CrossRef]

91. Cherepovitsyn, A.; Ilinova, A.; Evseeva, O. Stakeholders management of carbon sequestration project in the state-business-society system. *Zap. Gorn. Inst. /J. Min. Inst.* **2019**, *240*, 731. [CrossRef]

92. Nugent, D.; Sovacool, B.K. Assessing the lifecycle greenhouse gas emissions from solar PV and wind energy: A critical meta-survey. *Energy Policy* **2014**, *65*, 229–244. [CrossRef]

93. Tcvetkov, P.; Cherepovitsyn, A.; Makhovikov, A. Economic assessment of heat and power generation from small-scale liquefied natural gas in Russia. *Energy Rep.* **2020**, *6*, 391–402. [CrossRef]

94. VYGON. Consulting Global LNG Market: Illusory Glut. 2018. Available online: https://vygon.consulting/upload/iblock/486/vygon_consulting_lng_world_balance_en_executive_summary.pdf (accessed on 12 February 2019).

95. Dudley, B. *BP Statistical Review of World Energy 2019*; British Petroleum: London, UK, 2019.

96. Dudley, B. *BP Statistical Review of World Energy 2020*; British Petroleum: London, UK, 2020.

97. Savcenko, K.; Hornby, G. The future of European gas after Groningen. S&P Global Platts. 2020. Available online: https://www.spglobal.com/platts/plattscontent/_assets/_files/en/specialreports/naturalgas/groningen-european-gas-report.pdf (accessed on 12 February 2019).

98. Howell, N.; Pereira, R. LNG in Europe. Current Trends, the European LNG Landscape and Country Focus. Gaffney, Cline Associates. Available online: https://www.europeangashub.com/wp-content/uploads/2019/10/DM-_6044045-v1-Article_LNG_in_Europe_HOWELL_PEREIRA.pdf (accessed on 12 February 2019).

99. IGU. *Global Gas Report 2019*; International Gas Union: Barcelona, Spain, 2019. Available online: https://www.wgc2021.org/wp-content/uploads/2019/04/LNG2019-IGU-World-LNG-report2019.pdf (accessed on 12 February 2019).

100. GIIGNL. The LNG Industry. Annual Report 2019. Available online: https://giignl.org/sites/default/files/PUBLIC_AREA/Publications/giignl_annual_report_2019-compressed.pdf (accessed on 12 February 2019).

101. Belova, M.; Kolbikova, E.; Timonin, I. Russia's place ona global LNG map. *OilGas J.* **2019**, 74–81. Available online: https://vygon.consulting/upload/iblock/3c6/OGJR_2019_04_small.pdf (accessed on 12 February 2019).

102. S&P Global Platts. 2019 Review and 2020 Outlook. 2019. Available online: https://www.spglobal.com/platts/plattscontent/_assets/_files/en/specialreports/oil/platts_2020_outlook_report.pdf (accessed on 12 February 2019).

103. McKinsey. Global gas and LNG Outlook to 2035. 2018. Available online: https://www.mckinsey.com/solutions/energy-insights/global-gas-lng-outlook-to-2035/~{}/media/3C7FB7DF5E4A47E393AF0CDB080FAD08.ashx (accessed on 12 February 2019).

104. Henderson, J.; Yermakov, V. *Russian LNG: Becoming a Global Force. Working Paper*; Oxford Institute for Energy Studies: Oxford, UK, 2019; 37 p. [CrossRef]

105. Hashimoto, H. 2020 Gas Market Outlook. In *434th Forum on Research Works*; The Institute of Energy Economics: Tokyo, Japan. Available online: https://globallnghub.com/wp-content/uploads/2020/01/8787.pdf (accessed on 23 December 2019).

106. IEEJ. *EPRINC The Future of Asian LNG 2018 (The Road to Nagoya)*; The Institute of Energy Economics: Tokyo, Japan, 2018. Available online: https://eneken.ieej.or.jp/data/8140.pdf (accessed on 12 February 2019).

107. IGU. *Global Gas Report 2018*; International Gas Union: Barcelona, Spain, 2018. Available online: https://www.snam.it/export/sites/snam-rp/repository/file/gas_naturale/global-gas-report/global_gas_report_2018.pdf (accessed on 12 February 2019).

108. Althouse Group. Overview of the gas processing industry in Russia. 2019. Available online: https://althausgroup.ru/wp-content/uploads/2019/10/Analiz-rynka-produktov-gazopererabotki.pdf (accessed on 12 February 2019). (In Russian)

109. Bilfinger Tebodin. *The Russian Chemical Market*; Market study report; The Embassy of the Kingdom of the Netherlands; Bilfinger Tebodin: Moscow, Russia, 2018. Available online: https://www.pronline.ru/Handlers/ShowRelisesFile.ashx?relizid=6998&num=2 (accessed on 12 February 2019).

110. VYGON Consulting. Gas Chemistry of Russia. Part 1. Methanol: So far only plans. 2019. Available online: https://vygon.consulting/upload/iblock/f22/vygon_consulting_russian_methanol_industry_development.pdf (accessed on 12 February 2019). (In Russian).

111. Gautier, D.L.; Bird, K.J.; Charpentier, R.R.; Grantz, A.; Houseknecht, D.W.; Klett, T.R.; Sørensen, K. Assessment of undiscovered oil and gas in the Arctic. *Science* **2009**, *324*, 1175–1179. [CrossRef] [PubMed]

112. Viacheslav, Z.; Alina, I. Problems of unconventional gas resources production in arctic zone-Russia. *Espacios* **2018**, *39*, 42.

113. Brecha, R.J. Logistic curves, extraction costs and effective peak oil. *Energy Policy* **2012**, *51*, 586–597. [CrossRef]

114. Litvinenko, V.; Trushko, V.; Dvoinikov, M. Method of construction of an offshore drilling platform on the shallow shelf of the Arctic seas. Russian Patent 2704451 C1, 28 October 2019. (In Russian).

115. Bumpus, A.; Comello, S. Emerging clean energy technology investment trends. *Nat. Clim. Chang.* **2017**, *7*, 382–385. [CrossRef]

116. Kivimaa, P.; Kern, F. Creative destruction or mere niche support? Innovation policy mixes for sustainability transitions. *Res. Policy* **2016**, *45*, 205–217. [CrossRef]

117. Schmidt, T.S.; Sewerin, S. Measuring the temporal dynamics of policy mixes–An empirical analysis of renewable energy policy mixes' balance and design features in nine countries. *Res. Policy* **2019**, *48*, 103557. [CrossRef]

118. Pickl, M.J. The renewable energy strategies of oil majors–From oil to energy? *Energy Strategy Rev.* **2019**, *26*, 100370. [CrossRef]

119. Sovacool, B.K. How long will it take? Conceptualizing the temporal dynamics of energy transitions. *Energy Res. Soc. Sci.* **2016**, *13*, 202–215. [CrossRef]

120. Peters, G.P. The 'best available science' to inform 1.5 C policy choices. *Nat. Clim. Chang.* **2016**, *6*, 646. [CrossRef]

 *resources*

Article

Transformation of the Personnel Training System for Oil and Gas Projects in the Russian Arctic

Ekaterina Samylovskaya [1,*], Regina-Elizaveta Kudryavtseva [2], Dmitriy Medvedev [3], Sergey Grinyaev [3] and Alfred Nordmann [4]

[1] Faculty of humanities and sciences, St.-Petersburg Mining University, 199106 St Petersburg, Russia
[2] Institute of History, St.-Petersburg State University, 199034 St Petersburg, Russia; aethel@yandex.ru
[3] Faculty of Integrated Security of the Fuel and Energy Complex, I.M. Gubkin Russian State University of Oil and Gas, 119991 Moscow, Russia; medvedev.d@gubkin.ru (D.M.); gsn@gubkin.pro (S.G.)
[4] Institute of Philosophy, Darmstadt Technical University, 64289 Darmstadt, Germany; nordmann@phil.tu-darmstadt.de
*   Correspondence: samylovskaya_ea@pers.spmi.ru; Tel.: +7-921-367-37-91

Received: 19 October 2020; Accepted: 19 November 2020; Published: 22 November 2020

Abstract: This paper analyses the process of transforming specialist training systems for oil and gas projects in the Arctic, which has been taking place within the structure of education in Russia over the past decade. Using classical methods of analysis, synthesis, and classification, the authors studied the main global trends in training personnel for the Arctic and the manifestations of these trends in the system of training Russian specialists. To identify the qualitative characteristics of the educational system development, the authors applied the survey method and composed a list of leading universities in training personnel for the Russian Arctic, as well as the "Arctic professions of the future". As a result of the study, the authors came to the conclusion that global trends in training "Arctic personnel" show the need to develop an interdisciplinary approach, to form basic knowledge in Natural Sciences, to study the socio-cultural specifics of the region, to develop new educational standards, to implement the concept of 'Life Long Learning', to widely introduce digital technologies and to internationalize education. In general, the Russian personnel training system is adapting to changing conditions, in particular, some progress has been made in the formation of "digital" competencies and skills to work in a developed IT infrastructure. The introduction of "digital fields" has led to an increase in the demand for IT specialists in the Arctic oil and gas sector. With the help of an expert survey, it was revealed that in the future, the most popular professions, along with "drillers" and transport specialists, will be IT specialists who ensure the "digital fields" functioning. The leading Russian universities that train specialists for modern oil and gas projects in the Arctic have been identified. It is noted that not all leading industry universities in Russia are participating in international educational projects and organizations. There is skepticism about the internationalization of education.

Keywords: oil and gas education; personnel training; Arctic education system; oil and gas projects; international cooperation

---

1. Introduction

In the vector of development of both the global and Russian economy, the Arctic is the most important region with a huge potential for the Russian oil and gas industry. According to experts, the Arctic zone of the Russian Federation has 7.3 billion tons of oil reserves and 55 trillion cubic meters of natural gas [1], so Russia is developing an extensive program to develop production capacities and realize the potential of the region [2].

Arctic resources are of considerable interest to non-regional powers. Thus, according to American analysts, the most aggressive policy is led by China, which, according to their estimates, is the largest investor in the Arctic countries [3,4]. Other major economies (Japan, Germany, France, Great Britain, etc.) are also actively expanding their participation in developing the region [5]. To a greater extent, they are suppliers of technologies that ensure the extraction of hydrocarbons on the Arctic shelf, as well as of competencies for working with these technologies.

Indeed, along with the intensification of economic activity in the Arctic region, the issue of improving the personnel training system has become more urgent. At present, there is a shortage of highly qualified specialists in the oil and gas industry in several specialties [6], ranging from workers to engineers and managers. This deficit will only increase as large-scale projects are implemented [6,7].

According to experts, at the moment, the most popular in the Arctic are highly qualified specialists in the sphere of extractive industry, in particular, specialists developing offshore oil and gas fields, ice-class marine shipbuilding, port, and shipping infrastructure, and others [4]. At the same time, successful implementation of oil and gas projects in the Arctic requires qualified specialists not only with higher, but also with professional education, as well as service and support personnel. In turn, until recently, the number of graduates with higher education was estimated at dozens of graduates per year. Today, the situation has somewhat changed: We can talk about several hundred graduates of various specialties. However, for example, in the near future, only for the development of the Barents Sea fields, there is a need for 15 thousand specialists with higher education, more than 50% of whom have a marine oil and gas education and a qualification of "mining engineer" [8,9].

Authors set goals to see the process of the transformation of the specialists training system for oil and gas projects in the Arctic occurring in the education structure in Russia over the last decade, for which it seems necessary to examine the major global trends in training for the Arctic; to trace the manifestations of these trends in the system of Russian specialists' training; to identify the list of universities that are leaders in training for the Russian Arctic, preparing "Arctic jobs of the future".

2. Materials and Methods

Despite a fairly large number of studies in the sphere of education for Arctic training [10–18], there is no consolidated international or Russian statistics on training specialists for the oil and gas industry in the Arctic at the moment. In this regard, this study focuses on studying the directions of transformation of the personnel training system.

Based on the analysis of modern foreign and Russian scientific literature on the problem of training personnel for work in the Arctic region, the main current global trends in the training of Arctic specialists were identified.

To identify qualitative characteristics for developing the educational system, in August 2020, an expert survey was conducted among the heads of educational and scientific organizations engaged in training Arctic personnel, as well as among representatives of mining enterprises and businesses in the oil and gas industry of the Russian Federation. The expert survey was sent to 42 respondents from 23 institutions. As a result, 30 experts participated in the survey: Unfortunately, 12 respondents could not participate (Table 1).

Using the methods of systematization and classification, we analyzed the scientific and educational programs for training personnel for the Arctic suggested by those universities that were named by experts to be the leading centers for training Arctic specialists. Based on the method of comparative analysis, the authors analyzed the level of compliance of scientific and educational programs at these universities with the selected global trends.

Table 1. Expert group description [1].

The Institution Activity Specifics	Name of the Institution	Number of Participating Respondents	Number of those Who Refused
Scientific (directors of institutes/centers, heads of departments)	Kola Science Centre RAS	3	1
	E.M. Primakov National Research Institute of World Economy and International Relations RAS	1	-
	European Institute RAS	1	-
	Arctic and Antarctic Research Institute	1	-
	Institute of oil and gas RAS	1	-
	Karelia Science Centre RAS	-	1
	Komi Scientific Centre at the Ural branch of the Russian Academy of Sciences	-	1
	"Arctic" Research Centre of the Far Eastern branch RAS	-	1
Universities (deans and heads of departments of specialties)	Murmansk State Technical University	2	-
	Northern (Arctic) Federal University	2	-
	Gubkin Russian State University of oil and gas (2 respondents)	2	-
	Saint Petersburg Mining University	2	-
	Murmansk Arctic State University	2	1
	Petrozavodsk State University	-	1
	North-Eastern Federal University named after M. K. Ammosov	-	1
	Far-Eastern Federal University	-	1
Companies that specialize in oil and gas production (department heads and shift managers)	'Gazprom Neft' PJSC	3	-
	'Gazprom Neft shelf' LLC	2	-
	'Rosneft NK' PJSC	2	-
	'NOVATEK' PJSC	-	2
Institutions involved in the development and implementation of oil and gas projects (heads of departments)	'SPG' group of companies	3	-
	'Iceberg' Central Design Bureau	-	2
Expert center (head of the expert center and experts)	Expert center of the Arctic development project office	3	-

[1] Compiled by the authors.

3. Results of the Expert Survey

According to the results of the survey, the experts identified themselves as specialists working in the following areas of professional activity:

- Oil and gas complex—27%;
- Science—27%;
- Education—19%;
- Business (Economics and consulting)—27%.

It is noteworthy that 100% of the respondents who took part in the survey agree with the statement that there is currently a shortage of qualified personnel for implementing Russian oil and gas projects in the Arctic.

The respondents were asked to assess the qualitative characteristics of the development of the educational system for training personnel for oil and gas projects in the Arctic.

The absolute majority of respondents note that today, in the process of training "Arctic personnel", the importance of using digital technologies to solve professional problems is growing. They all emphasize the need to develop international cooperation in the sphere of training "Arctic personnel" and a practice-oriented approach to training specialists for Arctic projects. 13% of respondents in the sphere of economics/consulting (7% of respondents) and science (6% of respondents) question the statement that in the process of training modern specialists, the importance of mastering basic knowledge in meteorology, physics, biology, and ecology is growing, regardless of the direction of training. 28% of respondents also related to economics/consulting (20%) and education (8%) questioned the statement that the participation of universities from non-Arctic countries in training specialists for the Arctic contributes to increasing the effectiveness of Arctic projects (Table 2).

Thus, all the surveyed representatives of the oil and gas industry and business, as well as the vast majority of representatives of science and education, note that all the above quality characteristics are necessary for training qualified modern specialists for oil and gas projects in the Arctic, thus confirming the need to implement an interdisciplinary, integrated and international approach in the educational process. In turn, it is noteworthy that representatives of business (economics and consulting) underestimate the role of interdisciplinary knowledge (knowledge of meteorology, physics, biology, ecology) and an international approach in training specialists. We dare to assume that this is due to the professional specifics of respondents who are focused on developing narrowly focused competencies and specific skills.

Table 2. Qualitative characteristics of the development of the educational system for training personnel for oil and gas projects in the Arctic (30 experts interviewed) [1].

Qualitative Characteristics	Respondents' Answers
The importance of using digital technologies to solve professional tasks is growing	**Fully agree—73%:** oil and gas complex—27% science—9% education—10% business—27%
	Quite agree—27%: oil and gas complex—0% science—18% education—9% business—0%
There is a growing need to develop international cooperation in the sphere of training "Arctic personnel"	**Fully agree—47%:** oil and gas complex—12% science—11% education—12% business—12%
	Quite agree—53% oil and gas complex—15% science—16% education—7% business—15%
The importance of a practice-oriented approach to training specialists for Arctic projects is growing	**Fully agree—60%** oil and gas complex—18% science—12% education—14% business—16%
	Quite agree—40% oil and gas complex—9% science—15% education—5% business—11%

Table 2. *Cont.*

Qualitative Characteristics	Respondents' Answers
There is a growing importance of mastering basic knowledge in meteorology, physics, biology, and ecology, regardless of the sphere of study	**Fully agree—60%:** oil and gas complex—24% science—24% education—12% business—0%
	Quite agree—27%: oil and gas complex—3% science—3% education—4% business—17%
	Rather disagree—13%: oil and gas complex—0% science—0% education—3% business—10%
Participation of universities from non-Arctic countries in training specialists for the Arctic contributes to increasing the effectiveness of Arctic projects	**Quite agree—73%:** oil and gas complex—27% science—27% education—16% business—3%
	Rather disagree—27%: oil and gas complex—0% science—0% education—3% business—24%

[1] Composed by the authors basing on the experts' interview analyses.

As a result of the survey, the top five Russian universities that train qualified specialists in the oil and gas sector that meet modern requirements for developing the oil and gas industry include: I. M. Gubkin Russian State University of oil and gas (28.2% of the total number of responses), Saint Petersburg Mining University (25.6%), Saint Petersburg Polytechnic University (12.8%), Northern (Arctic) Federal University (10.2%), Murmansk State Technical University and Ufa State Oil Technical University (5.1% each) (Table 3).

Table 3. Russian universities that train qualified specialists in the oil and gas sector that meet modern requirements for developing the oil and gas industry in the Arctic (30 experts interviewed) [1].

Higher Educational Institution	Number of Responses (%)
I. M. Gubkin Russian State University of oil and gas	28.2
Saint Petersburg Mining University	25.6
Saint Petersburg Polytechnic University	12.8
Northern (Arctic) Federal University named after M. V. Lomonosov	10.2
Murmansk State Technical University	5.1
Ufa State Oil Technical University	5.1
Ukhta State Technical University	2.6
Northeast Federal University named after M. K. Ammosov	2.6
National Research Institute of Technology "MISiS"	2.6
Tomsk Polytechnic University	2.6
Saint Petersburg State University of Economics	2.6

[1] Composed by the authors basing on the experts' interview analyses.

It is worth noting that representatives of extractive enterprises and businesses in the oil and gas industry (i.e., representatives of potential employers interested in university graduates), in addition to the obvious leading universities, also mentioned the following universities: Northern (Arctic) Federal University, Ufa State Oil Technical University, Murmansk State Technical University, Ukhta State Technical University, Tomsk Polytechnic University and North-Eastern Federal University named after M. K. Ammosov.

To analyze the objectivity of the university assessment, respondents were asked to answer the question: "Which university did you graduate from?". The survey showed that four respondents (13%) indicated their "Alma mater" among the leading universities that provide training for Arctic oil and gas projects. Therefore, the psychological moment of commitment to "their own" university could play a role in the assessment, and this percentage can be considered a probable error. The high percentage of probable error can be explained by several reasons: First, the specifics of the expert survey, when a small number of professional participants take part in it; second, the specifics of the sphere the survey is aimed at.

The survey identified the most popular professions for Arctic oil and gas projects in the next 5–10 years. The experts named 21 professions, among which the most frequently mentioned were:

- drilling engineers, including specialists in offshore drilling (14.3% of the total number of responses);
- specialists in the sphere of marine transport—pilots, captains, security specialists in marine transport (14.3%);
- database specialists—programmers (11.9%);
- mining engineers (9.5%);
- geologists (7.1%);
- environmentalists (7.1%);
- construction workers (7.1%).

The data obtained on the professions that will be in demand in the Arctic in the future are quite correlated with the list of universities that, according to experts, train qualified specialists in the oil and gas sector that meet modern requirements for developing the oil and gas industry in the region: The leading universities mentioned above train personnel in these specialties.

Given the complexity of living and working conditions in the Arctic, the absolute majority of experts surveyed (87%) consider it necessary to conduct special psychophysical training for future Arctic specialists.

4. Discussion: New Trends in Training Specialists for the Arctic

Modern foreign and Russian experts in the field of education, analyzing the main global trends in the training of "Arctic" personnel, note that the transformation of the training system is influenced by the following global factors:

- increased inter-state competition that covers not only traditional markets for goods and capital, but also human potential, and consequently, the increasing role of human capital as the main factor of socio-economic development [6,19];
- digitalization and technological complexity of projects in the Arctic, the intellectualization of activities, resulting in the formation of qualitatively new qualification requirements and educational standards [6,7,19];
- internationalization of the educational process, increasing the importance of international network forms of education [6,19].

It is noteworthy that the presented trends overlap with the qualitative characteristics identified in the expert survey that are necessary for developing the educational system for training personnel for oil and gas projects in the Russian Arctic. Moreover, the characteristics allow us to more accurately reveal the manifestations of these trends in the Russian educational system.

4.1. Increased Competition, Development of New Standards

Analysis of the survey results and modern pedagogical and professional literature [6,19] showed that "Arctic cadres" should have interdisciplinary knowledge. So, regardless of the chosen direction of future professional activity, students should get a basic knowledge of meteorology, geography, geo-ecology, physics, chemistry, and biology to form a more complete understanding of the region's features. This is important because when solving professional tasks in Northern latitudes, it is necessary to understand the nature of natural and climatic factors and the associated risks and dangers to human life and the security of industrial facilities [20]. Environmental and socio-economic spheres of life are the most important areas of human activity in the Arctic [21], so it is no accident that the method of modeling environmental and economic risks in the region is increasingly used in the implementation of Arctic projects [22], in particular in the oil and gas industry [23,24].

In addition, extreme weather conditions, lack of convenient logistics, and a complex communication situation lead to the need to solve non-standard tasks in the course of performing professional activities. A modern specialist in the oil and gas industry should start implementing projects in a comprehensive manner, taking into account the level of scientific and technological progress in the industry, the specifics of the socio-economic life of the Arctic region, and should be psychologically ready to live and work in extreme Arctic conditions [25–27].

Accordingly, the specialist should be able to flexibly and creatively approach the solution of complex situations that do not fit into the "standard templates".

In this regard, the competencies of "problem solving skills" are of particular importance, and not just "knowing the classic ways to solve them". The focus on obtaining skills and abilities is emphasized by researchers who analyze the staff training system for the Arctic [12,20].

It should be emphasized that it is important to increase the adaptability of the national training system to ensure its sustainability in the face of rapid changes in the socio-economic sphere. Education is the most important factor in the sustainable development of the Arctic territories. When considering the features of the education system in the Arctic, several researchers emphasize the implementation of the concept of Arctic sustainable education (ASE) [28,29], which is provided in several domains, such as: Community, cooperation, problem-solving skills, and opportunities for active, local, and cultural training in sustainable development [28,29] through an inclusive approach that takes into account the cultural, linguistic, and other characteristics of the peoples living in the Arctic.

A relevant area of work is the preparation of new educational standards for training specialists for the Arctic. Work in this direction began in 2018–2019, when the "Arctic national scientific-educational consortium" Association conducted preliminary work to explore the possibilities of developing these standards. Now it regularly provides them to the Ministry of education, presents them at the meetings of the Council for the Arctic and Antarctic under the Federation Council of the Russian Federation, at the meetings of the Expert Council under the State Duma for legislative support of the Far North regions development, and so on [20].

So, in Russia today, programs are being implemented to train engineers for technological support of the oil and gas industry in the Arctic, geological exploration in the Northern latitudes. A special standard of educational programs is being developed for doctors who will work in the Arctic.

To date, 42 educational organizations of higher education, 12 of higher professional education, and 72 of secondary professional education operate in the Arctic zone of the Russian Federation [9]. At the same time, specialists with higher education for the Arctic are also trained in "non—Arctic" subjects of the Russian Federation—a total of 17 subjects of the Russian Federation (including six subjects of the Russian Arctic) [20]. In this regard, it becomes necessary to concord the network interaction of universities that train specialists for the Arctic. To this end, in 2016, the Ministry of education and science of the Russian Federation (today the Ministry of higher education and science of the Russian Federation) created "National Arctic scientific and educational consortium" Association.

The consortium is a voluntary association of universities, scientific organizations, and enterprises that implement training programs and scientific research of the region for the sustainable development of the Russian Arctic territories. In fact, the National Arctic scientific and educational consortium acts as the national counterpart of UArctic. As of September 2020, the Association includes 33 organizations, including 23 higher education institutions, of which only nine are located directly in the Russian Arctic (Table 4) [30]. Thus, it is clear that the Russian scientific and educational environment has clearly formed an opinion that network interaction is necessary for implementing Arctic projects.

Table 4. Higher educational institutions, the Association members [1].

Universities Located in the Arctic Zone of the Russian Federation	Universities out of the Arctic Zone of the Russian Federation
1. Northern (Arctic) Federal University named after M. V. Lomonosov 2. North-Eastern Federal University named after M. K. Ammosov 3. Murmansk Arctic State University 4. Murmansk State Technical University 5. Petrozavodsk State University 6. Northern State Medical University 7. Syktyvkar State University named after Pitirim Sorokin 8. Ukhta State Technical University 9. Ugra State University	1. National Research Tomsk State University 2. Siberian Federal University 3. Far Eastern Federal University 4. Tyumen State University 5. Ural Federal University named after the first President of Russia B. N. Yeltsin 6. Admiral Makarov State University of sea and river fleet 7. Novosibirsk State Technical University 8. Russian State Hydrometeorological University 9. Ryazan State University named after S. A. Yesenin 10. Saint Petersburg State University 11. Siberian Automobile and Road University 12. Tyumen Industrial University 13. Ufa State Aviation Technical University 14. Ufa State Oil Technical University

[1] Composed by the authors basing on National Arctic scientific and educational consortium site data analyses [30].

It is noteworthy that the Association does not include two leading universities that train specialists in the production of hydrocarbons in the Arctic—Gubkin Russian State University of Oil and Gas (Scientific Research Institute) and Saint Petersburg Mining University. In turn, the expert survey showed that these two universities are among the leaders in training qualified specialists in the oil and gas sector that meet modern requirements for developing the oil and gas industry in the Arctic. Along with them, the Association does not include the third leader of the expert survey—Saint Petersburg Polytechnic University.

Therefore, we can say that these universities somewhat ignore the domestic project of scientific and educational collaboration in the sphere of Arctic projects. This position is probably related to the broad scientific and educational ties of these universities directly with oil and gas companies and enterprises operating in the Arctic, as well as to the traditionally broad and direct ties with the leading scientific, educational, and industrial centers of the country. Saint Petersburg Mining University and Saint Petersburg Polytechnic University are the largest technical universities with a rich history: They have been creating technical, scientific-technical, and scientific-educational personnel for the entire country for decades and even centuries, thus creating entire scientific schools. In addition, these universities, despite many years of experience in training specialists for the oil and gas industry (including in the complicated conditions of the Arctic), relatively recently announced "their Arctic specialization". So, in 2019, the Centre of competence in the field of engineering and technology of field development in the Arctic conditions (CC "Arctic") was established at Saint Petersburg Mining University, which has already received a patent for invention No. 2704451 "Methods for constructing an offshore drilling platform on the shallow shelf of the Arctic seas" [31]. At the beginning of 2020, Saint Petersburg Mining University decided to start training specialists in the specialty "liquefied natural gas", in particular for 'Yamal SPG' project.

In 2019, the Centre for innovative Arctic technologies was established at Saint Petersburg Polytechnic University. At the same time, since 2015, the university has had a scientific and educational center "Gazpromneft-Polytech", which is engaged in scientific and practical developments and training of specialists in the sphere of developing mathematical models of oil and gas production processes by solving real problems of industry, i.e., directly in digital technologies that are so necessary for modern conditions of the oil and gas industry in the Arctic [32]. Thus, a scientific and educational "block" in the sphere of oil and gas projects in the Arctic is potentially cooperated and created in St. Petersburg between these two universities, which can create strong competition in the industry both in Russia and internationally.

As noted above, the expert survey showed that the leading university that traditionally trains specialists for oil and gas projects in the Arctic is I.M. Gubkin Russian State University of oil and gas. The university is a participant in major oil and gas projects in the Arctic, and also trains specialists for Arctic projects together with major oil and gas companies. So, in 2018, the university and 'NOVATEK' launched a master's degree program in the sphere of 'Cryogenic technologies and equipment for the gas industry' to train specialists in the LNG industry. Students enrolled in this program have an internship at the company's enterprises with the possibility of further employment.

In addition, the university trains masters in the sphere of "development and operation of offshore oil and gas fields". The research and education center "Underwater mining complexes" was created together with 'Gazprom' PJSC. To improve the skills, additional educational programs are being implemented, in particular, "Introduction to subsea hydrocarbon production systems" (initiated by 'Gazprom' PJSC).

In addition to Gubkin Russian State University, several universities provide personnel for implementing Arctic oil and gas megaprojects: Lomonosov Northern (Arctic) Federal University, Murmansk Technical State University, Novosibirsk State Technical University, Tyumen Industrial Institute, Siberian State University of Geosystems and Technologies, Marine State University named after Admiral G. I. Nevelsky, Far Eastern Federal University, Ukhta State Technical University, North-Eastern Federal University named after M. K. Ammosov, Tomsk Polytechnic University, Ufa State Oil Technical University, etc.

It is obvious that the title of the center for Arctic research and the main Arctic university, including in oil and gas projects, is claimed by the Northern (Arctic) Federal University, based on which NANOK was formed. The University trains bachelors in the profile "Operation and maintenance of oil and gas facilities of the Arctic shelf (full-time training) and masters in the program "Development of oil and gas fields of the Arctic shelf". Since 2011, the University has an Innovation and Technology Centre for Arctic oil and gas laboratory research. And in March 2019 it was decided to create an international center for integrated research of the shelf and coastal zone of the Arctic seas of Russia. In addition, since 2012, the university has been implementing the unique for Russia educational program "Arctic Floating University", which is an annual interdisciplinary scientific and educational marine expedition on board the "Professor Molchanov" research vessel. The program is focused on studying and monitoring the ecology, climate, and bioresources of the Arctic region, as well as studying the humanitarian side (the history of Arctic development, the Arctic in the system of international relations, and the legal space of the Arctic) [33].

Murmansk State University is a leading university that trains specialists in the sphere of Arctic sea transport and shipping, as well as oil and gas projects in the Arctic. The university annually recruits students in the profile of training "Operation and maintenance of oil and gas facilities of the Arctic shelf", as well as training specialists in the sphere of ice shipbuilding, navigation, construction of port and shipping infrastructure, human-made security [34].

According to experts, Tomsk Polytechnic University trains specialists in the sphere of oil and gas, and is also one of the leaders in comprehensive research of the Arctic shelf. The University has created an international scientific and educational laboratory for the study of carbon in the Arctic seas, which performs research work in the direction of "Siberian Arctic shelf as a source of greenhouse gases of planetary significance: Quantitative assessment of flows and identification of possible environmental and climate consequences" [35]. In addition, it should be noted that Tomsk State University implements the Master's program "Study of Siberia and the Arctic", which is aimed at training specialists in the field of ecology and environmental management in Siberia and the Arctic. Such specialists would know modern methods of environmental research and methodology for conducting scientific research to solve applied problems. In this regard, we can say that Tomsk is forming one of Russia's leading scientific and educational centers for training personnel for Arctic projects [36].

Based on a comparative analysis of the number of educational programs for training Arctic personnel, the following world leaders in training specialists for oil and gas projects can be identified: UiT The Arctic University of Norway, University of Copenhagen, University of Lapland, The Norwegian Institute of Bioeconomy Research (research Institute), University of Iceland, Luleå University of Technology. At the same time, we can note a significant increase in the number of programs in the Arctic direction, for example, "Arctic engineering", in all leading universities in Denmark, Iceland, Canada, Norway, the USA, Finland, and Sweden. At the same time, university programs are interdisciplinary in nature, for example: McGill University implements the "Arctic Field Studies Semester" training program [37], the University of Northern British Columbia implements the "Northern Studies" program [38], the University of Alaska-Fairbanks—"Arctic and Northern Studies" [39]; since 2019, Colorado School of Mines has opened several Arctic-themed programs.

It can be noted that the training system in the named countries, as well as in Russia, pays less attention to the Humanities, focusing more on the training of engineers, geologists, biologists, logisticians. At the same time, major Arctic universities, such as the University of Northern British Columbia, enroll students for arts, social sciences, and health sciences [40].

In addition, since years 2013–2014, training programs in near-Arctic countries have been actively developed. In particular, the UK began the first steps in training personnel for the Arctic in 2013 as part of the first British Arctic policy framework, "Adapting to Change" [41]. The UK has also established the UK government's Science and Innovation Network (SIN), which includes representatives of the Arctic powers and serves as a necessary tool for research collaborations. SIN's activities are aimed at providing training and student exchanges, as well as creating and building scientific potential [42]. Since 2018, the amount of financial support for the Scientific network from the UK government has increased significantly, which indicates that the country is interested in Arctic research.

A comparative analysis of leading Russian universities (according to experts) and world leaders in training personnel for the oil and gas industry according to the QS rating showed that Russian universities are significantly behind Western leaders. Moreover, the failure of specialized universities for the oil and gas industry is revealed (including Gubkin Russian State University of oil and gas and Saint Petersburg Mining University) both in the world ranking and in subject ratings. The situation is somewhat better with three universities (Peter the Great Saint Petersburg Polytechnic University, Tomsk Polytechnic University, "MISiS" National Research Technological University), but their achievements are lost against the background of obvious leaders—the University of Northern British Columbia, McGill University and the University of Copenhagen. The exceptions are several European universities that are not included in any rating at all: The University of Lapland and the University of Iceland. Table 5. [43] There is low competitiveness of Russian universities at the world level, and the nature of their training is catching up.

Table 5. Comparative analysis of Russian and world leaders in training specialists for oil and gas projects according to the QS University Rankings (QS UR) for 2020 [1,2].

Universities	QS WUR	EECA UR	QS WUR by Subject					
			Engineering— Mineral and Mining	Engineering— Petroleum	Engineering and Technology	Computer Science and Information Systems	Earth and Marine Science	Environmental Studies
Russian universities (in the order specified by experts)								
Gubkin Russian State University of Oil and Gas	-	182	-	-	-	-	-	-
St. Petersburg Mining University	-	173	18	-	-	-	-	-
Peter the Great Saint Petersburg Polytechnic University	401	47	-	-	191	301–350	-	-
Northern (Arctic) State University named after M. V. Lomonosov	-	231–240	-	-	-	-	-	-
Murmansk State Technical University	-	-	-	-	-	-	-	-
Ufa State Oil Technical University	-	301–350	-	-	-	-	-	-
Ukhta State Technical University	-	-	-	-	-	-	-	-
North-Eastern Federal University named after M. K. Ammosov	-	231–240	-	-	-	-	-	-
National Research Technology University 'MISiS'	428	45	46	-	247	-	-	-
Tomsk Polytechnical University	401	30	-	26	282	501–550	-	-
Saint Petersburg State University of Economics	-	241–250	-	-	-	-	-	-
Universities—world leaders (in order of appearance in the paper text)								
UiT The Arctic University of Norway	416	-	-	-	-	-	-	301–350
University of Copenhagen	76	-	-	51–57	-	101–150	51–100	42
University of Lapland	-	-	-	-	-	-	-	-
University of Iceland	-	-	-	-	-	-	-	-

Table 5. *Cont.*

Universities	QS WUR	EECA UR	QS WUR by Subject					
			Engineering—Mineral and Mining	Engineering—Petroleum	Engineering and Technology	Computer Science and Information Systems	Earth and Marine Science	Environmental Studies
Universities—world leaders (in order of appearance in the paper text)								
Luleå University of Technology	-	-	-	-	-	-		301–350
McGill University	31	-	6	-	47	51–100	27	29
University of Northern British Columbia	45	-	9	-	-	-	-	-
University of Alaska-Fairbanks	-	-	-	-	-	-	151–200	351–400
Colorado School of Mines	-	-	1	18	227		36	201–250

[1] The authors consider the subject areas (QS), which development is in demand in the implementation of oil and gas projects in the Arctic. [2] Composed by the authors basing on the QS University Rankings site analyses [43].

4.2. Life Long Learning

One of the ways to ensure "sustainable education" is to implement the concept of 'Life Long Learning' [29]. Speaking about the training of specialists for the oil and gas industry in Russia, first of all, we can talk about advanced training and retraining programs implemented in specialized universities, which the authors wrote about above.

To train highly qualified specialists, major Russian oil and gas companies implement continuing education programs in the "school-university-enterprise" system, which form an external personnel reserve. An example is the corporate program of 'Rosneft' Corporation. In 26 regions of the country, 117 'Rosneft' classes have been created (for students in grades 10 and 11) in 62 schools. In the process of training personnel with higher education, the company cooperates based on long-term agreements with 60 Russian and foreign universities. The company also works with students of partner universities, organizes the recruitment and selection of the best graduates to work at 'Rosneft' enterprises [44].

As an example of implementing this 'Life Long Learning' concept in the oil and gas industry through cooperation between the University and an oil company, we note the professional retraining program "Procurement and logistics of offshore projects in the oil and gas industry" of the Murmansk State Technical University and 'Gazprom Neft'. This program is practice-oriented for training specialists in the sphere of economics and management for the oil and gas industry: It is aimed at studying the specifics of working on offshore and Arctic fields, as well as building an effective supply chain for facilities located there and managing it [45]. A training platform is being organized on the basis of 'Gazprom Neft' PJSC, as well as practical training in the 'Gazprom Neft' group of companies [46].

Multinational oil and gas corporations are engaged in the process of training their own employees in new technologies and approaches in the industry. So, on the territory of the Arctic zone of the Russian Federation, there are training and multifunctional centers of large corporations, such as 'Gazprom', 'Rosneft', 'LUKOIL' and others (about 56 centers) [8]. In addition, these companies are implementing additional educational programs for university graduates to attract talented young professionals. For example, the practice-oriented additional educational program "Gazprom Neft" "At the start!", in which university graduates of 2019 or 2020 have the opportunity to implement their own projects in the development of "smart" energy, complete an internship at 'Gazprom Neft' enterprises, and then receive an invitation to a permanent job in the company [47].

4.3. Digitalization

In the last decade, the most important factor in changing the training process is the intensive digitalization of all areas of socio-economic development without exception.

Until the second half of the 2010s, the Russian oil and gas sector considerably lagged behind in the sphere of digitalization compared to other industries, as well as from world leaders—the USA, Norway, France, etc. [48] A qualitative leap occurred with the development of difficult hydrocarbons in the Arctic which demands a change in methods of exploration and production resources—using digital technology. So, in 2015, the RAS Institute of oil and gas problems and I. M. Gubkin Russian State University of Oil and Gas initiated the development of a program for digitalization and intellectualization of the Russian oil and gas industry, which was reflected in the departmental project "Digital energy" as part of implementing the national program "Digital economy of the Russian Federation", adopted in 2017 [49].

Digitalization is designed to reduce the costs associated with hydrocarbon exploration and production in extremely difficult regional conditions (climatic, geological, demographic, and technical), as well as with the development and production of special equipment [50]. It helps to reduce the environmental consequences of human presence in the region—in the future, to minimize as much as possible, and ideally to reduce its presence to zero, thereby ensuring technological and environmental safety by reducing the probability of deviations and transferring competencies to the level of robotic systems [51]. It is worth noting that the extreme conditions of the Arctic are an objective factor in accelerating the introduction of remote-control technologies and intelligent automation of production.

The use of digital technologies in the fuel and energy complex, which is the basis for the economic development of the Arctic region, provides intelligent automation of processes at facilities. Major oil and gas companies already use BigData (including BigGeoData), IoT, industrial Internet, blockchain, and artificial intelligence technologies [49,50] to solve applied problems and plan to expand this practice in the Arctic. Thus, 'Gazprom Neft' is ahead of 'Rosneft' in terms of innovation rates in its oil and gas programs, which indicates that 'Gazprom Neft' is more attentive to the digital transformation of implementing the large-scale RF governmental program "Digital economy of the Russian Federation": For example, the programs "Project management center", "Cognitive geologist", "Digital drilling" [52].

The use of digital technologies in the fuel and energy complex, in particular, the development of "digital deposits", provides intelligent automation of processes at facilities. According to experts, "digitalization of wells", "digital drilling", and modeling of technological processes [53,54] can reduce the cost of field operation by approximately 15–20% [51,55]. In the context of lower energy prices for Arctic hydrocarbon deposits, this factor plays a special role.

It is worth noting that digital technologies in the energy sector are used not only within the upstream sector. Static and dynamic analysis of processes allows you to adjust and reorganize related business processes, and make management decisions quickly.

These trends occur against the background of a perceived shortage of professional personnel, which causes the need to algorithmize the competencies of professional knowledge and skills and retrain specialists in new digital specialties [50]. In this regard, a special role in the future 5–10 years will be played by the skills and abilities to remotely control the technological processes, working with databases, and ensuring their security.

As you know, these technologies are end-to-end, which, on the one hand, opens up additional opportunities, and on the other, creates new risks and threats. In this regard, an important condition for effective training of a specialist is inevitably associated with obtaining a basic knowledge of Internet technologies and the complex nature of modern threats. All leading universities that train personnel for the Arctic have conducted training in the digital technologies, and often have separate divisions specializing in this area. For example, the aforementioned scientific-educational center "Gazpromneft-Polytech" (St.-Petersburg Polytechnic University), Teaching and research center of digital technologies (Saint-Petersburg Mining University), Department of integrated security TEK Russian State University of oil and gas (national research institute) named after I. M. Gubkin (which is the basis of the first in Russia laboratory study of detection of computer attacks with the example of virtual enterprises of oil and gas complex), and so on. [56,57].

The high knowledge intensity of Arctic projects underscores the need for in-depth monitoring of the technology market and the development of promising technical solutions and consideration when drawing up professional standards.

4.4. Internationalization of Arctic Education

Although the observed increase in international competition in the development and implementation of technological innovations in the Arctic, international cooperation in the sphere of Arctic education continues to develop steadily.

International exchange of training experience, the convergence of standards, and formation of international educational standards allows us to bridge the gap between educational processes, innovative solutions, and know-how applied in different parts of the Arctic.

A striking example of the internationalization of Arctic education is the international network project UArctic, created in 2001 on the initiative of the Arctic Council to create a unified scientific and educational network of organizations working in the spheres of higher education and research in the Arctic region (organizations located in the Arctic and implementing scientific and educational projects for the Arctic). If at the time of the project creation it included no more than 30 participants, then in 2020, it already had 153 participants from 11 countries. It is noteworthy that the majority of participants are from Russia, 27%—41 participants [58] (Figure 1). However, neither Gubkin Russian State University of Oil and Gas, nor Saint Petersburg Mining University, nor Saint Petersburg Polytechnic University are members of UArctic.

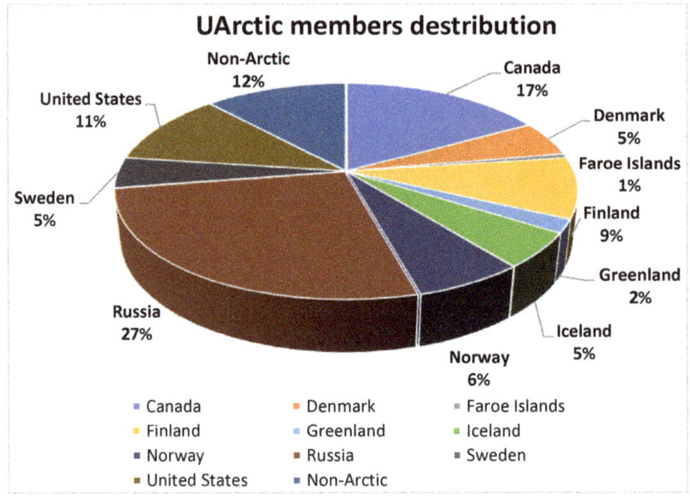

Figure 1. UArctic members distribution [58].

In turn, the analysis of the number of proposed scientific and educational programs showed that, despite their large number, Russian project participants are not inferior to their western counterparts in the number of implemented programs for training specialists.

Thus, within the framework of the project, out of 18 bachelor's degree programs (43 courses in total), only two programs and six courses are implemented by Russian organizations. While Norway, which has only 6% of the total number of participants in the UArctic Program, implements eleven undergraduate programs and six courses. Out of 81 master's degree programs, only eight are implemented by participants from Russia. PhD programs are not presented by participants Russian at all, but participants Russian offer seven PhD courses (Table 6) [58].

Table 6. The number of programs and courses implemented under the UArctic project [1].

Country	Number of Bachelor's Programs/ Courses	%	Country	Number of Master's Programs/ Courses	%	Country	Number of PhD Programs/ Courses	%
Canada	0/1	-/2.3%	Canada	1/1	1%/1%	Canada	0/1	-/2.9%
Denmark	0		Denmark	17/3	21%/3%	Denmark	0	
Faroe Islands	0		Faroe Islands	0	0	Faroe Islands	0	
Finland	3/7	17%/16%	Finland	17/5	21%/6%	Finland	1/4	50%/12%
Greenland	0		Greenland	0		Greenland	0	
Iceland	0/6	-/13.9%	Iceland	4/37	5%/43%	Iceland	0/3	-/8.8%
Norway	11/6	61%/14%	Norway	27/3	33%/3%	Norway	0/3	-/8.8%
Russia	2/6	11%/14%	Russia	8/6	10%/7%	Russia	0/7	-/21%
Sweden	0/7	-/16%	Sweden	1/20	1%/23%	Sweden	0/1	-/2.9%
United States	2/7	11%/16%	United States	5/8	6%/9%	United States	1/8	50%/24%
Non-Arctic	0/2	-/4.6%	Non-Arctic	1/2	1%/2%	Non-Arctic	0/6	-/18%

[1] Composed by the authors basing on the UArctic site data analyses [58].

Programs and courses are implemented in various forms—international summer and winter schools, training programs. In total, 24 different training programs presented by Russian universities are being implemented. Attention is drawn to the fact that the Summer school of the Saint Petersburg Mining University is marked among the programs, while the university is not an official member of UArctic (Table 7).

Table 7. Training programs implemented by Russian universities within the framework of UArctic [1].

No.	University	Educational Program
1	Far Eastern Federal University	School of Engineering "Ice Mechanics" Annual International Winter Course
2	Northern (Arctic) Federal University	• Arctic Floating University • Peoples and Culture of the Circumpolar World • International PhD School "Russia in the Arctic Dialogue: Local and Global Context" • Arctic Winter School 2020 • Arctic Summer School 2020 • Summer School in Arctic Law • Diverse Arctic: Local Challenges—Global Changes • Russian Studies Programme • Environmental Risks Management in the Arctic (ERMA) • European Studies: Arctic Focus • History and Culture of the Circumpolar World: comparative research • EU and Russia in the Arctic: History of Cultural and Political Interaction • Arctic Law
3	Saint-Petersburg Mining University	Summer schools at Saint-Petersburg Mining University
4	Arctic State Agrotechnological University	• Apiculture and bee-biology • Natural horse husbandry in Arctic and northern regions of Yakutia • Northern reindeer husbandry • Business planning and project management • Pathology of productive animals • Arctic food security: Traditional nutrition and Arctic food
5	Siberian Federal University	Biological Engineering
6	St. Petersburg University	Geology
7	Industrial University of Tyumen	Logistics and Supply Chain Management

[1] Composed by the authors basing on the UArctic site data analyses [58].

It is noteworthy that among all the training programs and courses offered by UArctic, the summer school of Saint Petersburg State University is the only program focused on developing the oil and gas complex in the Arctic: With a focus on technical and earth sciences. However, the project does not include Gubkin University Summer School "Development of offshore fields", which is important for training personnel for the Arctic. In turn, the vast majority of programs implemented by UArctic relate to the field of interdisciplinary research that combines social and earth sciences and covers such topics as land, environment, peoples, cultures, and politics in the Arctic and subarctic states—Circumpolar Studies, as well as economic problems of the region. (An exception is 'FEFU School of Engineering, 'Ice Mechanics' Annual International Winter courses, and the 'Geology' program implemented by Saint Petersburg University).

Thus, the analysis of UArctic training programs and courses showed that despite a significant number of participating Russian universities (including universities that train personnel for the oil and gas industry), in the vast majority of cases, their role is limited to providing students with courses, rather than implementing their own training programs. This seems to indicate that Arctic education in Russia is catching up and that the education system is lagging behind the general trends in training Arctic personnel: Russian universities are learning, not teaching. There is an interdisciplinary approach to UArctic programs with an emphasis on earth and environmental sciences, socio-humanitarian and economic courses.

The obvious flagship of this educational movement is the Northern (Arctic) University named after M. V. Lomonosov, which implements the majority of the training program and is the base of the UArctic research office in Russia.

5. Conclusions

The changes in the system of training specialists described in this work are aimed at minimizing the existing imbalances in the labor market in the Arctic zone of the Russian Federation.

The focus on high-tech technologies in the Arctic leads to an increase in the importance of human capital as a factor in the efficiency of economic development in the Arctic. It is obvious that as economic activity increases, the personnel deficit will increase. In this regard, based on existing forecasts, the training system should adapt to new challenges and adjust educational programs and directions. The future of the Arctic is in the hands of highly qualified specialists who have the skills to remotely control complex technological processes in oil and gas fields and marine transport.

In the twenty-first century, young specialists in the oil and gas industry who are being trained to work in the Arctic should acquire fluency in information technology and competence in working with complex robotics. Artificial Intelligence (AI) in the harsh natural conditions of the Arctic is already helping people, and in the near future, may even replace them.

At the same time, we can conclude that education is an important factor in preserving the unity of the social space of the Arctic. The development of international cooperation in the field of personnel training, in addition to practical educational tasks, makes it possible to strengthen cross-cultural ties that create the basis for sustainable development of the region.

As the results of the study showed, training personnel for work in the Arctic has a special specificity and often falls out of the general principles of ensuring the educational process. Global trends in training "Arctic personnel" show the need to develop an interdisciplinary approach, basic knowledge of natural sciences, to study the socio-cultural specifics of the region, to develop new educational standards, to implement the concept of 'Life Long Learning', to widely introduce digital technologies and internationalize education.

The analysis of Russian educational programs for training specialists allowed us to conclude that, in general, the domestic system of training is adapting to changing conditions, in particular, certain progress has been made in the formation of "digital" competencies and skills in the conditions of a developed IT infrastructure.

The introduction of "digital fields" has led to an increase in the demand for IT specialists in the Arctic oil and gas sector. With the help of an expert survey, it was revealed that in the future, the most popular professions, along with "drillers" and specialists in transport, will be IT specialists who ensure the functioning of "digital fields".

Leading universities that train specialists for modern oil and gas projects in the Arctic are Gubkin Russian State University of Oil and Gas, Saint Petersburg Mining University, Peter the Great Saint Petersburg Polytechnic University, Lomonosov Northern (Arctic) Federal University, and Murmansk State Technical University.

At the same time, it should be noted that not all leading Russian universities are included in international educational projects and organizations, and there is skepticism about the internationalization of education. In addition, there is a catch-up nature of Arctic education in Russia and the lag of the education system from the general trends in the training of Arctic personnel, and the low competitiveness of Russian universities at the world level. Probably, the processes of cooperation, primarily in the field of oil and gas projects, are constrained by the continuing tense relations between the largest states of the Arctic region.

Author Contributions: Conceptualization, E.S. and D.M.; methodology, D.M.; validation, S.G. and A.N.; formal analysis, E.S.; resources, R.-E.K.; writing—original draft preparation, E.S. and D.M.; writing—review and editing, S.G. and A.N.; visualization, R.-E.K.; project administration, E.S. All authors have read and agreed to the published version of the manuscript.

Funding: This research received no external funding.

Conflicts of Interest: The authors declare no conflict of interest.

References

1. Savard, C.; Nikulina, A.Y.; Mécemmène, C.; Mokhova, E. The Electrification of Ships Using the Northern Sea Route: An Approach. *J. Open Innov. Technol. Mark. Complex.* **2020**, *6*, 13. [CrossRef]

2. Litvinenko, V. The Role of Hydrocarbons in the Global Energy Agenda: The Focus on Liquefied Natural Gas. *Resources* **2020**, *9*, 59. [CrossRef]

3. Rosen, M.; Thuringer, C. *Unconstrained Foreign Direct Investment: An Emerging Challenge to Arctic Security*; CNA Corp.: Washington, DC, USA, 2017.

4. Zagorskii, A.V. China accepts conditions in the Arctic. *Mirovaya Ekon. I Mezhdunarodnye Otnos.* **2019**, *7*, 76–83. (In Russian)

5. Väätänen, V.; Zimmerbauer, K. Territory–network interplay in the co-constitution of the Arctic and 'to-be' Arctic states. *Territ. Politi- Gov.* **2019**, *8*, 372–389. [CrossRef]

6. Sokolova, I.; Smorchkova, V.; Yusov, A. Human resources of the Russian Arctic regions. *Public Adm.* **2017**, *19*, 86–93. (In Russian) [CrossRef]

7. Fadeev, A.M.; Larichkin, F.D.; Cherepovitsyn, A.E. Features of Training of Personnel for the Arctic Marine Hydrocarbon Fields Development. *J. Min. Inst.* **2011**, *194*, 332–338.

8. Fadeev, A.M.; Cherepovitsyn, A.E.; Larichkin, F.D.; Tsvetkova, A.Y. Personnel Support for the Implementation of Offshore Projects in the Arctic as an Effective Tool for Strategic Management of the Oil and Gas Complex. The North and the market: Shaping the economic order **2018**, *2*, 16–25. Available online: http://www.iep.kolasc.net.ru/journal/files/2018-2.pdf (accessed on 30 August 2020). (In Russian).

9. Fadeev, A.M.; Cherepovitsyn, A.E.; Larichkin, F.D. *Strategic Management of the Oil And Gas Complex in the Arctic*; KSC RAS: Apatity, Russia, 2019; pp. 217–225.

10. Keskitalo, P.; Määttä, K.; Uusiautti, S. Sámi education in Finland. *Early Child Dev. Care* **2012**, *182*, 329–343. [CrossRef]

11. Uusiautti, S.; Happo, I.; Määttä, K. Challenges and Strengths of Early Childhood Education in Sparsely Populated Small Provinces the Case of Lapland, Finland. *Br. J. Educ. Soc. Behav. Sci.* **2014**, *4*, 562–572. [CrossRef]

12. Määttä, K.; Uusiautti, S. Arctic Education in the Future. In *Human Migration in the Arctic*; Uusiautti, S., Yeasmin, N., Eds.; Palgrave Macmillan: Singapore, 2019; pp. 213–238.

13. Vasilyeva, O.E.; Dmitrieva, A.A.; Gdalin, D.A.; Ilyinsky, S.V.; Popov, M.I. High education and human resources in the Russian Arctic region: Problems and particular qualities of development. *IOP Conf. Series Earth Environ. Sci.* **2020**, *539*. [CrossRef]

14. Veretennikov, N.P.; Agarkov, S.A.; Kozmenko, S.Y.; Ulchenko, M.V. Features of Marine Education in the Arctic Region. In Proceedings of the 2018 XVII Russian Scientific and Practical Conference on Planning and Teaching Engineering Staff for the Industrial and Economic Complex of the Region (PTES), St. Petersburg, Russia, 14–15 November 2018; Saint Petersburg Electrotechnica University: Saint Petersburg, Russia, 2018; pp. 175–177.

15. Kudryavtseva, R.-E.A.; Guseva, N.A.; Koroleva, J. Arctic education in the field of polar research. *IOP Conf. Series Earth Environ. Sci.* **2020**, *539*. [CrossRef]

16. Nikolaeva, A.D.; Neustroev, N.D.; Neustroeva, A.N.; Bugaeva, A.P.; Shergina, T.A.; Kozhurova, A.A. Regional model of indigenous education: The case of the Sakha republic (Yakutia). *Int. Trans. J. Eng. Manag. Appl. Sci. Technol.* **2019**, *10*, 10A19M.

17. Pitukhina, M.; Tolstoguzov, O.; Privara, A.; Volokh, V. Transarctic cooperation potential evaluation of northern universities: Research performance of arctic universities' education indicators. In Proceedings of the 12th International Technology, Education and Development Conference, Valencia, Spain, 5–7 March 2018; IATED: Valencia, Spain, 2018; pp. 4103–4113.

18. Gaski, M.; Halvorsen, P.A.; Aaraas, I.J.; Aasland, O.G. Does the University of Tromso- The Arctic University of Norway educate doctors to work in rural communities? *Tidsskr. Den Nor. Legeforening* **2017**, *137*, 1026–1031. [CrossRef]

19. Arruda, G.M. *Sustainable Energy Education in the Arctic: The Role of Higher Education*; Routledge: London, UK, 2020; p. 278.

20. Kudryashova, E.V.; Nenasheva, M.V.; Saburov, A.A. Network Interaction of Universities in the Context of the Russian Arctic Development. *Vysshee Obraz. v Ross.* **2020**, *29*, 105–113. (In Russian) [CrossRef]

21. Djamilia, S.; Kuzaeva, A.; Glushkova, A. Formation of a database of indicators and analysis of the environmental and socio-economic vital activity spheres of Russian Federation Arctic zone municipalities. *Int. J. Syst. Assur. Eng. Manag.* **2019**, *11*, 1–25. [CrossRef]

22. Skripnuk, D.F.; Kikkas, K.N.; Safonova, A.S.; Volodarskaya, E.B. Comparison of international transport corridors in the Arctic based on the autoregressive distributed lag model. *IOP Conf. Series Earth Environ. Sci.* **2019**, *302*. [CrossRef]

23. Kruk, M.; Semenov, A.; Cherepovitsyn, A.; Nikulina, A. Environmental and economic damage from the development of oil and gas fields in the Arctic shelf of the Russian Federation. *Eur. Res. Stud. J.* **2018**, *2*, 423–433. (In Russian) [CrossRef]

24. Didenko, N.I.; Skripnuk, D.F.; Mirolyubova, O.V. Urbanization and Greenhouse Gas Emissions from Industry. *IOP Conf. Series Earth Environ. Sci.* **2017**, *72*, 012014. [CrossRef]

25. Sharok, V.; Iakovleva, I.; Vakhnin, N. Social and psychological aspects of individual adaptation in Arctic conditions. *IOP Conf. Series Earth Environ. Sci.* **2019**, *302*, 1. [CrossRef]

26. Tatianina, L.; Vakhnina, E. Personal determinants of trust in Arctic ships' crews. *IOP Conf. Ser. Earth Environ. Sci.* **2020**, *554*, 012007. [CrossRef]

27. Berge, S.T. Pedagogical Pathways for Indigenous Business Education: Learning from Current Indigenous Business Practices. *Int. Indig. Policy J.* **2020**, *11*, 1–20. [CrossRef]

28. Vizina, Y.; Williamson, K.J. *Indigenous Peoples and Education in The Arctic Region*; United Nations Department of Economic and Social Affairs: New York, NY, USA, 2017; pp. 41–64.

29. Määttä, K.; Hyvärinen, S.; Äärelä, T.; Uusiautti, S. Five Basic Cornerstones of Sustainability Education in the Arctic. *Sustainability* **2020**, *12*, 1431. [CrossRef]

30. Arctic National Scientific-Educational Consortium. Available online: http://arctic-union.ru (accessed on 30 August 2020). (In Russian).

31. Center of Competence in the Sphere of Engineering and Technology of Field Development in the Arctic. Available online: https://spmi.ru/rezultaty-nauchnoy-deyatelnosti (accessed on 30 August 2020). (In Russian).

32. 'Gazpromneft-Polytech' Research and Education Center. Available online: https://gpn.spbstu.ru/projects/ (accessed on 30 August 2020). (In Russian).

33. Arctic Floating University. Available online: https://narfu.ru/science/expeditions/floating_university/2020/ (accessed on 30 August 2020). (In Russian).

34. Areas of MSTU Training. Available online: http://abit.mstu.edu.ru/ (accessed on 30 August 2020). (In Russian).

35. Comprehensive Research of the Arctic Shelf. Available online: https://tpu.ru/research/fields/arctic (accessed on 30 August 2020). (In Russian).

36. Master's Program "Study of Siberia and the Arctic". Available online: http://tssw.ru/page/magisterskaya-programma-izuchenie-sibiri-i-arktiki/ (accessed on 5 November 2020). (In Russian).

37. Arctic Field Study Semester. Available online: https://www.mcgill.ca/arctic/academic (accessed on 13 September 2020).

38. Northern Studies. Available online: https://www.unbc.ca/northern-studies (accessed on 13 September 2020).

39. Arctic and Northern Studies Program. Available online: https://www.uaf.edu/arctic/ (accessed on 13 September 2020).

40. Programs and Courses of University of Northern British Columbia. Available online: https://www.unbc.ca/programs (accessed on 13 September 2020).

41. Adapting To Change UK Policy towards the Arctic. Available online: https://assets.publishing.service.gov.uk/government/uploads/system/uploads/attachment_data/file/251216/Adapting_To_Change_UK_policy_towards_the_Arctic.pdf (accessed on 13 September 2020).

42. Science and Innovation Network (SIN). Available online: https://www.gov.uk/world/organisations/uk-science-and-innovation-network (accessed on 13 September 2020).

43. QS Top. Universities. Available online: https://www.topuniversities.com/ (accessed on 13 September 2020).

44. Youth Policy of "Rosneft" Company. Available online: https://www.rosneft.ru/Development/personnel/young_specialists/ (accessed on 30 August 2020). (In Russian).

45. Kudryashova, E.V.; Sorokin, S.E.; Bugaenko, O.D. University-Industry Interaction as an Element of the University's «Third Mission». *Vysshee Obraz. v Ross.* **2020**, *29*, 9–21. (In Russian) [CrossRef]

46. Professional Retraining Program "Procurement and Logistics of Offshore Projects in the Oil and Gas Industry". Available online: http://www.mstu.edu.ru/press/ads/16-07-2019/zakupkiilogistikashelf.shtml (accessed on 30 August 2020). (In Russian).

47. Additional Educational Program "At the Start!". Available online: https://gpn-nastart.ru/ (accessed on 30 August 2020). (In Russian).

48. Makhovikov, A.B.; Katuntsov, E.V.; Kosarev, O.V.; Tsvetkov, P.S. Digital transformation in oil and gas extraction. In *Innovation-Based Development of the Mineral Resources Sector: Challenges and Prospects*; CRC Press: Boca Raton, FL, USA, 2018; pp. 531–538.

49. Dmitrievsky, A.N.; Eremin, N.A. Big geodata in the digital oil and gas ecosystem. *Energy Strategy* **2018**, *2*, 31–39.

50. Abukova, L.A.; Borisenko, N.Y.; Martynov, V.G.; Dmitrievsky, A.N.; Eremin, N.A. Digital modernization of the gas complex: Scientific research and personnel support. *Sci. J. Russ. Gas Soc.* **2017**, *4*, 3–12.

51. Kosarev, O.V.; Tcvetkov, P.S.; Makhovikov, A.B.; Vodkaylo, E.G.; Zulin, V.A.; Bykasov, D.A. Modeling of industrial iot complex for underground space scanning on the base of arduino platform. In *Topical Issues of Rational Use of Natural Resources 2018*; CRC Press: Boca Raton, FL, USA, 2019; pp. 407–412. (In Russian)

52. Gafurov, A.R.; Skotarenko, O.V.; Nikitin, Y.A.; Plotnikov, V.A. Digital transformation prospects for the offshore project supply chain in the Russian Arctic. *IOP Conf. Series Earth Environ. Sci.* **2020**, *554*, 012009. [CrossRef]

53. Litvinenko, B.C.; Dvoinikov, M. Methodology for determining the parameters of drilling mode for directional straight sections of well using screw downhole motors. *J. Min. Inst.* **2020**, *241*, 105. [CrossRef]

54. Chudinova, I.; Nikolaev, N.; Petrov, A. Design of domestic compositions of drilling fluids for drilling wells in shales. In *Youth Technical Sessions Proceedings*; CRC Press: Boca Raton, FL, USA, 2019; Volume 1, pp. 371–375.

55. "Gazprom Neft" on the Path of Comprehensive Digital Transformation. Available online: https://news.myseldon.com/ru/news/index/203832229 (accessed on 30 August 2020). (In Russian).

56. Tkacheva, V.L.; Grinyaev, S.N.; Pravikov, D.I.; Fatyanov, A.A.; Shushkevich, Y.A. *Fuel and Energy Complex in the Era of Digital Economy Formation: Application of Digital Economy Technologies in Power Engineering*; Gubkin Russian State University of Oil and Gas (SRI): Moscow, Russia, 2019; p. 234. (In Russian)

57. Gubkin University Has Created A Training Laboratory for Studying The Means of Detecting Computer Attacks. Available online: https://gubkin.ru/news/detail.php?ID=39680&sphrase_id=9160622 (accessed on 30 August 2020). (In Russian).

58. Study in Russia. UArctic. Available online: https://education.uarctic.org/universities/russia/ (accessed on 30 August 2020).

Publisher's Note: MDPI stays neutral with regard to jurisdictional claims in published maps and institutional affiliations.

Article

Scenario Modeling of Sustainable Development of Energy Supply in the Arctic

Yuriy Zhukovskiy [1], Pavel Tsvetkov [2,*], Aleksandra Buldysko [3], Yana Malkova [3], Antonina Stoianova [1] and Anastasia Koshenkova [4]

[1] Educational Research Center for Digital Technologies, Saint Petersburg Mining University, 2 21st Line, 199106 Saint Petersburg, Russia; spmi_energo@mail.ru (Y.Z.); Stoyanova_AD@pers.spmi.ru (A.S.)

[2] Department of Economics, Organization and Management, Saint Petersburg Mining University, 2 21st Line, 199106 Saint Petersburg, Russia

[3] Department of Electrical Engineering, Saint Petersburg Mining University, 2 21st Line, 199106 Saint Petersburg, Russia; Buldysko_AD@pers.spmi.ru (A.B.); Malkova_YaM@pers.spmi.ru (Y.M.)

[4] Department of Environmental Geology, Saint Petersburg Mining University, 2 21st Line, 199106 Saint Petersburg, Russia; s181735@pers.spmi.ru

[*] Correspondence: pscvetkov@yandex.ru

Citation: Zhukovskiy, Y.; Tsvetkov, P.; Buldysko, A.; Malkova, Y.; Stoianova, A.; Koshenkova, A. Scenario Modeling of Sustainable Development of Energy Supply in the Arctic. *Resources* **2021**, *10*, 124. https://doi.org/10.3390/resources10120124

Academic Editor: Elena Rada

Received: 28 October 2021
Accepted: 2 December 2021
Published: 7 December 2021

Publisher's Note: MDPI stays neutral with regard to jurisdictional claims in published maps and institutional affiliations.

Abstract: The 21st century is characterized not only by large-scale transformations but also by the speed with which they occur. Transformations—political, economic, social, technological, environmental, and legal-in synergy have always been a catalyst for reactions in society. The field of energy supply, like many others, is extremely susceptible to the external influence of such factors. To a large extent, this applies to remote (especially from the position of energy supply) regions. The authors outline an approach to justifying the development of the Arctic energy infrastructure through an analysis of the demand for the amount of energy consumed and energy sources, taking into account global trends. The methodology is based on scenario modeling of technological demand. It is based on a study of the specific needs of consumers, available technologies, and identified risks. The paper proposes development scenarios and presents a model that takes them into account. Modeling results show that in all scenarios, up to 50% of the energy balance in 2035 will take gas, but the role of carbon-free energy sources will increase. The mathematical model allowed forecasting the demand for energy types by certain types of consumers, which makes it possible to determine the vector of development and stimulation of certain types of resources for energy production in the Arctic. The model enables considering not only the growth but also the decline in demand for certain types of consumers under different scenarios. In addition, authors' forecasts, through further modernization of the energy sector in the Arctic region, can contribute to the creation of prerequisites that will be stimulating and profitable for the growth of investment in sustainable energy sources to supply consumers. The scientific significance of the work lies in the application of a consistent hybrid modeling approach to forecasting demand for energy resources in the Arctic region. The results of the study are useful in drafting a scenario of regional development, taking into account the Sustainable Development Goals, as well as identifying areas of technology and energy infrastructure stimulation.

Keywords: SDG-goals; Arctic; energy supply; scenario modeling; technological demand; energy scenarios; sustainable energy; hydrogen; renewable energy sources; sustainability

1. Introduction

A combination of external factors (political, economic, social, technological, environmental, and legal) has led to the emergence and consequently the aggravation of global challenges [1–4]. Among them are over-population and urbanization, decentralization of the world, environmental and climatic changes, increasing consumption of energy and resources. Delay in addressing these global challenges undermines the sustainability and security of the development of all mankind. Both for Russia and the whole world, the Arctic can become an answer to these challenges [5,6].

Quite a lot of research is directed towards the Arctic, this region and its features have always attracted the scientific community, much attention is paid to the social component. Recently, the region has attracted even more attention due to climate change and new opportunities, both predictable and unpredictable [7].

The Arctic zone has potential as a resource base of hydrocarbon, which is about 13 billion tons of oil and 86 trillion m^3 of gas. Their development is both a driver of economic sector progress and a lifeline for the implementation of the energy transition [7]. The value of the region is supported by the status of the guarantor of national security of the country and the cultural significance of the indigenous peoples of the North [6]. Mistakes made in the development of the potential of the Arctic can turn into both an ecological catastrophe and an economic foul—the cost of wrong decisions is several times higher [8,9]. Therefore, our study develops and complements the forecasting made earlier in order to reduce the risks of making decisions on development and focus on the resources, levers, and incentives for the necessary development.

According to the analysis of scientific works in the field of forecasting quality demand due to the growth of specific consumers, it was revealed that research is aimed at obtaining energy in certain conditions and in the short term. Many researchers resort to statistical and machine learning methods. In [10], the mixed integer programming method is used, which allows determining the state and power level of all generators to maximize the profit of the gas company. In [11], a support vector machine-based simulation is applied to predict solar and wind energy resources. The study [12] proposes a typical-load-profile-supported convolutional neural network for predicting plant-level electrical load. In [13], the forecast for natural gas consumption is based on a decomposition method by combining three different components: a trend-driven time series, a seasonal component based on a linear loop model, and a transit component to estimate daily fluctuations using explanatory variables.

In the field of research devoted directly to forecasting the growth of the Arctic region and demand in the Arctic, a small amount of work has been identified. Thus, in [14], the application of statistical models and neural networks to predict energy demand for two settlements was studied. Of course, in connection with the environmental agenda, attention is paid to renewable energy sources. In work [15], using econometric modeling, the potential impact of renewable sources in the Arctic on the sustainability of the region is estimated.

Thus, no potential models have been identified that could carry out strategic forecasting according to the risks and allow allocating a relevant set of resources in accordance with consumers. In connection with the goals of achieving carbon neutrality, the demand for hydrocarbons as a source of electricity is not predicted. Demand response activities are geared towards optimizing load schedules and lowering costs but do not take into account customer characteristics and do not take a long-term perspective. In the works on forecasting the growth of demand for electricity, the regional potential is not considered in terms of the development of tourism, science, the provision of medical services, and others.

Our proposed research is based on the growth of consumers, taking into account their qualitative characteristics, taking into account the opinions of experts, taking into account the analysis of interrelationships, and building the growth of demand for power, energy, and types of energy as a single structure. This will enable early preparation of energy infrastructure and long-term assessment of the dynamics of carbon footprint and climate change while maintaining the region's environmental sustainability and energy supply.

The study not only proposes a forecast of energy demand in various scenario conditions but also reveals the structure of this demand on the part of the consumer and the types of energy that will ensure sustainable energy supply in the Arctic.

The complex approach in the estimation of the needs of consumers of energy will allow to level gradually the naturally arising risks. It is necessary to create a solid foundation to ensure the development of resources, despite the specifics of the region. The basis for this should be infrastructure with a new logic of functioning, interaction, and sustainable development within the framework of the growing digitalization and transition to Industry 4.0.

Given the exhaustion of the continental resource base and taking into account the potential of the Arctic zone in terms of reserves, enhanced by the creation of transport and logistics hubs and the military-political aspect, the exploration, and development of the Arctic zone become a strategically important and economically beneficial step [16,17]. Hence, it is necessary to ensure the development of promising areas, maintain and modernize the existing, often outdated infrastructure, and create comfortable conditions for people to live in.

An indispensable condition for the development of the region is a reliable uninterrupted power supply to the territories, following global trends and sustainability requirements [18]. The progress in energy supply in the AZRF is determined by the global level of energy development, where a significant impact is made by the ongoing energy transition and digital transformation of energy. This is reflected in the 5D concept, which reveals the direction of current trends and tendencies and on the basis of which the analysis of current ways of possible development is carried out in Figure 1.

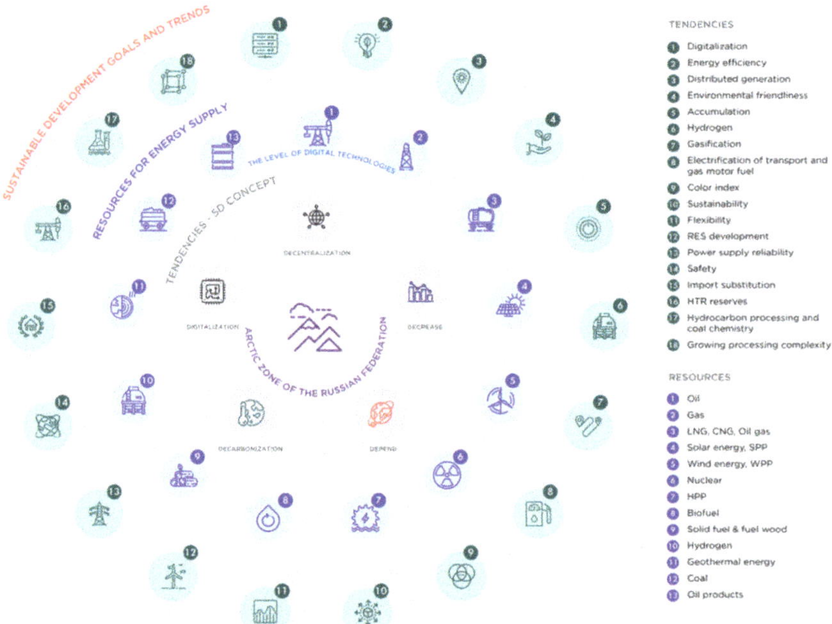

Figure 1. Linking the Sustainable Development Goals, technological trends and tendencies, and the resources under consideration for energy supply in the Russian Arctic.

Digitalization changes the work of the energy system; there is a new understanding of the use of electricity, power generation, and supply, the functioning of consumers and systems for the production, transporting, distribution, and storage of energy.

Decarbonization is related to climate change and primary fuel supply, electrification, and energy conservation and brings economic, environmental as well as social benefits.

Decentralization is related to changing the growth logic of energy systems and the distribution of energy consumption trends toward a larger number of small energy and economic nodes, which stimulates innovations from energy storage to intelligent data-based control systems.

Dependence is about the increasing information and technological dependence of production, transportation, and use of different types of energy on each other [19]. Integration of electricity, gas, water, heat, cold, and collaboration with all participants of the fuel and energy complex [20].

Decrease indicated a trend to reduce consumption not only of energy but also of all kinds of resources and materials, reduction of waste, and their increasing involvement in recycling [21]. This trend is facilitated by a qualitative change in consumer behavior and new production technologies that allow customizing products [22].

In its turn, each trend corresponds to a certain set of technological tendencies in the energy supply. Considering the current tendencies in energy supply, a certain set of necessary energy resources for regional development is formed [23,24]. Thus, it is possible to estimate the need for energy resources for this or that investment project. Given the need for sustainable development of the Arctic region [25], the development of technologies in the field of renewable energy and the achievement of carbon neutrality is a priority. However, to reduce the share of traditional hydrocarbon raw materials, it is necessary to properly plan the development of energy demand and its environmental friendliness in the short term until 2035 [26].

2. Materials and Methods

The model proposed in the article is based on the Delphi method. The Delphi method is designed to assess the aggregate expert opinion on complex problems, potential consequences and the effectiveness of the use of cost measures. The structured opinion of a group of experts in comparison with the opinion of one specialist is a tool for further and more accurate forecasting [18,19,27–29].

Following the basic structure of Delphi, the authors developed a questionnaire, the passage of which was allocated 1.5 weeks for 60 experts: employees of companies and universities of the mineral and fuel and energy complexes. The purpose of the survey was to determine the priority set of fuel resources for individual consumers, taking into account the political, economic, social, technological, environmental, and legal risks of using each resource. Risks were assessed on a nine-point scale, where 1 is the minimum risk impact; 9—the maximum impact of risk. Also in modeling, the influence of external factors was considered on the basis of scenario forecasting [18,30,31]. Given that many works noted the importance of risk assessment in the scenario forecasting based on expert evaluations, in our scenarios, we necessarily considered:

- Technologies and lines of their development of primary interest;
- Analysis of the current state and prospects for the development of the energy sector in the Arctic;
- Comprehensive analysis of global factors affecting the development of consumers [32,33];
- Consideration in modeling expert assessments based on the results of a survey of pro-field experts.

The study was divided into three stages, which are shown in Figure 2. The research methodology includes the following stages: first, a review of the region—the current state of the energy system and energy infrastructure, a study of the resource base, the fuel and energy balance, and related problems, including climatic. Then the study is divided into technological modeling of demand and scenario modeling. In the technological section, the list of key consumers is determined, and current and future centers of energy loads are highlighted. Based on an analysis of trends and tendencies, requirements are derived for each type of consumer, in terms of reliability category, required installed capacity, mobility, seasonality, and carbon footprint. Similarly, an assessment of possible resources by technical and economic criteria and CO_2 emissions is carried out. Scenario modeling generates scenarios according to the conditions of which and the results of a survey among experts the demand for energy consumption will change, as well as the probability of meeting the demand for different types of resources.

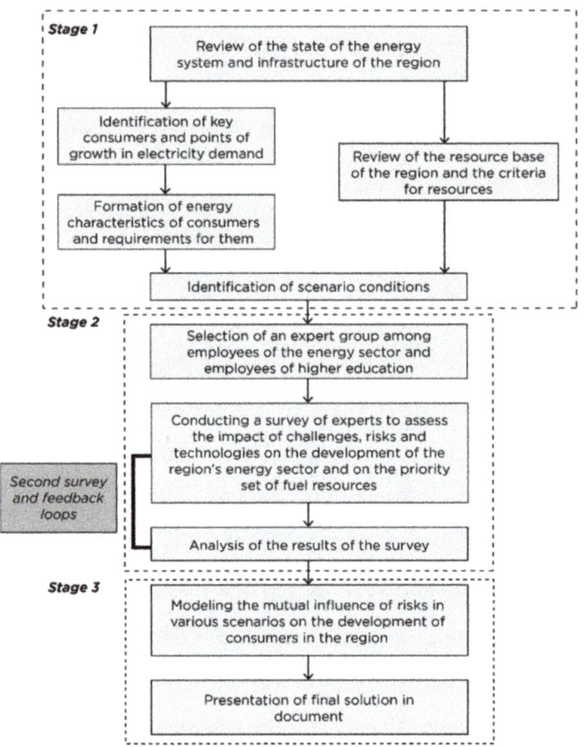

Figure 2. Research Methodology.

2.1. Analysis of the Development Potential of the Arctic

The Russian Arctic (AZRF) is a geographical area within the Russian Federation, located above the Arctic Circle. It includes Murmansk and Arkhangelsk Regions, the Komi Republic and Yakutia, Krasnoyarsk Territory, Nenets, Chukotka, Yamalo-Nenets Autonomous Districts, Belomorsk, Kemsk, and Lo-Ukhsk districts of Karelia, as well as lands and islands located in the Arctic Ocean and some uluses of Yakutia [34].

The study identified the following types of prospective consumers for the territory of the Arctic—they are also referenced points and points of growth (consumer agglomerations), on which the further development of the territory will be built:

1. Military bases;
2. Hydrocarbon deposits and rare-earth metal deposits;
3. Settlements (single-industry towns);
4. Scientific research bases;
5. Logistic clusters, hubs;
6. Medical bases;
7. Agricultural complexes;
8. Tourist bases;
9. Data centers (DPCs).

The study analyzes the development potential of all of these types of consumers, but in the final version of the work, the authors formulate conclusions only on some of them.

Hydrocarbon deposits and rare-earth metals. The scales of the Arctic resource potential are estimated as follows: about 70% of the total unexplored gas reserves in the Arctic are managed by the Russian Federation [35]; in the Arctic regions of the country, there are strategic and rare-earth metals—about 10% of the world reserves of nickel, 19% of the platinum group metals (PGM), 10% of titanium, over 3% of zinc, cobalt, gold, and silver; the largest coal and diamond deposits [36]. It is worth noting that 70% of the Arctic oil resources are concentrated in the shelf areas, and most of the explored fields are difficult to extract. Moreover, it is estimated that up to 25% of hydrocarbon reserves lie under the ocean strata on the Lomonosov Ridge [37].

Currently, coal and oil products have a significant advantage in the energy balance of the mainland Arctic, which negatively affects the fragile ecosystem of the region [38]. An essential role in reducing the negative impact and increasing the added value of the product will be played by such technologies as Clean Coal and the introduction of digital tools throughout the supply chain. Gas and gas-condensate potential can occupy a niche of gas-chemical production and ensure the production and export of liquefied natural gas (LNG) and hydrogen, which will not only make up for export losses due to the accelerating decline in demand for oil but also provide new investment and return flows [39].

Single-industry towns. The development of the Arctic's natural potential first of all requires a huge number of qualified specialists, but the actual living conditions [40] lead to a migration outflow of population. The difference in the socio-economic situation is due to the high diversification of the economy of single-industry towns and dependence on enterprises and is emphasized by the division of towns in the Arctic zone into categories [16,41–43].

Logistics clusters. One of the most important vectors of the development of the transport industry of the Russian Federation in the Arctic region is the establishment of permanent communication along the Northern Sea Route (NSR) [44], which requires the development of mainland infrastructure [45]. The Arctic zone is also characterized by a well-developed railway network; at present, there is the financing of the Northern Latitudinal Railway (NSR), Northern Latitudinal Railway-2, and Belkomur projects. The estimated traffic volume along the Northern Movement is 23.9 million tons of various cargoes per year [46]. The aviation infrastructure of the Arctic zone is currently represented by more than 160 airports and airfields.

The places of intersection of air, rail, and water hubs are the so-called transport hubs. In the complex development and development of the Arctic zone, they acquire enormous importance due to the tendency of formation of agglomerations near such clusters, transport accessibility, and available infrastructure. The work identified the following hubs: the currently existing ones (5) in Murmansk, Arkhangelsk, Sabetta, Vladivostok, Dudinka and the ones to be built (3) in Ust-Luga, Tiksi, Korsakov.

2.2. Energy Characteristics of Energy Consumers in the Arctic

Given the resource potential of the Arctic in the conditions of the gradual depletion of the continental base, the creation of transport and logistics hubs capable of giving a new impetus to the development of world trade, as well as the military and political aspect, the development and exploitation of the Arctic zone acquire a new strategic and economic momentum [47].

The development of the Arctic is a sequential and multistage process. It is necessary to ensure the development of promising areas, the maintenance and modernization of the existing infrastructure and the creation of comfortable living conditions for the population [48]. The remoteness from the main industrial centers of the country creates the need to build a large network of railways and roads: first, to maintain a high level of mining, and second, to supply the Arctic regions [49,50].

An indispensable condition for the development of the region is a reliable uninterrupted energy supply to existing and prospective consumers following global trends and requirements. Arctic states face a number of common problems in this area, among which are remoteness from central energy networks, the use of expensive diesel fuel for energy generation, the high level of tariffs for energy services, as well as the specifics of the life of small indigenous peoples. Arctic conditions require the development of affordable, reliable, and easy-to-operate technologies that can supply energy to remote areas under icing conditions, high humidity, and critically low temperatures, using environmentally friendly energy carriers.

As a result of research, the following characteristics of consumers have been revealed: the peculiarity of power supply of military bases—the necessity of independence from each other, energy sources working independently according to the special group of the first category of reliability and uninterruptibility of electric receivers. The power supply of military bases does not have to take into account its environmental and economic efficiency. In the structure of generating capacities of single-industry towns, there are a large number of boiler and diesel-generator sources. Low efficiency of heat and power networks and their significant wear and tear is observed. The Arctic region has great potential in the use of non-conventional renewable energy sources [51]. Wind energy has the greatest potential in the Murmansk region and can be used to create Power-to-X systems [52]. Power-to-X technology is a solution to the problem of the interconnection of energy sectors [53]. Steam-gas turbine units (SGU) are often used for distributed generation to obtain electric and thermal energy due to their high efficiency. The technology of converting electricity into heat (in this case Power-to-Heat) using heat pumps or heating rods is an innovative environmentally friendly way of heating buildings and even providing industrial enterprises with process heat. In the Arkhangelsk region, there is an active development of the industry of environmentally priority fuel-liquefied natural gas. The creation of an efficient support system for the development of small-scale power engineering, disclosure of floating nuclear power plants, and the increase of local resources efficiency will become the starting point of socio-economic development of North-West Russia. At present, scientific stations are mainly equipped with diesel power plants. For onshore research stations, the energy characteristics vary depending on the activity of use and the focus of research. It is perspective to use biofuel as a resource for the energy supply of agricultural complexes, which is received as a result of the cultivation of certain crops or animals' vital activity. In its turn, the transition to the use of biofuels will take a long time, so at the first stages, it is advisable to use wind-diesel installations, which are already in operation in the Murmansk Region [54]. The energy supply of data processing centers is characterized by reliability and uninterrupted operation. Most of the energy is consumed for cooling and operation of the equipment located in data centers. Currently, most data centers are powered by diesel generators, while large IT companies follow the trends and use renewable energy sources [55].

2.3. Requirements for Energy Consumers

For each of the allocated types of consumers, the requirements for energy supply were formed. For the convenience of perception, analysis, and application in the mathematical model, the selected requirements are summarized in Table 1.

Table 1. Requirements for consumers.

Type of Consumer	Reliability Category	Required Installed Capacity	Carbon Footprint	Mobility	Seasonality, m
Logistics hubs	First	Large	Regulations on reduction (international market)	Stationary	0–12
Sanitary unit Hospital	Second First special	Small Small	Voluntary agreements	Mobile Stationary	0–12 12
Scientific bases	Second	Small	Regulation on reduction	Variable mobility	0–12
Agricultural complexes	First (second)	Medium	Voluntary agreements	Stationary	0–12

Table 1. *Cont.*

Type of Consumer	Reliability Category	Required Installed Capacity	Carbon Footprint	Mobility	Seasonality, m
Military bases	First special	Large/Medium	No requirement for reductions	Stationary/mobile	12
DPCs	First special	Large	Voluntary agreements	Stationary/mobile	12
Single-industry towns	Third	Medium	Regulation on reduction	Stationary	12
Mining and oil & gas enterprises	Depends on the raw material. First, first special, second [56,57]	Large	Regulation on reduction	Stationary/mobile	12
Tourist bases	Third	Small	Regulation on reduction	Variable mobility	0–4

The transition to qualitative characteristics is due to the task of implicitly comparing the requirements of consumers and the criteria for energy sources: we resorted to this method so that it would be easier for experts to compare the resource and the consumer and assess the weight coefficients of their connection. In addition, due to a large variation in the installed capacity, we decided to assign such parameters as "capacity" qualitative characteristics. If you have close values of the installed capacity of the compared consumers, it may be more convenient for you to go to quantitative characteristics. Each of the qualitative characteristics presented by the authors is inherent for a certain primary resource. The method with their use emphasizes the influence of characteristics on the choice of consumers due to their specific's requirements for primary resources. The use of this connection when analyzing the compatibility of a resource with a consumer's requirement distinguishes our method from others.

2.4. Resources for Power Supply to Consumers

The main energy resources in all regions of the Arctic zone are fossil resources: natural gas, coal, diesel, fuel oil, gasoline, kerosene, associated petroleum gas (APG), coke, peat, and oil shale. The leading positions in the structure of the fuel and energy balance (FEB) of the Russian Arctic are taken by such resources as natural gas, coal, and oil products.

Natural gas. The priority in the use of this resource is given to the western regions of the Arctic zone (Yamal-Nenets Autonomous Okrug, Komi Republic, Arkhangelsk Oblast, Republic of Karelia, and Murmansk Oblast), where the infrastructure of local and main gas pipeline networks is developed. The YNAO and the Komi Republic are gas suppliers for the other western regions mentioned above. The Taimyr-Turukhan zone (Krasnoyarsk Krai) is the "greenest" of all the zones, with 60% of installed capacity coming from gas and the remaining 40% from water flow energy.

Coal. This resource is widespread in those Arctic regions where there is direct extraction, for example, in Chukotka AD. In other regions of the Arctic zone, coal is usually transported.

Oil products. Fuel oil and diesel oil are delivered to the territory of the Arctic in full. In the structure of fuel and energy mix of the Murmansk Region and the Republic of Karelia, they account for a significant part.

Nuclear fuel plays a significant role in the energy supply of the Murmansk Region and Chukotka AD. Further development of nuclear power in the Arctic is promising due to floating nuclear power plants. The energy of water streams is used in the energy balance of the Murmansk Region, and the Republic of Karelia has potential for the development of hydroelectric power. Associated petroleum gas is used in the energy regions where oil production takes place. In areas with a predominance of the woodworking industry, wood and municipal solid waste are used. Now, renewable energy sources do not make a significant contribution to the fuel and energy mix of Arctic consumers.

It is necessary to note the important role of LNG in the resource market. Considering the resource potential of the Eastern Arctic, that is, gold, tin, copper deposits, remote from fuel bases, it is necessary to have a resource that allows increasing the necessary capacities under requirements of carbon footprint reduction. Taking into account the course on the gasification of the Arctic region and development of the main transport route—the NSR, LNG has the prospects to take the leading position in the nearest decade not only on the external resource market but also on the domestic one.

Hydrogen is a potential resource, which in 2035–2050, with the development of appropriate technologies and the accumulation of sufficient experience in its use, could take significant positions.

Criteria for resources. All resources that are consumed in the region or are perspective from the point of view of use for energy supply are summarized in Table 2. They are evaluated by criteria, among which are economic and environmental: the capital cost of building a generation unit based on the selected resource (CAPEX), the average present value of electricity over the life cycle of the resource (LCOE), the net present value of building and operating a generation unit for the resource (NPV), as well as emissions, expressed in the equivalent of CO_2. Economic indicators are one of the most important assessed parameters of electrical engineering complexes, as a rule, they act as optimization criteria [58]. Quantitative indicators by criteria are given, which reflect in point expression the minimum (1 point) and maximum (5 points) level for the selected criterion.

Table 2. Criteria for resources.

Resource	Criteria			
	CAPEX	**CO$_2$**	**LCOE**	**NPV**
Fuel oil	4	5	1	2
Gas	5	3	2	5
Coal	4	5	3	3
LNG and CNG	5	2	2	4
Nuclear	5	2	4	5
ARES	4	2	3	4
Associated petroleum gas	4	4	3	4
Diesel	4	5	3	3
Hydrogen	5	2	3	4

All fuel resources (except for solid fuel and resources) at the stage of realization require considerable capital costs but getting energy from them entails consequences for the biosphere, which are caused by a different set of processes and the nature of their appearance. Thus, obtaining the end product from wind, solar, water, and land energy is less destructive to the environment in terms of disturbances and pollution due to the lack of transport processes and resource extraction [59]. Renewable energy sources are used locally and are not "transportable".

Hydrocarbons and coal are the most capacious fuels in terms of the number of processes and require a well-developed transportation infrastructure. Hydrogen fuel is a possible alternative to fossil fuels, which is worth considering as a source of clean energy and a means of storing it for Arctic consumers while developing technologies for the production, storage, transportation, and consumption of hydrogen. In its turn, natural gas has great potential in terms of exportable resources to the EU and APR countries due to the development of LNG production technologies.

We should also note the possibility of risks of LNG supply reduction to other countries through the use of hydrogen energy, the development of which generally helps to reduce the global dependence on gas. Petroleum products are ubiquitous and widely used fuels for all types of consumers.

Nuclear energy is also applicable to many consumers, is environmentally friendly due to minimal CO_2 emissions, and is convenient in terms of the long-term use of nuclear fuel, which is imported from other regions of Russia.

Using estimates for petroleum products as an example: CO_2 emissions for 2020 in the world amounted to more than 12 billion tons, or 30% of the total amount of emissions, respectively, the authors assigned a maximum rating of 5 points [60]. The LCOE for diesel installations varies according to specifications and fuel prices. On average, the LCOE will hover around \$50/MWh for small installations, while for solar panels, it will be over \$150/MWh, so an average rating of 3 was given [61,62]. CAPEX for oil fields continues to grow due to growing concerns about ESG and pressure on investors [63,64].

As a result of the analysis of review and analytical articles, a list of major global technological trends and related technologies was identified [65–70]. The existence of each trend is confirmed by examples of Russian and foreign company cases.

Against the background of global goals to reduce the carbon footprint of most industries, companies are striving to implement RESs and renewable energy, combining them with traditional sources under the control of IT technologies. Storage of produced energy is a no less urgent issue, which is solved by companies at the level of technology in the struggle for primacy in the creation of new storages [71]. Digitalization, implementation of microgrids, and active-adaptive networks are a number of additional technological trends caused by the global agenda.

At the moment, Hevel is building an autonomous hybrid power plant with a total capacity of 2.5 MW in the village of Tura. Hywind Scotland wind farm, built jointly by Equinor and Masdar, is operating in the Scottish waters. The total capacity of all wind turbines is 30 MW, the capacity factor is 50%. The development of hydrogen energy and transport should be singled out in the cases of RESs and RES companies: Airbus hydrogen gas turbines, Alstom hydrogen fuel cells [72]. For 6 years, Rubin Central Design Bureau has been preparing the Aisberg project, which includes the development of underwater production complexes of northern fields. The work includes the development of specialized drilling rigs, production stations, process control, and energy-saving equipment.

ROSATOM has developed a number of small nuclear power plants (SNPPs) designed for service in remote areas of the Arctic. Among them, there are transportable complexes, as well as complexes designed for offshore operation, so power units from 1 to >100 MW are represented [73]. The harsh conditions of the Arctic, in general, promote the development of autonomous technologies, as well as remote control technologies. Even today, there are a large number of Russian and foreign projects in the field of unmanned technologies, flying and underwater vehicles for remote control for transport, construction, research, and oil and gas industry needs. The Russian Helicopters Holding Company has developed the VRT-300 multifunctional unmanned helicopter for the Arctic and NSR development, capable of cargo transportation, environmental monitoring, search and rescue operations, as well as equipment and road infrastructure diagnostics. Kalashnikov Concern, Lazurite Central Design Bureau, and others are also engaged in the design of drones.

2.5. Scenario Development

The methodological approach in the study is based on several forecasting scenarios, which are broken down into a sequence of time intervals of 2020–2025; 2025–2030; 2030–2035. Planning or scenario analysis is a consolidated and structured process of creating future opportunities that have socio-economic, environmental, and technological implications.

Scenario planning was based on an analysis of the Arctic's external environment, followed by the identification of the main factors affecting consumer development, electricity demand, and capacity. As a result of the analysis of the external environment, a list of macro-environmental factors that have the greatest influence on the development of consumers in the Arctic in the period under consideration was compiled.

Based on the brainstorming method, the most significant factors were identified and optimized using cause-effect diagrams. This resulted in the selection of the most significant and independent factors. The brainstorming method has established itself as an effective way to generate creative and effective ideas when solving outstanding and complex problems that require the involvement of specialists from different specialties. The method is widely known and applicable, including in solving scientific and engineering problems [74–76]. The research work presented in this article implies a comprehensive study and the involvement of specialists and young scientists from different areas of the fuel and energy complex and MSC, which makes it necessary to organize work, including by the brainstorming method. Based on this method, the most significant factors were identified and optimized using cause-effect diagrams.

During the forecasting phase, several variants of different scenario outcomes were generated. The purpose of combining the most significant factors was to establish the interdependence between the predicted outcomes of the factors under consideration and to write scenarios.

The scenarios (Table 3) combine a variety of factors. For example, such as global economic growth, political factors, environmental issues, and technological development, that illustrate the relationship between the main driving forces. Scenario driving forces include various types of factors, some of them, such as the COVID-19 pandemic, arise spontaneously. Others represent sustained trends, such as Digitalization, Decarbonization, Decentralization, Dependence, and Decrease. Today's policies of companies and governments take into account the need to achieve sustainable development goals. Thus, the SDGs are also becoming one of the most important drivers for scenario planning.

Table 3. Description of scenarios.

Title 1	2025	2030	2035
Negative scenario "Cold Menace"	1. Virus development and mutation—not being able to financially overcome the vaccine race; 2. Reduction of energy consumption by 5% annually; 3. Oil price of $50–60/barrel; 4. A set of measures to support new fields at the level of 20% profit tax and 15% mineral extraction tax; 5. Lack of transparent regulation and certainty in the FEC and MRC; 6. Lack of international investment in Arctic development projects.	1. Reduction of the industry of offline culture and public events; 2. Minimum consumption of energy and services; 3. High volatility of prices for energy resources—lack of investments in projects to develop new fields; 4. Oil price $40–55/barrel; 5. Lack of possibility to enter foreign markets due to sanctions pressure; 6. World trade declines by 3–5% annually.	1. The energy poverty of the countries; 2. Lack of any investments in energy infrastructure, and their subsequent outflow; 3. Growing risks of man-made accidents and lack of funds to eliminate natural disasters; 4. Oil price 40–50 USD/bbl; 5. Increase in social tensions; 6. Low level of innovation, education, and culture.
Neutral scenario "Northern Outcast"	1. Containment of coronavirus infection without significant quarantine restrictions; 2. Electricity consumption increases by 40%; 3. Energy intensity of GDP does not change, specific consumption per capita grows by 1%; 4. Export restrictions—instability of mineral and energy supplies; 5. Stabilization of energy consumption at the same level, without regard to environmental and climatic situation; 6. Increase in global energy by 1–3% annually.	1. The lack of the former level of international trade in resources due to the import substitution race; 2. Inability to fix the carbon footprint; 3. Taking and holding leadership positions in creating international transport and logistics systems, and developing and using the Arctic is impossible; 4. Maximum oil price—$55–65/barrel; 5. Carbon footprint of energy resources is not a reference point for the energy supply of consumers; 6. Emergence of conflicts over resource shortages due to climate change; 7. Military build-up.	1. The development of territories is carried out at the expense of orientation on the domestic market; 2. The growth of investment in research and development is up to 8%; 3. Threat of development of Arctic territories because of cataclysms, caused by global warming; 4. Set of measures to support new mines at the level of 10% income tax and 10% mineral extraction tax.

Table 3. *Cont.*

Title 1	2025	2030	2035
Positive scenario "Energy Awakens"	1. Leveling the negative consequences of the crisis and quarantine measures; 2. Restoring the disrupted supply chain of energy, materials, and goods; 3. Sustainable development of the FEC and MRC in the Arctic on the basis of digital technologies [77,78]; 4. Oil price—$70+/barrel in all scenarios; 5. Renewal of fixed assets in the energy sector and network infrastructure; 6. Growth of global trade by 3–5% annually; 7. Set of support measures at the level of 0–3% profit tax and 4–5% mineral extraction tax.	1. Formation and development of ecologically and socially oriented points of growth; 2. Growth of demand for new technologies and equipment in the Arctic; 3. Infrastructural and legislative opportunities for small and medium businesses to locate in the AZRF; 4. Transparency and openness to internal and external consumer markets through digital technology; 5. Scientific and technological breakthrough at the global level through digital integration of stakeholders; 6. Emergence of a window of opportunity for companies supplying technologies; 7. Increase of R&D investments up to 20% relative to 2021 due to appearance of venture capital.	1. Coordinated development of the Arctic through international planning, funding, and regulatory frameworks; 2. Digital transformation in the management of the life cycle of energy and mineral resources in the Arctic; 3. The emergence of digital industries, smart factories, and high-tech spaces that operate through platform solutions; 4. Sustainability and reliability of energy supply to the Arctic consumers through new approaches to resource supply to consumers; 5. Innovative rebirth of the Arctic FEC and MRC into a high-tech and efficient infrastructure, providing quantitative and qualitative growth of the Russian economy.

Survey design methodology and results of the survey. To form scenarios for the development of electricity and energy supply of the Arctic consumers, a survey was conducted on the basis of expert evaluations. A poll was developed to get expert data to forecast Arctic energy development and the priority set of fuel resources for three time intervals: 2020–2025, 2025–2030, 2030–2035, taking into account the political, economic, social, technological, environmental and legal risks of each resource's use in the Arctic zone. The risks were assessed on a nine-point scale, where 1 is the minimum impact of risk; 9 is the maximum impact of risk. An expert assessment was made of the development of types of consumers and demand for energy resources.

Each region in the Arctic zone is characterized by its own set of fuel resources, which at the moment is conditioned by the presence of hydrocarbon or coal deposits and developed transport infrastructure of resources.

Since hydrocarbons and coal occupy priority positions in the structure of the region's fuel and energy balance, it was decided to divide the Arctic zone into three groups following the affiliation of hydrocarbons (gas, oil, and oil products) and coal to these territories:

I.	Regions where fossil resources are predominantly "imported", that is, the Murmansk Region, the Republic of Karelia, the Arkhangelsk Region, and the Komi Republic.

II.	Regions where fossil resources are "local", that is, are extracted in the regions in question—Republic of Karelia, Yamalo-Nenets AD.

III.	Regions where fossil resources are both "local" and "imported"—"mixed", that is, Krasnoyarsk Krai, Yakutia, and Chukotka AO.

Figure 3 presents a map of the fuel and energy balance of the Arctic region.

The time frame for the survey was 1 month. The expert group was selected from various structures, scientific and social schools. During this time, 64 people took part in the survey: employees of 5 educational institutions, employees of 5 companies of mineral complex, employees of 7 companies of the fuel and energy complex.

Thus, the forecasting of the energy sector development is based on a survey of a large number of professional workers and teachers of specialized institutions, close to the subject. As a result of this research, key consumers and the most sought-after resources for energy supply were identified.

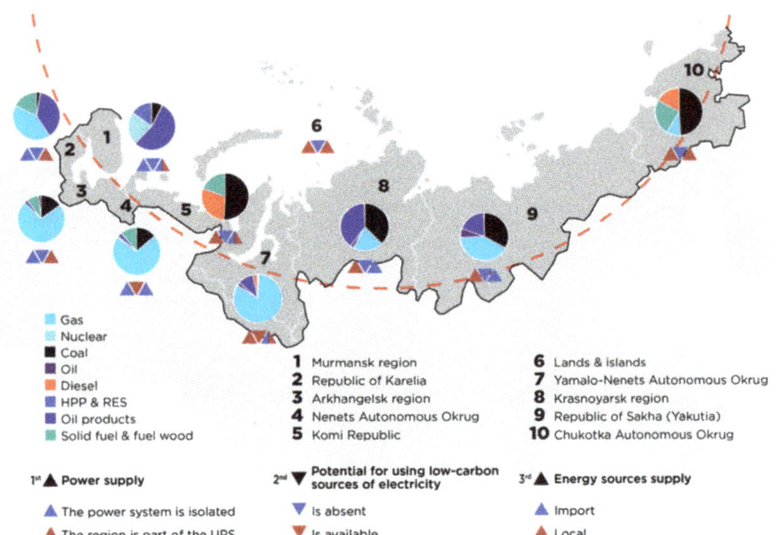

Figure 3. Structure of the fuel and energy balance of the Arctic region.

A complex step-by-step work was carried out: from assessment of the current situation in the energy sector of the Arctic to the choice of research method, the concept of questionnaire and experts, data analysis, and its use for forecasting in various scenarios.

Figure 4 shows a generalized methodology of building a mathematical model of scenario forecasting.

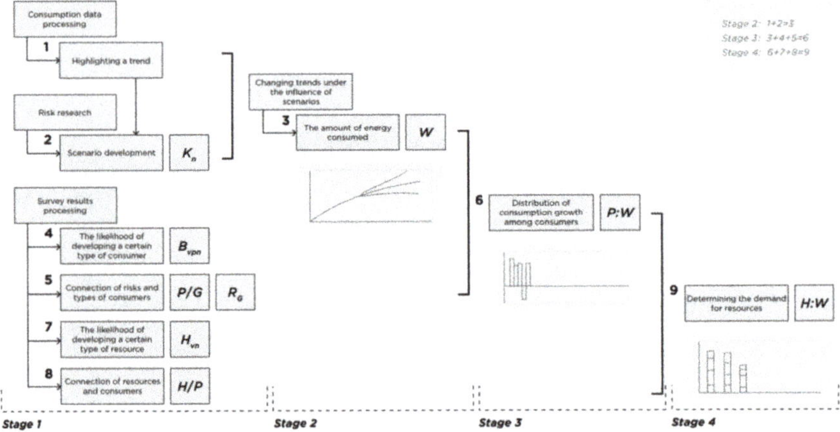

Figure 4. Methodology for constructing a mathematical model of scenario forecasting.

- Step 1: Calculation of the generalized risk impact factor K_n for each of the formed scenarios, based on the risk analysis and identified consumption trends from the processing of data on electricity consumption for the period 2010–2020;
- Step 2: Calculation of the cumulative impact of risk on the development of Arctic consumers. Calculation of energy consumption W in three scenarios at different time ranges based on an assessment of the impact of global challenges on consumption trends;

- Step 3: Calculation of energy consumption and distribution by types of consumers based on the calculation of the basic vector of the probability of development of a certain type of consumer B_{vpn} and the results of the calculation of the total weight coefficients of the connection between risks and types of arctic consumers R_G;
- Step 4: Calculation of the distribution of demand for resources between types of consumers based on the calculation of the basic vector of the probability of an increase in demand for resources H_{vpn} and the matrix of the relationship of weight coefficients of consumers with resources.

2.6. Mathematical Model of Scenario Forecasting

To formalize and establish the numerical values of the mutual influence of risks in different scenarios on the development of Arctic consumers, the interaction matrices compiled by the research participants were used [79]. The application of the method of interaction matrices makes it possible to identify the degree of mutual influence of the factors of the considered set and predict their behavior in the future. After analyzing and processing the results of the survey of the expert group, the risks were arranged on the plane (Figure 5) in accordance with the following axes:

- K_s is the axis of ordinates, the strength of the influence of risks on the development of arctic consumers; it takes values from -1 to 1 (strength decreases/increases);
- K_d is the abscissa axis, the influence of risks on the rate of change in the number of consumers; it takes values from -1 to 1 (inhibits/accelerates);
- S characterizes the size of the bubbles, which reflects the significance of the respective risk for the growth of energy consumption in the Arctic; takes values from 0 to 1.

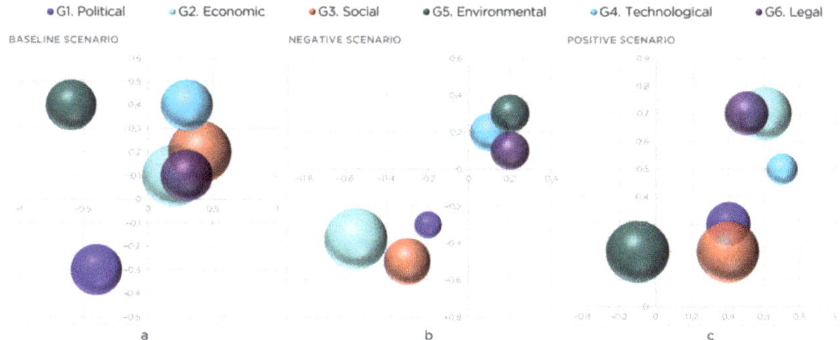

Figure 5. Results of PESTEL-risk analysis under different scenarios:(**a**) neutral scenario with $K = 0.095$; (**b**) negative scenario with $K = 0.21$; (**c**) positive scenario with $K = 0.585$.

Risk assessment varies depending on the scenario. The brainstorming session analyzed how risks would manifest themselves in negative, neutral, and positive scenarios.

It is important to note that we propose a methodology and demonstrate approaches to its implementation. When implementing the methodology in the future, the results will depend on the expert group; in this case, the forecast will be adjusted not in terms of the methods used, but in terms of expert opinion. The team of authors allows such a dependence, since the study is devoted to strategic planning, and it is necessary to use the knowledge of experts. The scenario conditions proposed by the authors are based on the study of risks and their impact, as well as on the assumptions of the development of certain risks for the worse or for the better, which is described in Table 3. The assessment and application of this approach were based on the literature, but specific assessments in the works are always different since they are formed on the basis of various scenarios. We propose a specific sequence of actions that allows for a more comprehensive assessment of future needs and the development of their sustainable provision.

It is also worth emphasizing that the matrix is dynamic; over time, it is possible to reassess the risks, since they are not static in nature, but undergo changes over the entire time interval considered in the study.

Based on the data of the interaction matrix, each group of risks (political, economic, social, technological, environmental, legal risks) is presented in the form of a generalized impact factor K_n, which is derived from the Formula (1):

$$K_n = \frac{(K_S + K_d)S}{2} \tag{1}$$

Based on these interaction matrices, each of the scenarios can be represented in the form of a generalized influence coefficient K_n.

Based on the results of the analysis and calculations for each of the scenarios, the final generalized impact factor (total) can be found, which determines the final degree of significance of the three indicators of each risk group for the development of Arctic consumers and, consequently, the growth of energy consumption and demand for the construction of energy sources and energy infrastructure.

The results of calculations of the generalized coefficients of influence of risk groups and scenarios are presented in Table 4 and Figure 5.

Table 4. Values of generalized coefficients of influence K_n.

Risk Group	K_n (Neutral Scenario)	K_n (Negative Scenario)	K_n (Positive Scenario)
G1. Political	−0.07	−0.025	0.07
G2. Economic	0.045	−0.25	0.195
G3. Social	0.09	−0.12	0.12
G4. Technological	0.07	0.03	0.06
G5. Environmental	−0.08	0.05	0.02
G6. Legal	0.04	0.03	0.12
Total	0.095	−0.285	0.585

2.6.1. Forecast Development of Consumer Types

The next stage of the study is to assess the impact of global challenges on the development of different types of consumers in the Arctic in order to further forecast the demand for energy by different consumers, taking into account the mutual influence of global challenges.

A total of nine types of consumers P1–P9 were identified (Table 5).

Table 5. Symbolic designation of Arctic consumers.

Type of Customer	P
Military bases	P1
Hydrocarbon deposits	P2
Settlements (single-industry towns)	P3
Scientific research bases	P4
Logistics clusters	P5
Medical bases	P6
Agricultural complexes	P7
Tourist centers	P8
Data Processing Centers (DPCs)	P9

Based on the survey, the influence of risk groups on the development of consumers for different time ranges was determined. At the same time, the processing of questions about the impact of risks at different time intervals was carried out. Coefficients take values from 0 to 1 in increments of 0.01. Taking into account the ranges, three effects were identified: 0.3 characterizes a weak effect; the range 0.3–0.6 corresponds to the medium effect conditions; 0.61 is in the context of a strong effect. It should be noted that these values were averaged for all experts, excluding observations with incomplete information about the main types of consumers. Thus, we obtained weighting coefficients, which allow assessing the degree of risk impact on the development of energy consumers (Table 6).

Table 6. Impact of risks in consumer development in 2020–2025.

Type of Customer	P	G1	G2	G3	G4	G5	G6	Cumulative Impact of Risk R_{Yi}
Military bases	P1	0.3	−0.2	0.07	0.1	0.01	−0.02	0.0433
Hydrocarbon deposits	P2	−0.28	0.17	0.11	0.14	−0.4	0.11	−0.043
Settlements (single-industry towns)	P3	0.01	−0.32	−0.21	0.12	−0.4	0.04	−0.127
Scientific research bases	P4	0.05	0.18	0.04	0.03	0.2	0.09	0.098
Logistics clusters	P5	−0.3	0.12	−0.1	0.2	−0.08	0.3	0.023
Medical bases	P6	0.07	0.16	−0.3	0.01	−0.1	0.06	−0.016
Agricultural complexes	P7	0.06	0.15	0.11	0.07	−0.22	0.05	0.036
Tourist centers	P8	−0.4	0.1	0.08	0.04	−0.06	0.04	−0.033
Data Processing Centers (DPCs)	P9	0.31	0.15	0.25	0.2	−0.35	0.12	0.113
Total								0.094

Further, the total weighting coefficients of the relationship between risks and types of Arctic consumers were summarized in Table 7.

Table 7. Total weighting coefficients of the relationship between risks and types of Arctic consumers.

	G1–G6 (2020–2025)	G1–G6 (2025–2030)	G1–G6 (2030–2035)	G1–G6 (2035+)
R_{Gj}	0.094	0.101	0.37	0.55

Based on the final weight coefficients of the connection between risks and types of consumers, the forecast of energy demand for consumers at different time ranges is determined.

The value of the base vector of the probability of development of a certain type of consumers, obtained from a survey of experts, is used. This vector is normalized internally by Formula (2):

$$B_{vpn} = \frac{B_{vpi}}{\sum_{i=1}^{n} B_{pi}} \cdot 100 \tag{2}$$

where B_{vpi} is the basic vector of the probability of development of the consumer species; B_{vpn} is the internal normalization of the basic vector of the probability of development of the consumer species.

The final forecast of energy demand growth for certain consumer types by years, taking into account scenarios, is calculated by Formula (3):

$$W_{iyn} = \frac{W_f(t) \cdot (1 \pm K_n \cdot R_{Gj}) \cdot B_{vpn_i}}{100} \tag{3}$$

where $W_f(t)$ is the allocated trend of electricity consumption for the period preceding the forecast one. The available data on the consumption of the period from 2010 to 2020 on the territory of the Arctic were taken as the basis.

Table 8 shows the results of calculations of energy demand up to 2025. For other time intervals, similar calculations were made.

Table 8. Forecast of energy consumption by type of consumer by 2025.

Type of Customer	P	Basic Vector of Probability Based on the Survey	Neutral Scenario (2020–2025), Billion kW·h	Negative Scenario (2020–2025) Billion kW·h	Positive Scenario (2020–2025) Billion kW·h
Military bases	P1	0.760606	5.041147651	4.849507195	5.288262977
Hydrocarbon deposits	P2	0.912121	6.045359407	5.815543595	6.341700847
Settlements (single-industry towns)	P3	0.693939	4.599291828	4.424448628	4.824747533
Scientific research bases	P4	0.657576	4.358284983	4.192603717	4.571926615
Logistics clusters	P5	0.690909	4.579209581	4.405129813	4.803680861
Medical bases	P6	0.475758	3.153230877	3.033360036	3.307801171
Agricultural complexes	P7	0.409091	2.711375053	2.60830147	2.844285727
Tourist centers	P8	0.342424	2.26951923	2.183242903	2.380770283
Data Processing Centers (DPCs)	P9	0.457576	3.032724141	2.917434393	3.181387236
Total			35.79014275	37.120479	36.59403675

On the basis of the expression, which takes into account the energy consumption change on the time interval and the normalized total risk influence, the distribution of the energy consumption increase (decrease) by consumer types was obtained.

Thus, at this stage, the prognosticated development of consumer types and the associated scenario change in energy consumption are justified.

2.6.2. Forecast for Resource Use Development

For the scenario study, the main resource types were considered: fuel oil, gas, coal, LNG and compressed natural gas (CNG), nuclear, nontraditional, and renewable energy sources, associated petroleum gas, and hydrogen H1–H9 (Table 9).

Table 9. Symbolic designation of resources.

Type of Resource	H
Fuel oil	H1
Gas	H2
Coal	H3
LNG and CNG	H4
Nuclear	H5
ARES	H6
APG	H7
Diesel	H8
Hydrogen	H9

The energy requirements (Table 1) of consumers determine the strength of the connection with the types of resources based on their criteria (Table 2). The table of weight coefficients is compiled based on the results of processing the experts' evaluation of the connection. According to the results of the analysis and comparison of consumers' requirements and characteristics of resource types, the connection matrix is compiled (Table 10). The highest value indicates a more appropriate choice and compliance with the resource to provide the given type of consumers, taking into account the fullest satisfaction of requirements. The coefficients take values from 0 to 1 in increments of 0.001 and, with this in mind, range as 0.001–0.3, a weak relationship; 0.301–0.6, a medium degree relationship; 0.601–1, a strong relationship.

Table 10. Matrix of connection of weight coefficients of consumers with resources.

H/P	P1	P2	P3	P4	P5	P6	P7	P8	P9
H1	0.86	0.5	0.55	0.57	0.1	0.3	0.1	0.1	0.1
H2	0.8	0.94	0.87	0.65	0.87	0.67	0.67	0.77	0.64
H3	0.86	0.76	0.56	0.45	0.58	0.34	0.22	0.1	0.1
H4	0.89	0.3	0.67	0.34	0.56	0.34	0.78	0.82	0.34
H5	0.67	0.92	0.45	0.34	0.87	0.32	0.45	0.34	0.89
H6	0.76	0.56	0.78	0.88	0.45	0.67	0.78	0.89	0.68
H7	0.3	0.8	0.5	0.4	0.76	0.45	0.65	0.43	0.78
H8	0.89	0.32	0.45	0.54	0.23	0.34	0.21	0.2	0.12
H9	0.78	0.56	0.67	0.45	0.67	0.56	0.78	0.67	0.9

As a result of a survey of experts, it was proposed to put the strength of influence and the direction of the influence of risks on the development of bases, then the assessment was averaged and entered into the appropriate cell. The expert assessments in the questionnaire were processed in the standard way adopted for this approach [79].

Changes in demand for certain types of consumers in the Arctic will determine changes in the demand for energy resources. For this purpose, based on the table of weight coefficients of connection between consumers and resources, the forecast of energy demand from a certain type of resource was determined.

Forecasting the increase in demand for energy resources (4) was carried out taking into account the basic vector of the probability of an increase in demand for resources (H_{vi}) and its internal normalization (H_{vn}). This vector was obtained on the basis of survey data. Then we modeled the connection matrix of electricity consumers and resources in the form of normalized vector (L_{vn}), which corrects the forecast of demand for certain types of resources and reflects the competitive distribution as a result of the scenario conditions but does not change the value of total demand.

$$H_i = \Delta W_{iyn} f(t) \cdot \frac{H_{vin} \cdot L_{vin} \cdot 100}{\sum_{i=1}^{n} H_{vin} \cdot L_{vin}} \tag{4}$$

where $\Delta W_{iyn} f(t)$—changes in energy consumption, taking into account the scenario conditions in the allocated time range.

Table 11 shows the results of the calculations of scenario forecasting of an increase or decrease in demand for resources in the time interval 2020–2025. Similar calculations were conducted for other time intervals.

Table 11. Results of calculations of scenario forecasting of demand for resources by 2025.

Type of Resource	H	H_v Base Probability Vector Based on the Survey	H_{vn} Internal Normalization of Basic Probability Vector	L_{vn} Normalization with Connection to Consumer Types	Neutral Scenario (2020–2025), Billion kW·h	Negative Scenario (2020–2025), Billion kW·h	Positive Scenario (2020–2025), Billion kW·h
Fuel oil	H1	0.760606	14.0853	7.038513	3.155958122	3.037160292	3.309144796
Gas	H2	0.912121	16.89113	15.22798	8.18813799	7.879916847	8.585581042
Coal	H3	0.693939	12.85072	8.787074	3.594644762	3.459333716	3.769124795
LNG and CNG	H4	0.657576	12.17733	11.15538	4.324347858	4.161569039	4.534246863
Nuclear	H5	0.690909	12.79461	11.62019	4.732867717	4.55471123	4.962595818
ARES	H6	0.475758	8.810333	14.27623	4.003962935	3.853244171	4.198310814
APG	H7	0.409091	7.575759	11.22178	2.706276212	2.604405488	2.83763583
Diesel	H8	0.342424	6.341185	7.304117	1.47442394	1.418923088	1.545990828
Hydrogen	H9	0.457576	8.47363	13.36875	3.606155464	3.470411128	3.781194214
	Total		100%	100%	35.786775	34.439675	37.523825

3. Results and Discussion

Figure 6 presents the results of the forecast of energy consumption by the Arctic consumers based on the scenario conditions and risks, where three scenarios of development of electricity consumption and the process of change in the fuel and energy balance in the Arctic for the period from 2021 to 2035 are considered.

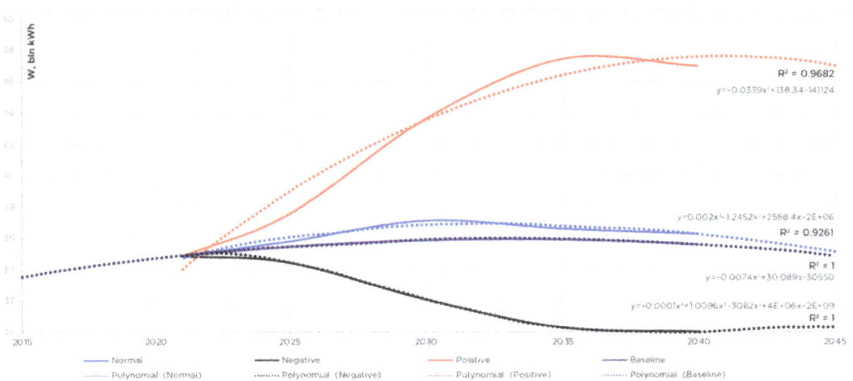

Figure 6. Forecast of energy consumption by Arctic consumers, billion kW·h.

The R^2 value—the coefficient of determination is the difference between the unit and the proportion of unexplained variance. This coefficient is applicable to determine the degree of correspondence between one random variable and many others. R^2 can be calculated automatically using, for example, standard MS Excel tools, as was done by the authors of the article.

The results of the calculated determination coefficients prove that the obtained mathematical model for predicting energy consumption using the developed hybrid method corresponds to the data from [80] sufficiently and does not contradict them. Relying on the available data on energy consumption by consumers in the Arctic, made it possible to carry out the initial iterations of the model tuning. In turn, this made it possible to impose on the collected database the influence of risks migrating over time.

Figure 7 shows the results for the distribution of the projected increase in energy consumption by type of consumers, which takes into account the energy consumption change on the time interval and the normalized total risk influence. The results indicate that the increase in electricity will be mainly due to data processing centers, hydrocarbon deposits and logistics clusters in neutral and positive scenarios. The growth will peak in the 2030s. Following the same scenarios there will be a degeneration of settlements based on industry. During the period under consideration, from 2021 to 2035, the volume of energy consumption will change mainly due to an increase or decrease in the demand for gas depending on the scenario variant of the Arctic zone development. The high potential of gas use is due to the large reserves of this resource in the Arctic and the developed infrastructure of gas pipelines in its territory. Thus, natural gas will occupy a leading position in the structure of the Arctic fuel and energy balance.

Accepting the positive development scenario as the best in terms of sustainability, a rather high contribution to the increase of energy consumption will be made by unconventional energy sources, such as atomic and hydrogen energy. The use of nuclear energy in the Arctic has proven its validity and effectiveness in the case of small and floating nuclear power plants. The results of the prognosis point to the development of hydrogen fuel as a new perspective energy source, which is planned to be produced based on the Kola NPP in the Murmansk Region and Yakutia, as well as on pilot sites in the Yamal-Nenets Autonomous District; and to start supplying LNG in nuclear tankers to remote Arctic areas. If the scenario is positive, there will be an increase in demand for LNG use, although this increase will not have a significant impact on the fuel and energy balance of the Arctic. Following the trend of decarbonization, the development of RESs will continue, but the increase in these resources is not comparable in volume with other resources under consideration, so it is not reflected in the simulation results. The largest increase in energy consumption will be achieved in 2030 due to gas, which is due to the planned gasification of the regions. The share of petroleum products will decrease and become equal to the share of hydrogen fuel. Such diversification of the resource mix will allow not to disturb the sensitive ecosystem of the Arctic and ensure an innovative breakthrough in Russia's energy industry, as well as the country's competitiveness in the market of clean fuels.

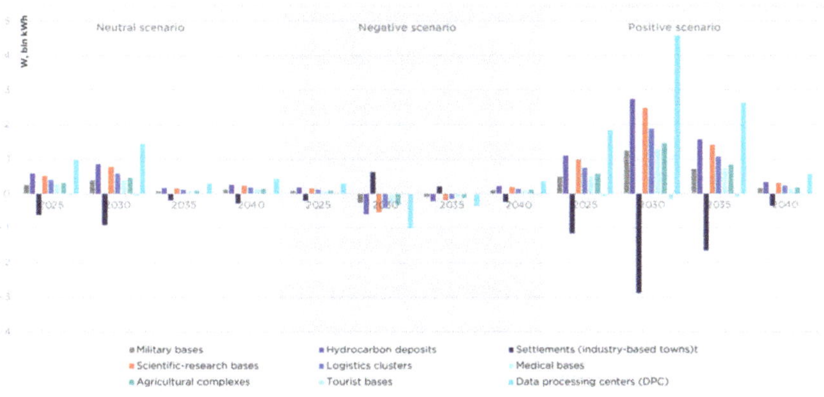

Figure 7. Distribution of the projected increase in energy consumption by type of consumers.

The model allowed taking into account not only the growth but also the decline of certain types of consumers under different scenarios. There is a noticeable increase in energy consumption of data centers, which is due to the emergence of new strategic and economic facilities, which will appear by 2025–2030 (supply of the NSR, SSH, military bases, research bases). Similarly, the demand for hydrocarbon deposits will increase, as the Arctic has a large potential of natural resources, which are already being exploited, and the development and emergence of technologies specialized for the Arctic conditions will have a high demand for energy consumption. Investment in the NSR, SSH, and the development of ports, railways, and highways will also be an impetus to increase the energy consumption of logistics clusters in the Arctic. The model clearly shows a drop in energy demand in single-industry towns. This can be explained by their low attractiveness to their current state and the prospect of attracting residents to live in these cities permanently is not observed. Consequently, we can conclude that mono-cities with the existing infrastructure and economy are a dead end, and in the Arctic, such development of territories is ineffective. Figure 8 shows the results of scenario forecasting of an increase or decrease in demand for resources at the corresponding time intervals.

The results showed that demand for hydrogen will increase over time due to loyalty and investment in hydrogen by large companies, as well as the emergence of infrastructure and research centers adapted to hydrogen fuel. Also, demand for gas, APG, LNG, and CNG will increase as LNG and CNG transportation campaigns will continue to roll out. The emergence of small NPPs and the prospect and approval by the Russian government of floating NPPs will be a step toward the growing demand for nuclear power in the Arctic. Demand for petroleum products will decrease due to the risks of spills leading to large fines, sanctions, and restrictions, which reduces the competitiveness of these fuels (fuel oil, diesel) compared to others. In the next 5–10 years, coal will still be in demand as a fuel for the Arctic, but gradually the demand for its use will start to fall.

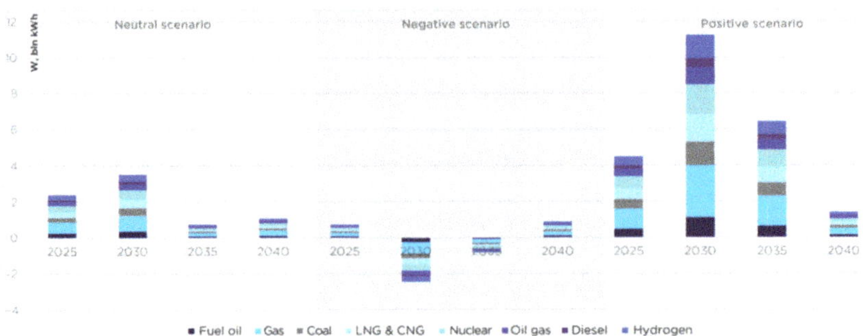

Figure 8. Results of resource demand forecasting.

Thus, scenario modeling of the impact of risks due to global challenges on the development of consumers in the Arctic, and the assessment of changes in demand for energy resources will allow tracing the mutual influence of risks, taking into account the scenario conditions on energy supply technology. This, in its turn, makes it possible to identify, through resources, the key technologies necessary for energy supply to Arctic consumers, taking into account all efficiency requirements while reducing the negative impact on the environmental situation in the region to achieve the goals of sustainable development.

4. Conclusions

Energy supply to consumers must be built on the principles of reliability, availability, and flexibility. Our research follows these principles and focuses on the consumer, considering his characteristics and needs. Energy supply also depends on the availability of resources, and our study answers this question using the potential of the region under consideration. The process of energy supply in the Arctic region should be considered as a single interconnected structure, where different types of consumers and energy sources are codependent. We have proposed a strategic planning approach based on the classification and analysis of the above variables, considering the opinion of experts, and scenario modeling of factors. The application of the relationship between consumer characteristics and the criteria for resources distinguishes our method from others. We understand that authors may encounter some limitations when repeating this technique. The risk groups identified in this study are applicable to the designated growth zones of the Russian Arctic. However, the technique is universal, which makes it possible to apply it in other studies, where their own risk groups and an expert group will be distinguished according to the specific request of the authors. Among the limitations, it is also worth highlighting that this approach does not consider the limitations caused by the infrastructure necessary to create conditions for the supply of energy with one or another resource in certain regions. The exact location of the growth in needs is also currently not determined by this method, which does not allow creating the necessary incentives for development at the regional level.

We would also like to emphasize that the study has great potential for development, among a number of urgent tasks for future work, the following can be distinguished: presentation of factor analysis of risks, which will allow us to assess the dynamics of changes in risks over time, as well as to highlight the group of the most significant risks and their relationship with each other; solution of the optimization problem: regional optimal distribution of resources with the least carbon dioxide emissions and financial investments. In addition, research on limiting the use of various resources, especially hydrogen, is promising.

Author Contributions: Conceptualization, A.B. and Y.M.; methodology, Y.M.; validation, A.K.; formal analysis, A.B. and Y.M.; investigation, A.S. and A.K.; writing-original draft preparation, A.B., A.K. and Y.M.; writing—review and editing, Y.Z. and P.T.; visualization, A.B. and Y.Z.; supervision, data curation P.T.; project administration, Y.Z. All authors have read and agreed to the published version of the manuscript.

Funding: The research was carried out with the financial support of the grant by the President of the Russian Federation for the state support of leading scientific schools of the Russian Federation, the number of the project NSh-2692.2020.5 "Modelling of ecological-balanced and economically sustainable development of hydrocarbon resources of the Arctic".

Conflicts of Interest: The authors declare no conflict of interest.

References

1. Cherepovitsyn, A.; Rutenko, E.; Solovyova, V. Sustainable Development of Oil and Gas Resources: A System of Environmental, Socio-Economic, and Innovation Indicators. *J. Mar. Sci. Eng.* **2021**, *9*, 1307. [CrossRef]
2. Tcvetkov, P. Climate Policy Imbalance in the Energy Sector: Time to Focus on the Value of CO2 Utilization. *Energies* **2021**, *14*, 411. [CrossRef]
3. Delakhova, A.M.; Grigoriev, E.P. Analysis of peculiarities of the food supply of the population of the northern regions. *Vector Econ.* **2021**, *11*, 244–248.
4. President of Russia. Decree of the President of the Russian Federation of October 26, 2020. "Strategy of Development of the Arctic Zone of the Russian Federation and Ensuring National Security for the Period Up to 2035". Available online: http://www.kremlin.ru/acts/bank/45972 (accessed on 18 October 2021).
5. NAVY. Northern Fleet. Operational-Strategic Association of the Russian Navy. Available online: https://flot.com/nowadays/structure/north/ (accessed on 18 October 2021).
6. TASS. ROSATOM Plans to Start Producing Hydrogen at Kola NPP in 2023. Available online: https://rosatom.ru/journalist/smi-about-industry/rosenergoatom-planiruet-v-2023-godunachat-proizvodstvo-vodoroda-na-kolskoy-aes/ (accessed on 18 October 2021).
7. Chan, F.T.; Stanislawczyk, K.; Sneekes, A.C.; Dvoretsky, A.; Gollasch, S.; Minchin, D.; David, M.; Jelmert, A.; Albretsen, J.; Bailey, S.A. Climate change opens new frontiers for marine species in the arctic: Current trends and future invasion risks. *Glob. Chang. Biol.* **2019**, *25*, 25–38. [CrossRef] [PubMed]
8. Tolvanen, A.; Eilu, P.; Juutinen, A.; Kangas, K.; Kivinen, M.; Markovaara-Koivisto, M.; Naskali, A.; Salokannel, V.; Tuulentie, S.; Similä, J. Mining in the arctic environment—A review from ecological, socioeconomic and legal perspectives. *J. Environ. Manag.* **2019**, *233*, 832–844. [CrossRef]
9. Treharne, R.; Bjerke, J.W.; Tømmervik, H.; Stendardi, L.; Phoenix, G.K. Arctic browning: Impacts of extreme climatic events on heathland ecosystem CO_2 fluxes. *Glob. Chang. Biol.* **2019**, *25*, 489–503. [CrossRef]
10. Abahussain, M.M.; Christie, R.D. Optimal scheduling of a natural gas processing facility with Price-based Demand Response. In Proceedings of the 2013 IEEE Power & Energy Society General Meeting, Vancouver, BC, Canada, 21–25 July 2013. [CrossRef]
11. Zendehboudi, A.; Baseer, M.; Saidur, R. Application of support vector machine models for forecasting solar and wind energy resources: A review. *J. Clean. Prod.* **2018**, *199*, 272–285. [CrossRef]
12. Walser, T.; Sauer, A. Typical load profile-supported convolutional neural network for short-term load forecasting in the industrial sector. *Energy AI* **2021**, *5*, 100104. [CrossRef]
13. Sánchez-Úbeda, E.F.; Berzosa, A. Modeling and forecasting industrial end-use natural gas consumption. *Energy Econ.* **2007**, *29*, 710–742. [CrossRef]
14. Foldvik Eikeland, O.; Bianchi, F.M.; Apostoleris, H.; Hansen, M.; Chiou, Y.-C.; Chiesa, M. Predicting Energy Demand in Semi-Remote Arctic Locations. *Energies* **2021**, *14*, 798. [CrossRef]
15. Brazovskaia, V.; Gutman, S.; Zaytsev, A. Potential Impact of Renewable Energy on the Sustainable Development of Russian Arctic Territories. *Energies* **2021**, *14*, 3691. [CrossRef]
16. Selina, V.S.; Skufyina, T.P.; Bashmakova, E.P.; Toropushina, E.E. North and the Arctic in the new paradigm of world development: Current problems, trends, prospects. In *Scientific and Analytical Report*; KCN RAS: Apatity, Russia, 2016; p. 420.
17. Official Information Portal of the Republic of Sakha (Yakutia). Decree of the Head of the Republic of Sakha (Yakutia) "On the Strategy of Social and Economic Development of the Arctic Zone of the Republic of Sakha (Yakutia) for the Period Up to 2035" of August 14, 2020 y. №1377. Available online: https://www.sakha.gov.ru/news/front/view/id/3204989 (accessed on 19 October 2021).
18. Zhukovskiy, Y.L.; Batuyeva, D.E.; Buldysko, A.D.; Gil, B.; Starshaia, V.V. Fossil Energy in the Framework of Sustainable Development: Analysis of Prospects and Development of Forecast Scenarios. *Energies* **2021**, *14*, 5268. [CrossRef]
19. National Research Council. *Hidden Costs of Energy: Unpriced Consequences of Energy Production and Use*; The National Academies Press: Washington, DC, USA, 2020. [CrossRef]
20. Van der Roest, E.; Snip, L.; Fens, T.; van Wijk, A. Introducing Power-to-H3: Combining renewable electricity with heat, water and hydrogen production, and storage in a neighborhood. *Appl. Energy* **2020**, *257*, 114024. [CrossRef]
21. Olkuski, T.; Suwała, W.; Wyrwa, A.; Zyśk, J.; Tora, B. Primary energy consumption in selected EU Countries compared to global trends. *Open Chem.* **2021**, *19*, 503–510. [CrossRef]

22. IEA. Material Efficiency in Clean Energy Transitions. 2019. Available online: https://iea.blob.core.windows.net/assets/52cb578 2-b6ed-4757-809f-928fd6c3384d/Material_Efficiency_in_Clean_Energy_Transitions.pdf (accessed on 20 October 2021).

23. Tcvetkov, P. Small-scale LNG projects: Theoretical framework for interaction between stakeholders. *Energy Rep.* **2022**, *8* (Suppl. 1), 928–933. [CrossRef]

24. Dvoynikov, M.; Buslaev, G.; Kunshin, A.; Sidorov, D.; Kraslawski, A.; Budovskaya, M. New Concepts of Hydrogen Pro-duction and Storage in Arctic Region. *Resources* **2021**, *10*, 3. [CrossRef]

25. Cherepovitsyn, A.; Evseyeva, O. Parameters of Sustainable Development: Case of Arctic Liquefied Natural Gas Projects. *Resources* **2021**, *10*, 1. [CrossRef]

26. Litvinenko, V. The Role of Hydrocarbons in the Global Energy Agenda: The Focus on Liquefied Natural Gas. *Resources* **2020**, *9*, 59. [CrossRef]

27. Worku, G.; Teferi, E.; Bantider, A.; Dile, Y.T. Prioritization of watershed management scenarios under climate change in the Jemma sub-basin of the Upper Blue Nile Basin, Ethiopia. *J. Hydrol. Reg. Stud.* **2020**, *31*, 100714. [CrossRef]

28. Jefferson, M. Scenario planning: Evidence to counter 'Black box' claims. *Technol. Forecast. Soc. Chang.* **2020**, *158*, 120156. [CrossRef]

29. Megan, M.G.; George, W. Delphi Method. *Wiley StatsRef Stat. Ref. Online* **2016**, 1–6. [CrossRef]

30. Linzenich, A.; Zaunbrecher, B.; Ziefle, M. "Risky transitions?" Risk perceptions, public concerns, and energy infrastructure in Germany. *Energy Res. Soc. Sci.* **2020**, *68*, 101554. [CrossRef]

31. Shove, E. Time to rethink energy research. *Nat. Energy* **2020**, *6*, 118–120. [CrossRef]

32. Chen, K.; Ren, Z.; Mu, S.; Sun, T.; Mu, R. Integrating the delphi survey into scenario planning for China's renewable energy development strategy towards 2030. *Technol. Forecast. Soc. Chang.* **2020**, *158*, 120157. [CrossRef]

33. Jones, A.W. Perceived barriers and policy solutions in clean energy infrastructure investment. *J. Clean. Prod.* **2015**, *104*, 297–304. [CrossRef]

34. Cherepovitsyn, A.E.; Lipina, S.A.; Evseeva, O.O. Innovative approach to the development of mineral and raw material potential of the Arctic zone of Russia. *Notes Min. Inst.* **2018**, *232*, 438–444. [CrossRef]

35. Filatova, I.; Nikolaichuk, L.; Zakaev, D.; Ilin, I. Public-private partnership as a tool of sustainable development in the oil-refining sector: Russian case. *Sustainability* **2021**, *13*, 5153. [CrossRef]

36. Volkov, A.V.; Bortnikov, N.S.; Lobanov, K.V.; Galyamov, A.L.; Chicherov, M.V. Deposits of strategic metals of the Arctic region. In Proceedings of the Fersman Scientific Session of the Institute of the KSC RAS 2019, Apatity, Russia, 7–10 April 2019; Volume 16, pp. 80–84. [CrossRef]

37. Nikishin, A.M.; Petrov, E.I.; Cloetingh, S. Arctic Ocean Mega Project: Paper 3—Mesozoic to Cenozoic geological evolution. *Earth-Sci. Rev.* **2021**, *217*, 103034. [CrossRef]

38. Litvinenko, V.S.; Dvoynikov, M.V.; Trushko, V.L. Elaboration of a conceptual solution for the development of the Arctic shelf from seasonally flooded coastal areas. *Int. J. Min. Sci. Technol.* **2021**, *1*. [CrossRef]

39. Fedoseyev, S.V.; Tsvetkov, P.S. Key factors in public perception of carbon dioxide capture and disposal projects. *Proc. Min. Inst.* **2019**, *237*, 361–368. [CrossRef]

40. GoArctic. The Arctic of Megaprojects: What's Changing in the Regions. Available online: https://goarctic.ru/society/arktika-megaproektov-chto-menyaetsya-v-regionakh/ (accessed on 20 October 2021).

41. Chelnokova, I. The Arctic. Northern contradictions. *Kommersant* **2020**, *224*, 25.

42. Institute for Applied Political Solutions. Analytical Report "Single-Industry" Towns in the Arctic Zone of the Russian Federa-tion: Problems and Opportunities for Development". Available online: http://www.arcticandnorth.ru/Encyclopedia_Arctic/monogoroda_AZRF.pdf (accessed on 19 October 2021).

43. Ranking of Sustainable Development of the Regions of the Russian Arctic. Available online: https://www.econ.msu.ru/sys/raw.php?o=73806&p=attachment (accessed on 19 October 2021).

44. Khramchikhin, A.A. Military and political situation in the Arctic and possible prospects for its development. *Vestnik MSTU* **2014**, *3*, 606–615.

45. Government of Russia. Decree of the Government of Russian Federation from 24.05.2021 № 1338-p "On Equipping the Arctic Aviation Units of EMERCOM of Russia with Aviation Equipment". Available online: http://government.ru/news/42301/ (accessed on 20 October 2021).

46. Ministry of the Russian Federation for the Development of the Far East and the Arctic. Development of Distributed Geo-Nergy in FEFD and Arctic: The Profile Committee of the State Duma Supported the Proposals of FERC. Available online: https://minvr.gov.ru/press-center/news/32116/ (accessed on 18 October 2021).

47. Young, R.O. Arctic Futures–Future Arctics? *Sustainability* **2021**, *13*, 9420. [CrossRef]

48. Chanysheva, A.; Kopp, P.; Romasheva, N.; Nikulina, A. Migration Attractiveness as a Factor in the Development of the Russian Arctic Mineral Resource Potential. *Resources* **2021**, *10*, 65. [CrossRef]

49. Carayannis, E.G.; Ilinova, A.; Cherepovitsyn, A. The Future of Energy and the Case of the Arctic Offshore: The Role of Strategic Management. *J. Mar. Sci. Eng.* **2021**, *9*, 134. [CrossRef]

50. Chanyasheva, A.; Ilinova, A. The Future of Russian Arctic Oil and Gas Projects: Problems of Assessing the Prospects. *J. Mar. Sci. Eng.* **2021**, *9*, 528. [CrossRef]

51. Corell, R.; Kim, J.D.; Kim, Y.H.; Moe, A.; VanderZwaag, D.L.; Young, O.R. (Eds.) Arctic Resource Development: Economics and Politics. In *The Arctic in World Affairs: A North-Pacific Dialogue on Global-Arctic Interactions*; Korea Maritime Institute: Busan, Korea; East-West Center KMI and EWC: Honolulu, HI, USA, 2019; pp. 205–224.

52. Sterner, M.; Specht, M. Power-to-Gas and Power-to-X—The History and Results of Developing a New Storage Concept. *Energies* **2021**, *14*, 6594. [CrossRef]

53. Gurieff, N. Power-to-X Renewable Resource Ecosystems. *Sustainability* **2020**, *12*, 8554. [CrossRef]

54. Roscongress. Healthcare in the Arctic: Results of Two Years and New Goals. Available online: https://roscongress.org/news/zdravoohranenie-arktiki-itogi-dvuh-let-i-novye-tseli/ (accessed on 18 October 2021).

55. Roscongress. Arctic Tourism in Russia Has a Chance. Available online: https://roscongress.org/materials/u-arkticheskogo-turizma-v-rossii-poyavilsya-shans (accessed on 17 October 2021).

56. Sychev, Y.A.; Zimin, R.Y. Improving the quality of electricity in the power supply systems of the mineral resource complex with hybrid filter-compensating devices. *J. Min. Inst.* **2021**, *247*, 132–140. [CrossRef]

57. Lavrenko, S.A.; Shishljannikov, D.I. Performance evaluation of heading-and-winning machines in the conditions of potash mines. *Appl. Sci.* **2021**, *8*, 3444. [CrossRef]

58. Al-falahi Monaaf, D.A.; Jayasinghe, S.D.G.; Enshaei, H. A review on recent size optimization methodologies for standalone solar and wind hybrid renewable energy system. *Energy Convers. Manag.* **2017**, *43*, 252–274. [CrossRef]

59. Turysheva, A.; Voytyuk, I.; Guerra, D. Estimation of Electricity Generation by an Electro-Technical Complex with Pho-toelectric Panels Using Statistical Methods. *Symmetry* **2021**, *13*, 1278. [CrossRef]

60. Our World in Data. CO_2 Emissions by Fuel. Available online: https://ourworldindata.org/emissions-by-fuel (accessed on 20 October 2021).

61. EIA. Levelized Costs of New Generation Resources in the Annual Energy Outlook 2021. Available online: https://www.eia.gov/outlooks/aeo/pdf/electricity_generation.pdf (accessed on 20 October 2021).

62. IEA. Projected Costs of Generating Electricity. Available online: https://www.oecd-nea.org/upload/docs/application/pdf/2020-12/egc-2020_2020-12-09_18-26-46_781.pdf (accessed on 20 October 2021).

63. Bloomberg. Cost of Capital Spikes for Fossil-Fuel Producers. Available online: https://www.bloomberg.com/news/articles/2021-11-09/cost-of-capital-widens-for-fossil-fuel-producers-green-insight (accessed on 21 October 2021).

64. IEA. The Oil and Gas Industry in Energy Transitions. Available online: https://www.iea.org/reports/the-oil-and-gas-industry-in-energy-transitions (accessed on 23 October 2021).

65. IEA. Energy Technology Perspectives. Available online: https://iea.blob.core.windows.net/assets/7f8aed40-89af-4348-be19-c8a67df0b9ea/Energy_Technology_Perspectives_2020_PDF.pdf (accessed on 20 October 2021).

66. Phebe, A.O.; Samuel, A.S. A review of renewable energy sources, sustainability issues and climate change mitigation. *Co-Gent Eng.* **2016**, *3*, 1–14. [CrossRef]

67. Gartner. Emerging Technology Roadmap for Large Enterprises 2020–2022. Available online: https://emtemp.gcom.cloud/ngw/globalassets/en/information-technology/documents/benchmarks/emerging-tech-roadmap-le-2020-2022.pdf (accessed on 19 October 2021).

68. McKinsey Digital. The Top Trends in Tech. Available online: https://www.mckinsey.com/business-functions/mckinsey-digital/our-insights/the-top-trends-in-tech (accessed on 20 October 2021).

69. BloombergNEF. New Energy Outlook 2021. Available online: https://about.bnef.com/blog/getting-on-track-for-net-zero-by-2050-will-require-rapid-scaling-of-investment-in-the-energy-transition-over-the-next-ten-years/ (accessed on 20 October 2021).

70. Gartner. Top Strategic Technology Trends for 2021. Available online: https://emtemp.gcom.cloud/ngw/globalassets/en/information-technology/documents/insights/top-tech-trends-ebook-2021.pdf (accessed on 20 October 2021).

71. Hannan, M.A.; Al-Shetwi, A.Q.; Begum, R.A.; Ker, P.J.; Rahman, S.A.; Mansor, M.; Mia, M.S.; Muttaqi, K.M.; Dong, Z.Y. Impact assessment of battery energy storage systems towards achieving sustainable development goals. *J. Energy Storage* **2021**, *42*, 103040. [CrossRef]

72. Hannan, M.A.; Faisal, M.; Ker, P.J.; Begum, R.A.; Dong, Z.Y.; Zhang, C. Review of optimal methods and algorithms for sizing energy storage systems to achieve decarbonization in microgrid applications. *Renew. Sustain. Energy Rev.* **2020**, *131*, 110022. [CrossRef]

73. Tung, K.-K.; Zhou, J. Using data to attribute episodes of warming and cooling in instrumental records. *Proc. Natl. Acad. Sci. USA* **2012**, *110*, 2058–2063. [CrossRef] [PubMed]

74. Kohn, N.W.; Smith, S.M. Collaborative fixation: Effects of others' ideas on brainstorming. *Appl. Cogn. Psychol.* **2010**, *25*, 359–371. [CrossRef]

75. Anthony, J.; Gibson, A. *Sustainability in Engineering Design*; Academic Press: Waltham, MA, USA, 2014.

76. Al-Samarraie, H.; Hurmuzan, S. A review of brainstorming techniques in higher education. *Think. Ski. Creat.* **2018**, *27*, 78–91. [CrossRef]

77. Vasilyeva, N.V.; Boikov, A.V.; Erokhina, O.O.; Trifonov, A.Y. Automated digitization of radial charts. *J. Min. Inst.* **2021**, *247*, 82–87. [CrossRef]

78. Koteleva, N.; Buslaev, G.; Valnev, V.; Kunshin, A. Augmented Reality System and Maintenance of Oil Pumps. *Int. J. Eng.* **2020**, *33*, 1620–1628. [CrossRef]

79. Shabalov, M.Y.; Zhukovskiy, Y.L.; Buldysko, A.D.; Gil, B.; Starshaia, V.V. The influence of technological changes in en-ergy efficiency on the infrastructure deterioration in the energy sector. *Energy Rep.* **2021**, *7*, 2664–2680. [CrossRef]
80. Rosstat. Available online: https://rosstat.gov.ru (accessed on 20 October 2021).

Article

Ensuring the Sustainability of Arctic Industrial Facilities under Conditions of Global Climate Change

George Buslaev, Pavel Tsvetkov *, Alexander Lavrik, Andrey Kunshin, Elizaveta Loseva and Dmitry Sidorov

Arctic Competence Center, Saint Petersburg Mining University, 199106 Saint Petersburg, Russia;
Buslaev_GV@pers.spmi.ru (G.B.); Lavrik_AYu@pers.spmi.ru (A.L.); Kunshin_A2@pers.spmi.ru (A.K.);
Loseva_ES@pers.spmi.ru (E.L.); Sidorov_DA2@pers.spmi.ru (D.S.)
* Correspondence: pscvetkov@yandex.ru

Abstract: Global climate change poses a challenge to the mineral development industry in the Arctic regions. Civil and industrial buildings designed and constructed without consideration of warming factors are beginning to collapse due to changes in the permafrost structure. St. Petersburg Mining University is developing technical and technological solutions for the construction of remote Arctic facilities and a methodology for their design based on physical and mathematical predictive modeling. The article presents the results of modeling the thermal regimes of permafrost soils in conditions of thermal influence of piles and proposes measures that allow a timely response to the loss of bearing capacity of piles. Designing pile foundations following the methodology proposed in the article to reduce the risks from global climate change will ensure the stability of remote Arctic facilities located in the zone of permafrost spreading.

Keywords: Arctic shelf; permafrost; global warming; ground thawing; modular pile foundation; physical and mathematical modeling; temperature stabilization

Citation: Buslaev, G.; Tsvetkov, P.; Lavrik, A.; Kunshin, A.; Loseva, E.; Sidorov, D. Ensuring the Sustainability of Arctic Industrial Facilities under Conditions of Global Climate Change. *Resources* **2021**, *10*, 128. https://doi.org/10.3390/resources10120128

Academic Editor: Ben McLellan

Received: 30 October 2021
Accepted: 13 December 2021
Published: 15 December 2021

Publisher's Note: MDPI stays neutral with regard to jurisdictional claims in published maps and institutional affiliations.

1. Introduction

Combating climate change is the United Nations' 13th Sustainable Development Goal [1]. According to the UN Intergovernmental Panel on Climate Change (IPCC) [2], the Arctic is warming faster than the rest of the world, which is highlighted in every report. For example, a global warming has already reached 1 °C above pre-industrial period (1850–1900). At the same time, the measured temperature increase in the Arctic was two to three times higher, and there were significant differences between Arctic regions [3]. IPCC climate models predict that this trend will continue: a 2 °C increase by 2100 globally is projected to result in a 4–7 °C increase in Arctic temperatures.

There are a number of studies on climate change on Earth that link warming processes with anthropogenic greenhouse gas emissions [4,5]. Global anthropogenic greenhouse gas emissions increased by 1.7% in 2017 and by about 2.7% in 2018 [6]. But even with all current national commitments to reduce greenhouse gas emissions, an average global average annual temperature increases of 3 °C is projected, corresponding to an average nighttime temperature increase of 7–11 °C in the Arctic [3].

Over the past decades, global warming has led to a widespread decrease in the cryosphere with loss of ice sheet and glacier mass, reduction in snow cover, and an increase in the area and thickness of Arctic sea ice, as well as an increase in permafrost temperatures [7]. Permafrost temperatures have risen to record-high levels (from 1980 to the present), including a recent increase of 0.29 ± 0.12 °C from 2007 to 2016 in the average polar and high-altitude regions of the world. Widespread thawing of permafrost is expected with high confidence this century and in future years [2].

The relevance of developing solutions aimed at ensuring the stability of foundations of objects located in the zone of permafrost spreading has recently increased in light of global climate change on planet Earth. For example, the accident at CHPP-3 of Norilsk-Taimyr

Energy Company on 29 May 2020, can be attributed to the consequences of the loss of bearing capacity of the foundations located in the permafrost spreading zone [8]. The incident spilled about 21,000 tons of oil products, of which 6000 tons ended up in the ground, and the rest in the Ambarnaya River and its tributary Daldykan, which flows into the large Pyasino Lake. From this lake flows the Pyasina River, which flows into the Kara Sea.

1.1. Increase of CO_2 Emissions as Permafrost Melts

A number of studies have noted the acceleration of climate change processes in the Arctic region, while an inverse relationship is observed in the Antarctic [9,10]. It is predicted that melting permafrost in the Arctic will lead to the emission of "additional" 25–85 billion tons of greenhouse gases per year (in terms of carbon), given that all mankind emits about 13 billion tons of carbon [11]. As a result, tundra soils will not absorb, but release "extra" carbon dioxide and methane. Currently, the tundra and other areas of permafrost are among the absorbers of greenhouse gases—areas in which natural systems absorb more greenhouse gases, including CO_2 and methane, than they are formed in this area. A large proportion is deposited in peat or soil, some of which is in a state of permafrost.

Due to higher temperatures and CO_2 concentrations, plants will be able to absorb more carbon dioxide—their "productivity" will increase from 69 to 88 billion tons of carbon. On the other hand, the melting of permafrost will cause organic deposits in the tundra soil to "unfreeze" and begin to rot, releasing carbon dioxide and methane [12].

By 2100, according to projections, the near-surface (within 3–4 m) permafrost area will decrease by 24 ± 16% (probable range) for RTC2. 6 (scenario with warming of 1.1–2.0 °C during 2031–2050 and 0.9–2.4 °C during 2081–2100) and 69 ± 20% (probable range) for RTC8.5 (1.5–2.4 °C during 2031–2050 and 3.2–5.4 °C during 2081–2100), Figure 1 [2,6,7,13,14].

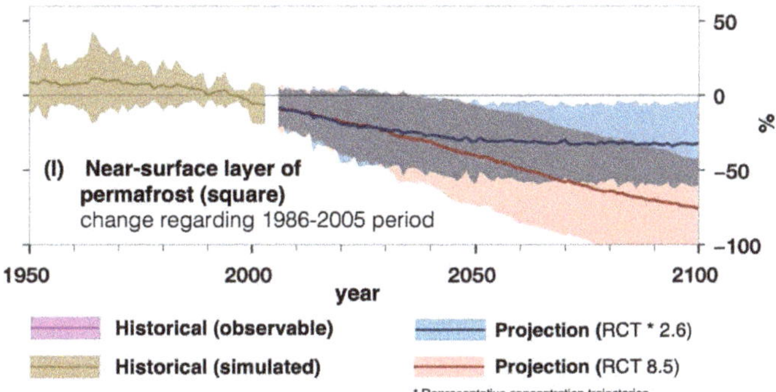

Figure 1. Historical observations and projections from RTK2.6 and RTK8.5 for near-surface permafrost [14].

1.2. Rising Global Sea Levels and Risk to Infrustructure

For coastal regions, including the Arctic, sea level rise poses a threat to the stability of industrial structures [15,16]. The risk of seasonal flooding of such territories increases, including that caused by the formation of a surge wave in the water area [17–19].

Even if the world goes down the path of low greenhouse gas emissions, the global sea level is likely to rise at least 0.3 m above 2000 levels by 2100. If we go down the high emission pathway, we cannot rule out a worst-case scenario where the 2100 level will exceed the one of 2000 by 2.5 m [14,20].

About 60% of the territory of the Russian Federation is in the permafrost zone with major mineral reserves in it, as shown in Figure 2 [21–26]. One of the important tasks of ensuring sustainable functioning of facilities in the Arctic zone of the Russian Federation is to prevent defrosting and thawing of permafrost [27,28].

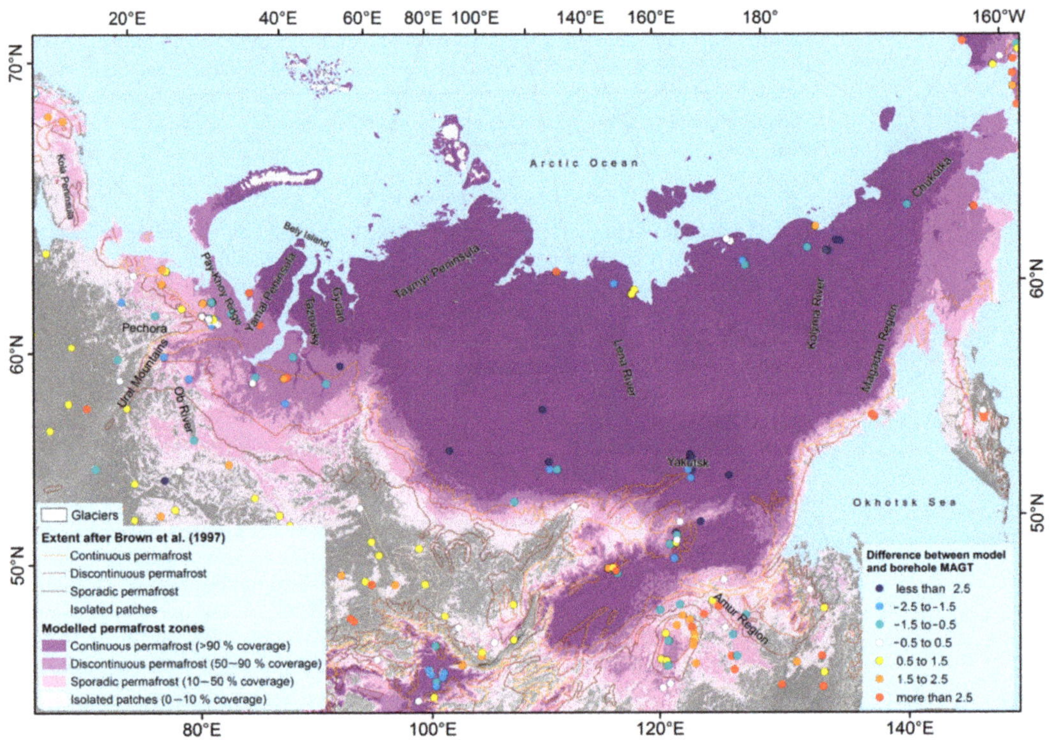

Figure 2. Global Permafrost zonation for Russia [26].

The wide distribution of permafrost over the entire Arctic shelf and the presence of an extremely harsh climate pose enormous difficulties for construction [29–31]. There are many ways of building objects in the Arctic conditions. For example, ventilated basement, which provides heat removal from the building and prevents its penetration into the ground, an injection consolidation technology or improvement of soil properties using Jet Grouting [32].

The main part of the shelf area under consideration is shallow, with prevailing depths of 1–3 m. Only in the extreme southern part of the area, the sea depth reaches 5 m and more. The presence of a weak subsoil base carries a risk associated with the insufficient stability of structures. For such conditions, the possibility of forced freezing with subsequent thermal stabilization of weak bottom sediments becomes topical in order to locate remote Arctic objects, both on land and at sea. Anyway, since the environmental impact of infrastructure facilities is known, it is necessary to apply the best technologies for the construction and operation of such facilities [33,34].

The aim of this paper is to search for answers to modern challenges arising from global climate change: thawing of permafrost and loss of stability of pile foundations, sea level rise and, as a consequence, an increase in the intensity of seasonal flooding of coastal Arctic territories.

The paper solves the problems related to design of modular pile foundations, modeling the consequences of global warming and its impact on the pile foundations bearing capacity, development a methodology for predicting changes [35] and monitoring changes during the operation of remote Arctic objects, development of measures for saving the bearing capacity of modular pile foundations.

2. Materials and Methods

2.1. Proposed Solution for Remote Arctic Oil and Gas Facilities

The proposed solution for the creation of industrial infrastructure facilities is a pile foundation, mounted in wintertime with pre-strengthening of soils, see Figure 3.

The concept of creating modular pile foundations developed at Mining University [36,37] implies optimal placement and year-round operation of equipment production infrastructure facilities in the allocated technological zones. To implement these solutions, new types of piles or traditional foundations, which are widely used in construction work in areas of permafrost soils, can be used.

Figure 3. Prototyping of production infrastructure facilities on a pile foundation.

The Arctic Research Center of Mining University developed design documentation for modular pile foundations (Figure 4) for placing drilling rigs in zones of permafrost spreading, subject to seasonal flooding. As a result of the research and development, it was found that the cost of constructing modular pile foundations is on average 50% less than the cost of constructing and maintaining sand dumps [38].

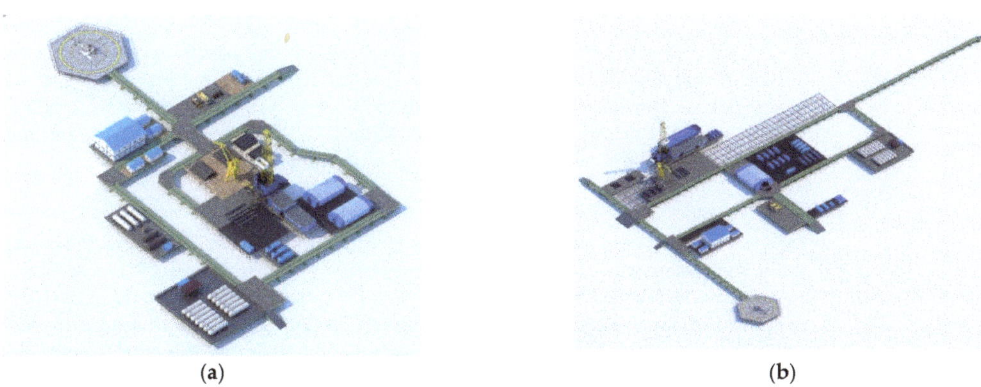

(**a**)   (**b**)

Figure 4. The modular pile base of a drilling rig. (**a**) for exploration drilling, (**b**) for production drilling.

2.2. Physical and Mathematical Modeling of Geotechnical Solutions for the Location of Arctic Oil and Gas Facilities under Climate Change

To select the best technological solutions, it is first necessary to understand how climate change in the Arctic zone affects the bearing capacity of the piles. For this purpose, according to statistical data, three warming scenarios for the period from 2031 to 2050 were developed as initial characteristics for modeling: (1) positive with a temperature increase of 2.2 °C (0.1 °C per year); (2) neutral with a temperature increase of 3 °C (0.16 °C per year); (3) negative with a temperature increase of 4.8 °C (0.24 °C per year); (4) locally negative with a temperature increase of 9.6 °C (0.5 °C per year). The initial data for modeling were obtained as a result of engineering surveys at a remote field in the Russian Arctic zone. The characteristics of soils for modeling the bearing capacity of foundations are summarized in Table 1.

Table 1. Soil characteristics for modeling the bearing capacity of foundations.

Layer	Type of Soils	Cut Open Thickness, m	Density of Dry Soil ρ_d, g/cm^3	Total Humidity W_{tot}, d.e.
Layer-1	Brown peat, malleable, highly porous	0.5	0.14	4.82
Layer-2	Sand, fine, solid frozen	3.5	1.58	0.23
Layer-3	Clay loam, dark gray hard frozen	4.5	1.26	0.31
Layer-4	Sandy silt	7.0	1.56	0.24

The physical and mechanical properties of soils are determined from laboratory data. Additional parameters such as heat capacity and thermal conductivity of the material in the thawed and frozen state were calculated according to the Russian standardization document [39]. The average active layer, according to the surveys, was 0.5 m.

The initial meteorological data are average monthly and annual air temperature and wind speed. The data are shown in Table 2.

Table 2. Air temperature and wind speed (average monthly and annual data).

						Air Temperature, °C						
I	II	III	IV	V	VI	VII	VIII	IX	X	XI	XII	per Year
−23.1	−24.6	−20.4	−15.3	−6.8	−0.7	6.0	5.8	2.6	−5.8	−14.0	−18.7	−9.5
						Wind Speed, m/s						
I	II	III	IV	V	VI	VII	VIII	IX	X	XI	XII	per Year
6.4	6.2	6.3	6.0	6.3	5.9	5.4	5.8	6.4	6.8	6.8	7.0	6.3

Modeling the distribution of soil temperatures around the pile and calculating the bearing capacity of the pile foundations was carried out in the Frost 3D software. In this researce we considered the air temperature increase to model active layer depth. However, there are other factors such as content of water [40–42], air content [42], organic matter [43], etc. Taking these factors into account is a direction for further research.

2.3. Proposed Solution for the Stability of Pile Foundations in Permafrost

Thermosyphons are an effective means of temperature stabilization of permafrost soils, the main elements of which are an evaporator and a condenser. In cold seasons, ground cooling is performed by natural convection of low-boiling refrigerant (ammonia, refrigerant, carbon dioxide, etc.) in a thermosyphon, with heated coolant flowing from the buried evaporator to the condenser, which dissipates the heat into the atmosphere, and cooled coolant back to the evaporator. During the warm season, when the ground temperature is lower than the atmospheric air temperature, the thermosiphon does not work. Thus, capacity of the ground temperature stabilization system should be calculated so that accumulated winter "cold" was enough to maintain the necessary ground temperature until the new cold season comes.

However, the accelerated global warming in recent years complicates the task of thermal stabilization of pile foundations. At the construction stage it is required to install more productive thermosiphons or a larger number of them. Also at the stage of operation it's required to provide for the possibility of installing new thermosiphons, taking into account the global warming. In this regard, there are developments providing not only passive cooling of the soil by thermosiphons, but also its active cooling. For example, the thermosiphon can have a second circuit, closed to the cooling machine, which allows to cool the soil adjacent to the pile in the warm season [44]. In another variant, the thermosiphon can have one circuit; however, when the warm season comes, the liquid refrigerant should be pumped out, and then the cold air should be circulated by means of an air turbo-cooling machine [45]. In [46], a combined thermal stabilization unit is proposed, which in addition to passive ground cooling by thermosiphon provides for active cooling, carried out by injection into the thermosiphon, cooled by throttling the refrigerant through a special removable nozzle of the thermosiphon. The method of year-round ground cooling by means of thermosiphon and compressor-condenser unit is also known, according to which the thermosiphon condenser is simultaneously a refrigerating machine evaporator [47]. Issues of using thermoelectric modules for year-round ground stabilization based on the Pelte effect are covered in [22]; however, the results of pilot operation by enterprises Fondamentproekt and Gazprom VNIIGAZ showed the complexity of providing the required ground temperature and high-power consumption of this technology.

At the same time, if the object is far away from the centralized power supply, which is quite common in relation to objects of the oil and gas sector, the organization of power supply of the refrigeration machine may present certain difficulties. In addition, it is possible that the fight against the effects of global warming actually aggravates the climate problem—if the refrigeration machine uses installations based on fossil fuels, especially diesel, fuel oil, etc. For these reasons, the idea of using renewable energy sources (RESs) to stabilize permafrost has been developed. The possibility of using RESs for active thermal stabilization is mentioned in [45–48].

Mining University proposed to use a combined system based on a two-circuit thermosiphon and a refrigeration machine connected to its second circuit. Power supply of the refrigeration machine is provided by RESs, for example—wind power or photovoltaic station, also there are storages of electricity and a reserve source to increase the reliability of power supply. The peculiarity of the proposed device is the placement of electric power storage, backup source, control system and other devices, sensitive to the temperature regime, in a thermally insulated container. The container is heated in the cold season at the expense of the heat removed from the thermosiphon condenser with an additional circuit with a coolant. During the warm season, when the thermosiphon is not in operation, heating of the thermally insulated container is usually not required, otherwise its own thermostat system can be used. The functional diagram of the proposed system is shown in Figure 5.

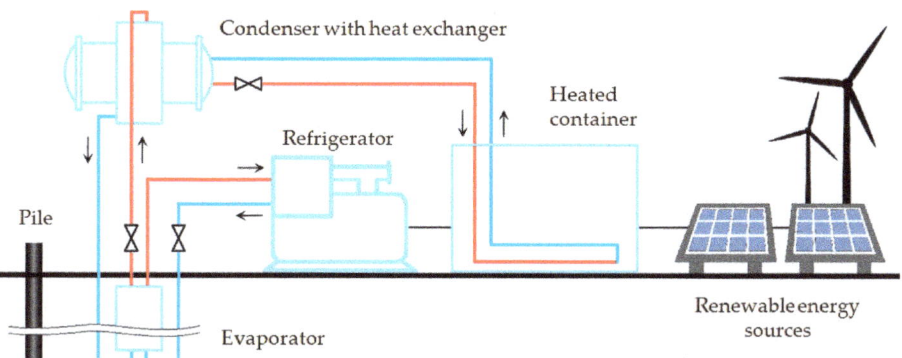

Figure 5. Functional diagram of an autonomous thermal stabilization unit with a thermosiphon and a refrigerating machine powered by renewable energy sources and energy storage system inside the heated container [49].

The heat capacity required to heat the equipment container was determined by Formula (1):

$$P = \frac{V \cdot \Delta T \cdot K}{860},$$ (1)

where V is the volume of the heated room, ΔT is the difference between the ambient air temperature and the desired temperature in the heated room, K is the coefficient of heat losses, 860 is the coefficient to convert kcal/h to kW.

For the calculation, the following geometric dimensions of the insulated container were taken: length 6 m, width 2.5 m, height 2.6 m. Such dimensions of the container allow to place electric energy accumulators, reserve source, control unit, electric energy conversion devices and circulation pump of the container heating system. The coefficient of heat losses is taken equal to 2, which corresponds to an average level of thermal insulation. So, in order to have 0 °C (the lower temperature limit for Russian-made lithium-ion energy storage units) inside the container in the coldest winter month (according to Table 2, February, temperature is −24.6 °C), heat power supply of about 2.23 kW is needed. It is also necessary to take into account thermal losses, which in each case will differ depending on the design of the system. Assuming the coefficient, taking into account heat losses by pipelines of hot water supply systems, equal to 1.15 kW, we obtain that in the coldest winter month heat capacity of 2.57 kW is required.

In the above mentioned Frost 3D software package the heat removed from the thermosiphon condensers is calculated, which can be used for heating the insulated container with equipment in the cold season.

Figure 6 shows a graph of the power of 14 thermosiphons under the considered conditions, a graph of the required thermal power to heat the container and a graph of the ambient air temperature.

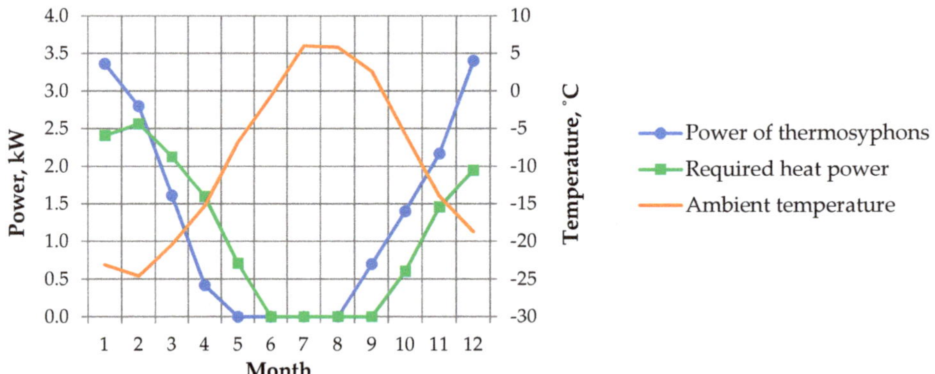

Figure 6. Annual graph of monthly averaged capacity of thermosyphon without application of jet grouting technology.

In accordance with the results of simulation of cooling capacity in the program Frost 3D and numerical simulation it was found that under the considered conditions, when using 14 thermosiphons, the required air temperature inside the container is maintained 9 months out of 12, and is not provided in March, April and May. The possibility of heat recovery when using a backup power generation source was not considered. For 3 out of 12 months of the year, the temperature inside the container must be maintained by other means of thermostatting. It should be noted that it is possible to use smaller containers.

The issues of thermostatting the bases of remote Arctic objects become even more relevant when placing gas-chemical complexes [50] with high energy flows in the areas of permafrost spreading. In any case, the number of thermosiphons is determined based on the year-round provision of the bearing capacity of piles, and the possibility of effective use of thermal energy of thermosiphon condensers is determined after fulfilling the conditions of mechanical stability of pile foundations.

3. Results and Analysis

In order to ensure the reliability of a building being erected it is necessary not only to choose a reliable foundation technology, but also to be able to predict the performance of the structure. With the help of Frost 3D simulation software, it is possible to obtain scientifically based predictions of the thermal regimes of permafrost soils in conditions of thermal influence of piles as well as of erected buildings and structures. This problem of heat distribution in the foundation over time in conditions of construction on permafrost soils is of primary importance.

According to the results of modeling for the year 2050, the following values and figures were obtained. Under the positive scenario with the temperature increase by 2.2 °C for the period from 2031 to 2050. (0.1 °C per year) the active layer will be 0.595 m. The modeling results of active layer changes due to warming up to 2050 under a positive scenario are shown in Figure 7.

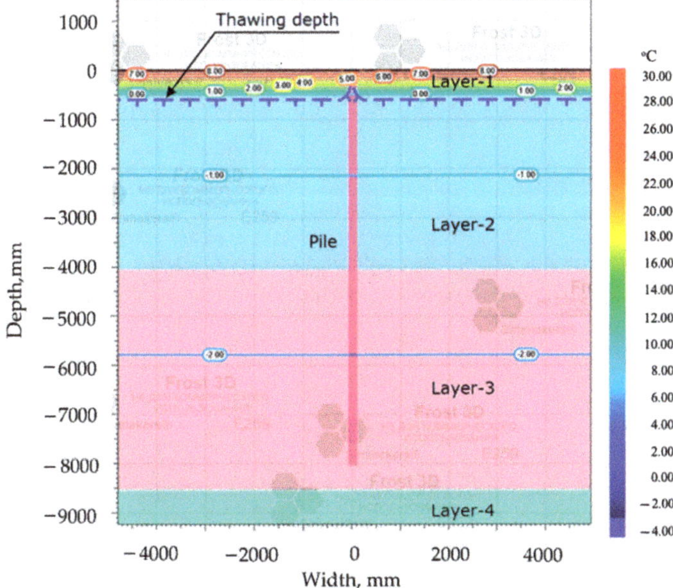

Figure 7. Temperature distribution for 2050 under a positive warming scenario.

Under the neutral scenario with a temperature increase by 3 °C for the period from 2031 to 2050. (0.16 °C per year) the active layer will be 0.673 m. The modeling results of active layer changes due to warming up to 2050 under a neutral scenario are shown in Figure 8.

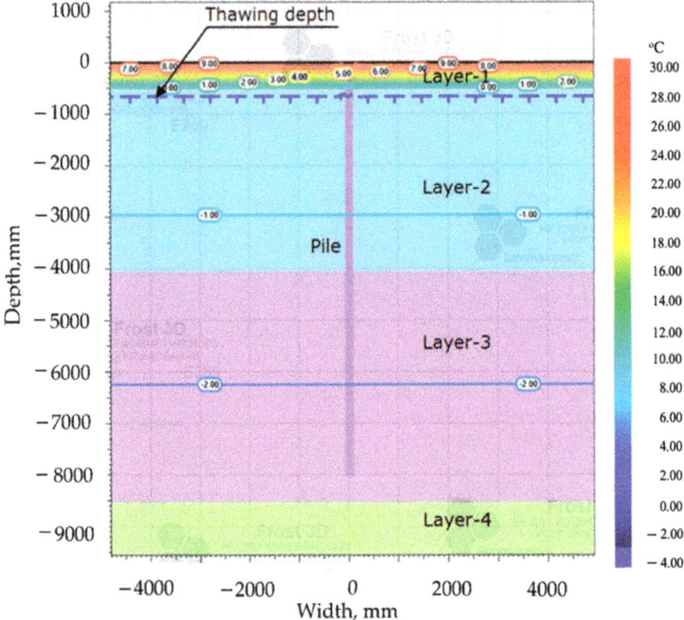

Figure 8. Temperature distribution for 2050 under a neutral warming scenario.

Under the negative scenario with an increase in temperature by 4.8 °C for the period from 2031 to 2050. (0.24 °C per year) the active layer will be 0.840 m. The modeling results of active layer changes due to warming up to 2050 under a negative scenario are shown in Figure 9.

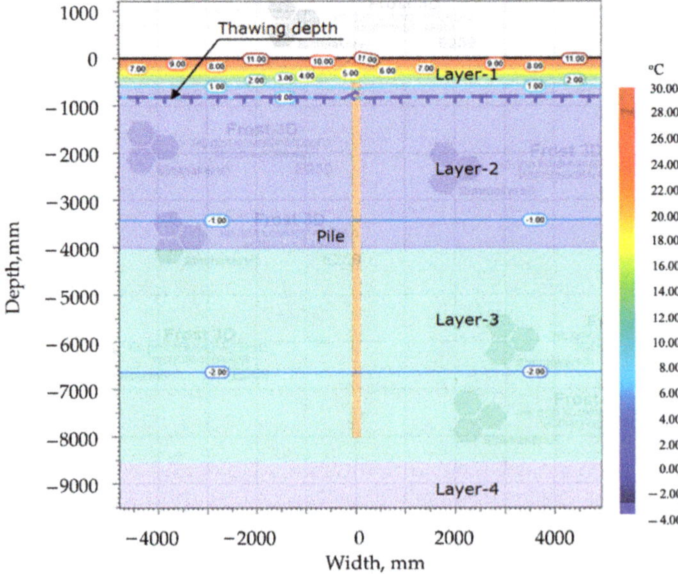

Figure 9. Temperature distribution for the year 2050 under a negative scenario of warming.

Under the local negative scenario with temperature increase by 9.6 °C for the period from 2031 to 2050. (0.5 °C per year) the active layer will be 1.868 m. The modeling results of active layer changes due to warming up to 2050 under a local negative scenario are shown in Figure 10.

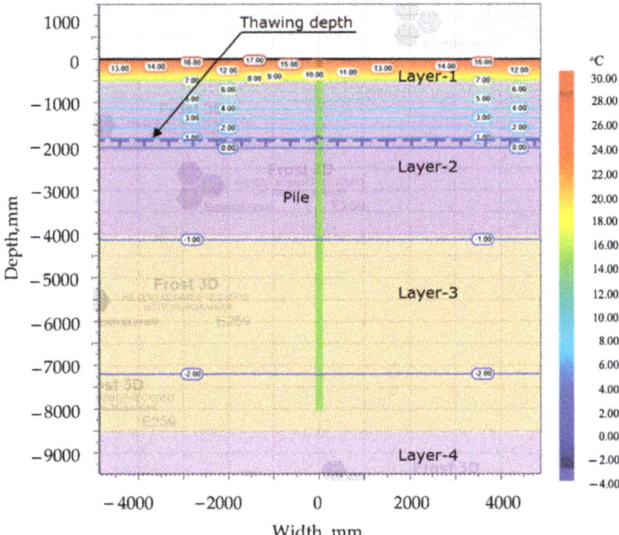

Figure 10. Temperature distribution for 2050 under a local negative warming scenario.

The simulation results (Table 3) showed that under a positive warming scenario (with an increase of temperature by 0.11 degrees per year), the active layer will be 0.595 m, the bearing capacity of the pile will be 81.93 tons in 2050. Under a neutral warming scenario (with an increase of temperature by 0.16 degrees per year), the active layer will be 0.673 m, and the bearing capacity of the pile will be 75.84 tons. In negative and locally negative scenarios, the active layer will be 0.840 and 1.868 m, respectively. The bearing capacity of the pile will be 71.32 and 63.79 tons. Analyzing the results, it can be concluded the active layer by 2050 may increase threefold, while the bearing capacity of the piles will decrease by more than 20%. In order to avoid emergency situations and to ensure further safe operation of construction facilities, it is necessary to provide additional measures.

Table 3. Results of modeling the bearing capacity of the soil for various scenarios.

Characteristics of Piles	Warming Scenario	Increase of Temperature per Year, °C	Active Layer by 08.2050, m	Pile Bearing Capacity as of 08.2050, tons
	Positive	0.11	0.595	81.93
L = 8 m, Ø = 0.2 m	Neutral	0.16	0.673	75.84
	Negative	0.24	0.840	71.32
	Locally negative	0.50	1.868	63.79

Besides, there was carried out the ground temperature state modeling when using seasonally acting cooling devices. The distribution of ground temperature when using thermosiphons for the winter period of 2050 under the local negative scenario is shown in Figure 11.

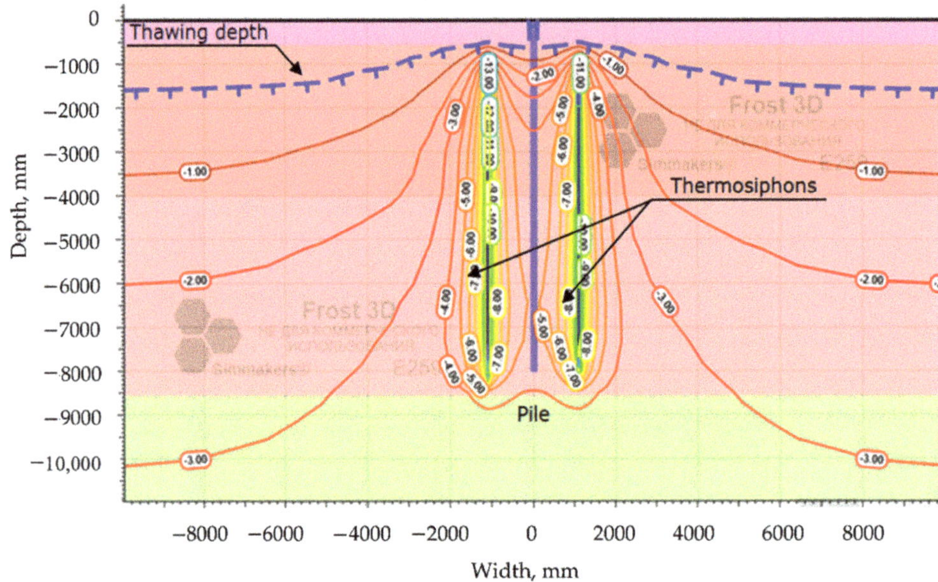

Figure 11. Modeling results of the winter ground temperature distribution.

The distribution of ground temperature when using thermosiphons for the summer period of 2050 under the local negative scenario is shown in Figure 12.

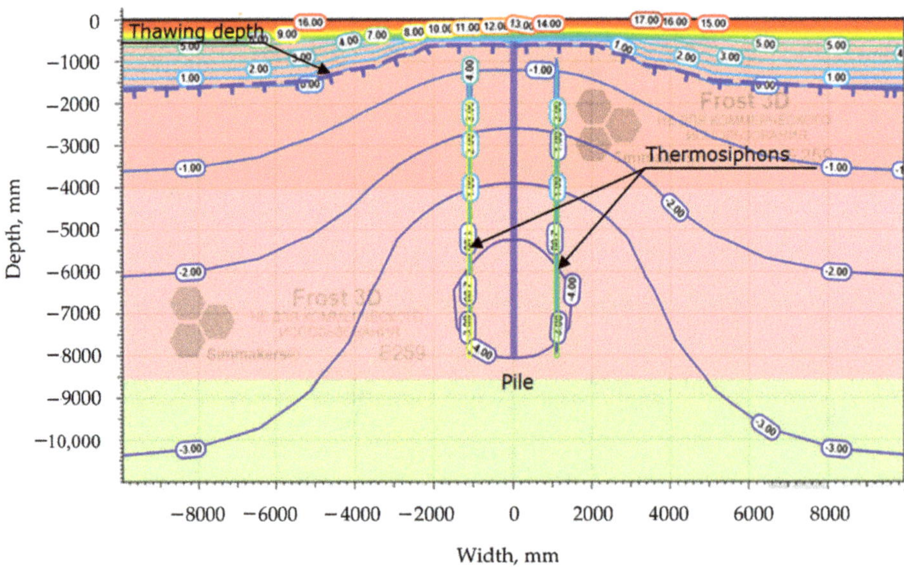

Figure 12. Modeling results of ground temperature distribution in the summer period.

4. Discussion

In continuation of the study [38] of scientists of the Saint Petersburg Mining University, the authors of this article propose the design of objects taking into account the impact of thawing of permafrost on the stability of the pile foundation.

Geotechnical design of industrial infrastructure in remote Arctic fields is proposed to be carried out in accordance with the steps of Figure 13.

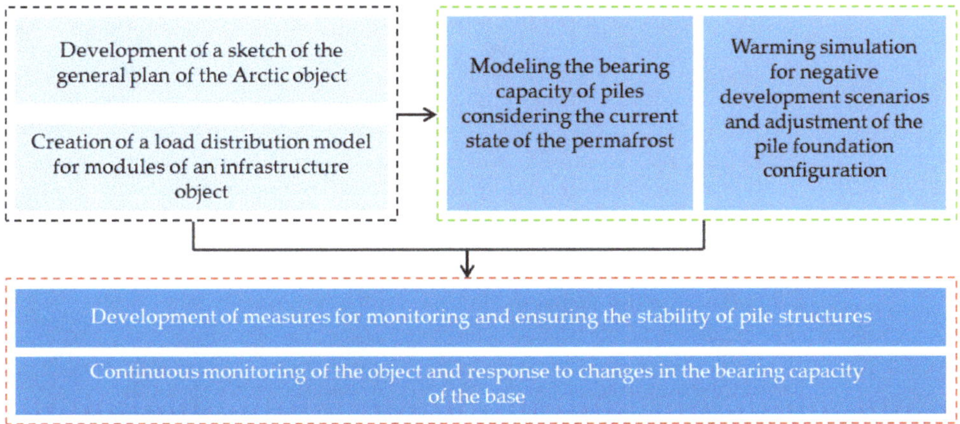

Figure 13. The proposed methodology for pile foundation design on a base.

The 1st block includes:

- determination of the list of necessary initial data for modeling the bearing capacity of piles for individual modules (taking into account concentrated and distributed loads on the pile foundation);
- development of a sketch of the general plan of the object.

The 2nd block is the modeling stage:

- modeling of bearing capacity and selection of characteristics of pile fields (type of pile and its geometric characteristics, distance between piles and method of installation) according to engineering survey data;
- modeling of thawing of frozen soils with different dynamics of changes in average annual temperatures.

The final block includes:

- selection of a list of measures to preserve the bearing capacity of piles, depending on the results presented in the second block;
- year-round monitoring of the stability of the facility during the entire period of its operation.

To ensure the stability of the pile foundations of remote industrial Arctic objects, it is proposed to carry out modeling of heat transfer processes in order to predict the rate of thawing of permafrost, taking into account climate change. In case of revealing a significant loss of the bearing capacity of the piles, the design documentation should be adjusted taking into account negative scenarios. During the operation of the pile foundation, it is necessary to monitor the bearing capacity of the piles and respond in a timely manner to changes occurring in the cryolitic zone.

5. Conclusions

Global climate change poses new challenges to the industry involved in the development of mineral reserves in the Arctic regions. Civil and industrial structures designed and built without taking into account the warming factor begin to collapse due to changes in the structure of permafrost. The situation may be aggravated by the rising level of the world ocean, leading to an increase in the area of flooded Arctic territories. Mining University is developing technical and technological solutions for the construction of remote Arctic facilities and a methodology for their design based on physical and mathematical predictive modeling [51]. The solutions proposed by the authors will make it possible to ensure the sustainability of infrastructure facilities in remote Arctic territories.

Author Contributions: Conceptualization, G.B., P.T.; methodology, A.K. and D.S.; software, A.L. and E.L.; validation, A.K. and D.S.; formal analysis, G.B.; investigation, A.K., D.S. and A.L.; resources, P.T.; data curation, G.B.; writing—original draft preparation, A.K. and D.S.; writing—review and editing, A.L., D.S. and A.K.; visualization, E.L. and A.L.; supervision, G.B.; project administration, G.B.; funding acquisition, P.T. All authors have read and agreed to the published version of the manuscript.

Funding: The framework for the sustainable development of Arctic facilities was created with the financial support of the grant by the President of the Russian Federation for the state support of leading scientific schools of the Russian Federation, the number of the project NSh-2692.2020.5 "Modelling of ecological-balanced and economically sustainable development of hydrocarbon resources of the Arctic".

Data Availability Statement: The data presented in this study are available on request from the corresponding author.

Acknowledgments: The framework for the sustainable development of Arctic facilities was created with the financial support of the grant by the President of the Russian Federation for the state support of leading scientific schools of the Russian Federation, the number of the project NSh-2692.2020.5 "Modelling of ecological-balanced and economically sustainable development of hydrocarbon resources of the Arctic".

Conflicts of Interest: The authors declare no conflict of interest.

References

1. Zhukovskiy, Y.; Tsvetkov, P.; Buldysko, A.; Malkova, Y.; Stoianova, A.; Koshenkova, A. Scenario Modeling of Sustainable Development of Energy Supply in the Arctic. *Resources* **2021**, *10*, 124. [CrossRef]
2. The Intergovernmental Panel on Climate Change. Global Warming of 1.5 C. Available online: www.ipcc.ch (accessed on 23 November 2021).
3. Masson-Delmotte, V.; Zhai, P.; Pörtner, H.-O.; Roberts, D.; Skea, J.; Shukla, P.R.; Pirani, A.; Moufouma-Okia, W.; Péan, C.; Pidcock, R.; et al. *IPCC Special Report: Global Warming of 1.5 C—Summary for Policymakers*; World Meteorological Organization: Geneva, Switzerland, 2018; 32p.
4. Stocker, T.F.; Qin, D.; Plattner, G.-K.; Tignor, M.; Allen, S.K.; Boschung, J.; Nauels, A.; Xia, Y.; Bex, V.; Midgley, P.M. IPCC. Summary for Policymakers. In *Climate Change 2013: The Physical Science Basis. Contribution of Working Group I to the Fifth Assessment Report of the Intergovernmental Panel on Climate Change*; Cambridge University Press: Cambridge, UK; New York, NY, USA, 2013; 28p. Available online: http://www.ipcc.ch/pdf/assessment-report/ar5/wg1/WG1AR5_SPM_FINAL.pdf (accessed on 23 November 2021).
5. Meinshausen, M.; Smith, S.J.; Calvin, K.; Daniel, J.S.; Kainuma, M.L.T.; Lamarque, J.F.; Matsumoto, K.; Montzka, S.A.; Raper, S.C.B.; Riahi, K.; et al. The RCP greenhouse gas concentrations and their extensions from 1765 to 2300. *Clim. Chang.* **2011**, *109*, 213–241. [CrossRef]
6. Le Quéré, C.; Andrew, R.M.; Friedlingstein, P.; Sitch, S.; Hauck, J.; Pongratz, J.; Pickers, P.A.; Korsbakken, J.I.; Peters, G.P.; Canadell, J.G.; et al. Global Carbon Budget 2018. *Earth Syst. Sci. Data* **2018**, *10*, 2141–2194. [CrossRef]
7. Biskaborn, B.K.; Smith, S.L.; Noetzli, J.; Matthes, H.; Vieira, G.; Streletskiy, D.A.; Schoeneich, P.; Romanovsky, V.E.; Lewkowicz, A.G.; Abramov, A.; et al. Permafrost is warming at a global scale. *Nat. Commun.* **2019**, *10*, 264. [CrossRef]
8. Rajendran, S.; Sadooni, F.N.; Al-Kuwari, H.A.-S.; Oleg, A.; Govil, H.; Nasir, S.; Vethamony, P. Monitoring oil spill in Norilsk, Russia using satellite data. *Sci. Rep.* **2021**, *11*, 3817. [CrossRef]
9. Meredith, M.; Sommerkorn, M.; Cassotta, S.; Derksen, C.; Ekaykin, A.; Hollowed, A.; Kofinas, G.; Mackintosh, A.; Melbourne-Thomas, J.; Muelbert, M.M.C.; et al. Chapter 3: Polar regions. In *The Ocean and Cryosphere in a Changing Climate: Summary for Policymakers*; Intergovernmental Panel on Climate Change; 2019; pp. 3-1–3-173. Available online: https://www.ipcc.ch/srocc/chapter/chapter-3-2/ (accessed on 23 November 2021).
10. Moon, T.A.; Overeem, I.; Druckenmiller, M.; Holland, M.; Huntington, H.; Kling, G.; Lovecraft, A.L.; Miller, G.; Scambos, T.; Schädel, C.; et al. The expanding footprint of rapid Arctic change. *Earths Future* **2019**, *7*, 212–218. [CrossRef]

11. Koven, C.D.; Ringeval, B.; Friedlingstein, P.; Ciais, P.; Cadule, P.; Khvorostyanov, D.; Krinner, G.; Tarnocai, C. Permafrost carbon-climate feedbacks accelerate global warming. *Proc. Natl. Acad. Sci. USA* **2011**, *108*, 14769–14774. [CrossRef]
12. Timofeev, A.V.; Piirainen, V.Y.; Bazhin, V.Y.; Titov, A.B. Operational Analysis and Medium-Term Forecasting of the Greenhouse Gas Generation Intensity in the Cryolithozone. *Atmosphere* **2021**, *12*, 1466. [CrossRef]
13. Vincent, W.F. Arctic Climate Change: Local Impacts, Global Consequences, and Policy Implications. In *The Palgrave Handbook of Arctic Policy and Politics*; Springer International Publishing: Cham, Switzerland, 2020.
14. Pörtner, H.-O.; Roberts, D.C.; Masson-Delmotte, V.; Zhai, P.; Tignor, M.; Poloczanska, E.; Mintenbeck, K.; Alegría, A.; Nicolai, M.; Okem, A.; et al. Summary for Policymakers. In *IPCC Special Report on the Ocean and Cryosphere in a Changing Climate*; 2019. Available online: https://www.ipcc.ch/site/assets/uploads/sites/3/2019/11/03_SROCC_SPM_FINAL.pdf (accessed on 23 November 2021).
15. Church, J.A.; White, N.J. Sea-Level Rise from the Late 19th to the Early 21st Century. *Surv. Geophys.* **2011**, *32*, 585–602. [CrossRef]
16. Domingues, R.; Goni, G.; Baringer, M.; Volkov, D. What Caused the Accelerated Sea Level Changes along the U.S. East Coast during 2010–2015. *Geophys. Res. Lett.* **2018**, *45*, 13367–13376. [CrossRef]
17. Leuliette, E. *The Budget of Recent Global Sea Level Rise 2005–2013*; National Oceanic and Atmospheric Administration: Washington, DC, USA, 2014. Available online: https://www.star.nesdis.noaa.gov/socd/lsa/SeaLevelRise/documents/NOAA_NESDIS_Sea_Level_Rise_Budget_Report_2014.pdf (accessed on 23 November 2021).
18. Sweet, W.V.; Kopp, R.E.; Weaver, C.P.; Obeysekera, T.; Horton, R.M.; Thieler, E.R.; Zervas, C. Global and Regional Sea Level Rise Scenarios for the United States. In *NOAA Technical Report NOS CO-OPS 083*; National Oceanic and Atmospheric Administration: Washington, DC, USA, 2017; 75p. Available online: https://tidesandcurrents.noaa.gov/publications/techrpt83_Global_and_Regional_SLR_Scenarios_for_the_US_final.pdf (accessed on 23 November 2021).
19. Sweet, W.V.; Park, J.; Marra, J.J.; Zervas, C.; Gill, S. Sea level rise and nuisance flood frequency changes around the U.S. In *NOAA Technical Report NOS CO-OPS 73*; National Oceanic and Atmospheric Administration: Washington, DC, USA, 2014; 53p. Available online: https://tidesandcurrents.noaa.gov/publications/NOAA_Technical_Report_NOS_COOPS_073.pdf (accessed on 23 November 2021).
20. Church, J.A.; Clark, P.U.; Cazenave, A. Sea Level Change. *CUP* **2013**, *13*, 1137–1216.
21. Romasheva, N.; Dmitrieva, D. Energy Resources Exploitation in the Russian Arctic: Challenges and Prospects for the Sustainable Development of the Ecosystem. *Energies* **2021**, *14*, 8300. [CrossRef]
22. Tcvetkov, P.; Cherepovitsyn, A.; Makhovikov, A. Economic assessment of heat and power generation from small-scale liquefied natural gas in Russia. *Energy Rep.* **2020**, *6*, 391–402. [CrossRef]
23. Tretyakov, N.; Cherepovitsyn, A.; Komendantova, N. Technology predictions for arctic hydrocarbon development: Digitalization potential, 5th International Conference on Technological Transformation: A New Role for Human, Machines and Management. *Technol. Transform.* **2020**, *157*, 241–251. [CrossRef]
24. Dmitrieva, D.; Romasheva, N. Sustainable Development of Oil and Gas Potential of the Arctic and Its Shelf Zone: The Role of Innovations. *J. Mar. Sci. Eng.* **2020**, *8*, 1003. [CrossRef]
25. Carayannis, E.G.; Ilinova, A.; Cherepovitsyn, A. The Future of Energy and the Case of the Arctic Offshore: The Role of Strategic Management. *J. Mar. Sci. Eng.* **2021**, *9*, 134. [CrossRef]
26. Obu, J.; Westermann, S.; Bartsch, A.; Berdnikov, N.; Christiansen, H.H.; Dashtseren, A.; Zou, D. Northern Hemisphere permafrost map based on TTOP modelling for 2000–2016 at 1 km^2 scale. *Earth Sci. Rev.* **2019**, *193*, 300–316. [CrossRef]
27. Tokarev, I.V. Use of isotope data (δ2H, δ18O, 234U/238U) in the study of permafrost degradation processes as a result of long-term climate variations. *J. Min. Inst.* **2008**, *176*, 191.
28. Vasiliev, G.G.; Dzhaljabov, A.A.; Leonovich, I.A. Analysis of the causes of engineering structures deformations at gas industry facilities in the permafrost zone. *J. Min. Inst.* **2021**, *249*, 377–385. [CrossRef]
29. Vernigor, V.M.; Morozov, K.V.; Bobrovnikov, V.N. On approaches to designing of thermal regime at ore mines under permafrost conditions. *J. Min. Inst.* **2013**, *205*, 139–140.
30. Potapov, A.I.; Pavlov, I.V. Acoustic emission diagnostics buildings mining companies in the North. *J. Min. Inst.* **2014**, *209*, 128–132.
31. Vasiltsov, V.S.; Vasiltsova, V.M. Strategic planning of arctic shelf development using fractal theory tools. *J. Min. Inst.* **2018**, *234*, 663. [CrossRef]
32. Mangushev, R.; Sakharov, I. *Foundations and Foundations: Textbook for Bachelors of Construction and Specialists in the Field of Construction of Unique Buildings and Structures*; Publishing house ASV: Moscow, Russia, 2019; 468p.
33. Jorgenson, M.T.; Grosse, G. Remote Sensing of Landscape Change in Permafrost Regions. *Permafr. Periglac. Process.* **2016**, *27*, 324–338. [CrossRef]
34. Korobkov, G.E.; Yanchushka, A.P.; Zakiryanov, M.V. Numerical modeling of a stress-strain state of a gas pipeline with cold bending offsets according to in-line inspection. *J. Min. Inst.* **2018**, *234*, 643. [CrossRef]
35. Islamov, S.; Grigoriev, A.; Beloglazov, I.; Savchenkov, S.; Gudmestad, O.T. Research Risk Factors in Monitoring Well Drilling—A Case Study Using Machine Learning Methods. *Symmetry* **2021**, *13*, 1293. [CrossRef]
36. Litvinenko, V.S.; Trushko, V.L.; Dvoinikov, M.V. Method of Ice-Resistant Drilling Platform Construction on Shallow Shelf of Arctic Seas. Patent for Invention RU 273739, 27 November 2020.
37. Litvinenko, V.S.; Trushko, V.L.; Dvoinikov, M.V. Method for the Construction of an Offshore Drilling Platform on the Shallow Shelf of the Arctic Seas. Patent for Invention RU 2704451, 28 October 2019.

38. Litvinenko, V.S.; Trushko, V.L.; Dvoinikov, M.V. Elaboration of a conceptual solution for the development of the Arctic shelf from seasonally flooded coastal areas. *Int. J. Min. Sci. Technol.* **2021**. [CrossRef]
39. Set of Rules SP 25.13330.2020. *Bottoms and Foundations on Permafrost Soils*; Standartinform: Moscow, Russia, 2020; Approved by Order of the Ministry of Construction and Housing and Communal Services of the Russian Federation No. 915/pr.
40. Weismüller, J.; Wollschläger, U.; Boike, J.; Pan, X.; Yu, Q.; Roth, K. Modeling the thermal dynamics of the active layer at two contrasting permafrost sites on Svalbard and on the Tibetan Plateau. *Cryosphere* **2011**, *5*, 741–757. [CrossRef]
41. Kane, D.; Hinkel, K.; Goerin, D.; Hinzman, L.; Outcalt, S. Non-conductive heat transfer associated with frozen soils. *Glob. Planet Chang.* **2001**, *29*, 275–292. [CrossRef]
42. Liu, G.; Xie, C.; Zhao, L.; Xiao, Y.; Wu, T.; Wang, W.; Liu, W. Permafrost warming near the northern limit of permafrost on the Qinghai–Tibetan Plateau during the period from 2005 to 2017: A case study in the Xidatan area. *Permafr. Periglac. Process.* **2020**, *32*, 323–334. [CrossRef]
43. Liu, G.; Wu, T.; Hu, G.; Wu, X.; Li, W. Permafrost existence is closely associated with soil organic matter preservation: Evidence from relationships among environmental factors and soil carbon in a permafrost boundary area. *Catena* **2021**, *196*, 104894. [CrossRef]
44. Andreyev, M.A.; Andreev, I.A.; Mironov, A.V.; Terentyev, M.A. Arrangement of bases and foundations of oil reservoirs in complicated conditions of Polar region. *Ind. Civ. Eng.* **2006**, *9*, 40–41.
45. Svetlov, L.P.; Vedrashko, E.M.; Voronoj, V.A.; Biryukov, O.R.; Ozornin, A.A.; Letin, E.V. Method of Forced Reduction of Permafrost Soil Temperature in the Bases of Pile Foundations of the Operating Bridge Supports. Patent for the Invention RU 2731343, 1 September 2020.
46. Bocharov, A.G.; Bocharov, M.E.; Kozhevnikov, S.A. Method of Cold Accumulation in the Ground. Patent for the Invention RU 2650005, 6 April 2018.
47. Prokopenko, I.F.; Shtefanov, Y.P.; Shishov, I.N. Year-round thermal stabilization of a building. In Proceedings of the Fifth Conference of Russian Geocryologists, Moscow, Russia, 14–17 June 2016; pp. 291–296. (In Russian)
48. Trushevsky, S.N.; Strebkov, D.S. Method and Device for Year-Round Cooling, Freezing of Foundation Soil and Heat Supply of Structures on Permafrost Soil in Cryolithozone Conditions. Patent for the Invention RU 2519012, 10 June 2014.
49. Lavrik, A.Y.; Buslaev, G.V.; Dvoinikov, M.V.; Zhukovsky, Y.L. Method of Combined Year-Round Temperature Stabilization of Soil. Application for the Invention RU 2021109950, 12 April 2021.
50. Dvoynikov, M.; Buslaev, G.; Kunshin, A.; Sidorov, D.; Kraslawski, A.; Budovskaya, M. New Concepts of Hydrogen Production and Storage in Arctic Region. *Resources* **2021**, *10*, 3. [CrossRef]
51. Litvinenko, V.S.; Leitchenkov, G.L.; Vasiliev, N.I. Anticipated sub-bottom geology of Lake Vostok and technological approaches considered for sampling. *Chem. Erde/Geochem.* **2020**, *80*, 125556. [CrossRef]

Article

Parameters of Sustainable Development: Case of Arctic Liquefied Natural Gas Projects

Alexey Cherepovitsyn * and Olga Evseeva

Economics, Organization and Management Department, Saint-Petersburg Mining University, Vasilievsky Island, 21 line 2, 199106 Saint-Petersburg, Russia; s185097@stud.spmi.ru
* Correspondence: Cherepovitsyn_AE@pers.spmi.ru; Tel.: +8-921-919-54-55

Abstract: Effective management of the social and economic development of the Arctic zone of the Russian Federation is today a significant scientific and practical task. It requires an integrated approach to meet the expectations of the state, business and society. The main drivers of growth for remote Arctic territories are large investment projects, which not only create production and sectorial results, but also stimulate the development of related sectors of the economy. Additionally, they contribute to the formation of modern infrastructure in the region and create conditions for the broad introduction of innovative technologies. The current problem with territorial development strategic planning is the assessment of the results that have been achieved. This includes approved lists of indicators that do not allow for a full assessment of the impact of the implemented projects. Assessment on the achievement of the region goals is also murky. This indicates a lack of consistency in regional development management. This article defines the importance of the indicators for an assessment of sustainable development management. The model of achieving external effects in project activities is described. The concept of sustainability of large-capacity complexes for the production of liquefied natural gas (LNG) is also formulated. Based on the needs of micro- and macro-environment projects, a list of indicators for assessing the sustainability of LNG projects has been proposed. On the basis of the proposed indicator list, a sustainability analysis of three Arctic LNG projects was carried out. Based on the example of LNG production, it was concluded that approaches to assessing the sustainable socio-economic development of the Arctic region and its industrial systems are interrelated, but there are differences between them.

Keywords: sustainability; sustainable development; socio-economic development of the northern territories; innovation; projects; Arctic; liquefied natural gas

Citation: Cherepovitsyn, A.; Evseeva, O. Parameters of Sustainable Development: Case of Arctic Liquefied Natural Gas Projects. *Resources* **2021**, *10*, 1. https://dx.doi.org/10.3390/resources10010001

Received: 20 October 2020
Accepted: 17 December 2020
Published: 23 December 2020

1. Introduction

The implementation of Russian strategic goals in the field of geopolitics, the economic development of Northern territories and a surge in hydrocarbon and other mineral raw materials production today is closely related to the development of the Arctic. The Arctic region has enormous energy and mineral resources concentrated in large and unique deposits [1,2]. Experts estimate that 20–25% of the world hydrocarbon resources are located in the Russian Arctic zone and today about 80% of gas and 60% of oil from the total production of the country are produced there [3,4]. The Arctic mining complex is represented by deposits of iron, apatite, phosphorus, titanium, tungsten, copper, nickel, antimony, mercury, cobalt, gold, silver, platinum, rare metals and rare-earth elements [5,6].

The resource potential of the Russian Arctic is a national priority. The implementation of Arctic commodity projects will allow for the launch of innovative processes aimed at testing unique technological solutions and developing organizational and project management tools [7,8]. Among the most important tasks of Arctic hydrocarbon projects is the socio-economic development of northern territories, extended reproduction, and effective use of mineral and raw materials resources. Furthermore, the technological development of industrial systems, ensuring the sustainability of natural ecosystems, and ensuring energy and national security of the country as a whole, are significant goals [9].

There are serious challenges and requirements for the implementation of oil and gas projects in the coordinated system, especially the area of region-industrial systems-ecology [10,11]. Often, promising areas for oil and gas production are removed from points of consumption. In this case, the lack of transport and industrial infrastructure may delay making investment decisions on projects. In addition, such areas may be located in the territories of indigenous peoples. There is a small degree of geological exploration of existing deposits, which increases the risk for new projects. Difficult operational conditions require the introduction of advanced technological solutions [12]. It is also expedient to consider the environmental factor since the activities of oil and gas companies have a significant impact on the environment, including the fact that new projects should leave a low carbon footprint [13–15]. The existing institutional environment must also be optimized and must contribute to the intensification of investment activities in the region. Legislative initiatives need to be supported by long-term strategic planning that reflects an effective set of measures for the development of businesses, social environments, integration aspects and environmental balance mechanisms [16].

The long-term goals of the state on the development of the Arctic are connected with the creation of large-scale transport and logistics, power, information and communication systems, safety and environmental protection complexes. There is a need to develop social infrastructure facilities, stimulate R&D and increase demand for domestic technologies and equipment. It is expected to increase the geological study of the subsurface and increase the production of Arctic raw materials [17]. The indicated effects are planned to be obtained through the implementation of large investment projects, including the use of public-private partnership mechanisms. In this regard, sustainable regional development requires project participants to not only accumulate productive and financial resources, but also to take a specific approach. This approach allows for the development of industrial and infrastructural mineral-raw-material systems in conditions of high instability in energy markets and the objective complexity of solving technological and socio-economic problems [18,19].

An important element of effective regional development management is indicative planning and monitoring aimed at quantitative measurement and control of target results. It is obvious that the relationship of new investment projects with regional goals, which has been repeatedly noted in strategic planning documents, requires an appropriate list of indicators. These indicators take into account not only the technological characteristics of industrial development, but also the impact on the regional economy, population, and the state of the environment. It is noted that the idea of sustainable development is acquiring particular relevance in the Arctic [20].

One of the main directions of the Russian Arctic fuel and energy complex development is the implementation of liquefied natural gas (LNG) projects. The importance of LNG production today is largely determined by the need to diversify Russian gas export markets [21]. The implementation of large-scale Arctic LNG projects creates substantial opportunities for using the mineral resource potential of the region with a course for comprehensive infrastructure development of remote Northern territories, inter-industry interaction, ecological, innovative and technological development [22].

The feasibility of the development of the LNG industry in the Arctic is determined by a number of factors. These factors include large gas reserves in the coastal zone, low average temperatures in the region, the relatively low cost of natural gas production, and a good geographical location relative to key markets. Combined, these variables determine the high competitiveness of Arctic LNG compared to the leading countries of the LNG market.

Currently, there is one large-capacity LNG plant operating in the Arctic region—Yamal LNG. Two more LNG projects are planned for implementation: Arctic LNG-2 and Ob LNG. Their main characteristics are shown in Table 1.

Table 1. Characteristics of Arctic LNG projects. Based on open data of operating companies. Budget efficiency is determined on the basis of discounting tax revenues to the regional budget for the period of project implementation.

	Yamal LNG	Arctic LNG-2	Ob LNG
Capacity, million tons	17.5	19.8	4.8
Starting year of the 1st lines	2017	2023	2022
Number of production lines	4	3	3
Resource base	The South-Tambeyskoye field	Utrenneye field	Verkhnetiuteyskoye and Zapadno-Seyakhinskoye fields
CAPEX, $ billion	27.5	21.3	5.7
Project participant	Novatek, Total, China National Petroleum Corporation (CNPC), Silk Road Fund (SRF)	Novatek, Total, CNPC, China National Offshore Oil Corporation (CNOOC), Japan Arctic LNG	Novatek
Regional infrastructure objects that are driven by the project	Sabetta seaport, Sabetta airport, shift settlement, mobile phone tower network	Utrenniy terminal	No data
Job creation	32,000	No data	No data
Budget efficiency at the regional level, $ million	1228	1664	409

The experience of the Yamal LNG project has shown that its results not only show the beginning of development of the South Tambey gas condensate field and an increase in the share of Russian LNG in the global market, but also the creation of a major transport hub in the town of Sabetta, including a seaport and international airport, a shift settlement, a fleet of gas tankers and icebreakers and the development of the first Russian liquefaction technology. Also, the contribution to the development of the Northern Sea Route is paramount [23]. The construction of communication lines has provided access to high-speed data transmission in several cities of the Far North, and the project has created a significant number of jobs. The functioning asset is a stable source of revenue to the federal and regional budgets. According to experts of the SKOLKOVO Energy Center, the development of the LNG industry in the Arctic has already had a great synergistic effect in the preparation of some projects in the field of coal, gold, non-ferrous and rare metal ore production, which indicates the impact of increasing the investment attractiveness of the region especially for foreign investors [24].

Therefore, it can be concluded that the implementation of LNG projects has an impact on both the development of the region and the development of related industries through the formation of demand for related products and services, which indicates the presence of pronounced external effects. Given the previously noted relationship between the results of large-scale industrial projects and the development goals of the Arctic region, such effects need to be quantified and systematized, creating the possibility of managing their achievement.

The purpose of this study is to develop and test a list of sustainability indicators for large-scale LNG production complexes in accordance with the needs of the region and business interests. To achieve this goal, it is necessary to answer the following research questions:

(1) What is the value of a quantitative assessment of the results in sustainable development management and why is it needed for Arctic industrial systems?
(2) What is the concept of sustainability for LNG projects?
(3) What is the potential of Arctic LNG projects to generate positive economic, social, and environmental outcomes?

(4) What should the sustainability indicators in a project approach be, and how will it differ from a corporate level assessment?

2. Theoretical Background for Defining the Concept of Project Sustainability

The term "sustainable development" became widely used after the "Our Common Future" report was presented by the Brundtland Commission [25]. The report provided a classic definition of sustainable development, which refers to "a process of change in which the exploitation of resources, the direction of investments, the orientation of technological development, and institutional change are all in harmony and enhance both current and future potential to meet human needs and aspirations" [26].

Russia approved the "Concept of the Russian Federation's Transition to Sustainable Development", presenting its own vision of the idea of sustainable development. According to the document, sustainable development means "stable socio-economic development that does not destroy its natural basis". The purpose of the gradual transition was determined "to ensure in the long term, a balanced solution to the problems of socio-economic development and preservation of a favorable environment and natural resource potential, meeting the needs of present and future generations of people".

Sustainable development is seen as a paradigm for thinking about the future, in which environmental, social and economic aspects are balanced in an effort to improve the quality of human life [27]. At the same time, the development itself testifies to a certain dynamic process. The term "sustainability" is used to describe the state of the system and its general target vision [28].

The most common description of sustainability involves three interconnected pillars, encompassing economic, social and environmental factors [29]. The sustainability model in scholarly writing is often depicted through three intersecting circles: society, environment and economy, with sustainability at the intersection of these spheres [30]. A similar relationship underlies the idea of the Triple Bottom Line (TBL) proposed by John Elkington and is often identified with the acronym "3P" (People, Planet, Profit) [31,32].

The characteristics of the interpretation and approach to sustainability assessment depend on the level of economic activity within which the evaluation object is considered. The Schukina L.V. study suggested that the following levels of sustainable development should be identified [33]:

- International (global)
- National
- Regional
- Sectorial
- Corporate

Despite the close relationship between the levels, each level has its own target trajectory of development. Sustainable development at the global level is focused on international partnerships to fight poverty and hunger, protect health and human rights, address climate change, preserve the biodiversity of the planet and its natural resources, prevent hostilities and protect the world's oceans [34].

At the national level, using the example of the Russian Federation, sustainable development involves ensuring national and environmental security, geopolitical interests, balanced development of economic sectors, resource availability, promotion of the well-being of the nation and realization of citizens' rights.

The components of sustainable development at the regional level include the stable functioning of industrial complexes, socio-economic and ecological systems of individual entities, comprehensive improvement of territories and settlements, provision of housing and communal services. These components create an impact on the population, industry, social, energy and transport infrastructure, improvement of well-being and the quality of local people's life, which ensures the preservation of culture and traditions [35].

Sustainable development of industry is determined by its competitiveness in domestic and foreign markets, innovation and technological potential, balanced functioning of production and economic units and their safety for the environment, efficiency of activity and the ability to provide necessary intra-industry proportions and connections [15,36].

Sustainable development at the level of economic entities (corporate level) includes the creation of effective economic results while respecting the safety of production cycles, ensuring a high level of quality of produced products, minimization of a negative impact on the environment, development, social support and protection of workers' rights, implementation of CSR programs and implementation of advanced resource management practices [37–39].

Consequently, sustainable development goals and targets are largely defined by key challenges, opportunities, and constraints at each specific level. Sustainable development is closely linked to stakeholder theory, which is based on the principle of the harmonization of interests and expectations of direct and indirect participants in relation to ongoing processes [40,41]. Sustainable development of global and local systems is the result of interaction of the state, business and society in economic, social and environmental spheres [42]. Table 2 shows the main expectations of each group of stakeholders in relation to the three areas of sustainability.

Table 2. Challenges in social, ecology and economic spheres in the system of regional authorities –business-society.

	State (Region)	**Business**	**Society**
Social sphere	Ensuring humanitarian security and social stability in the region.	Human capital development.	Universal access to quality social services.
Ecology	Preservation of a favorable environment, biodiversity, reproduction of natural resources.	Rational use of resources, minimization of anthropogenic impact on the environment.	Safe state of the environment and ubiquitous access to natural resources according to basic needs.
Economy	Achieving maximum welfare of the region's population, stable growth of the territory's economy, integration and penetration into other local and global markets.	Stable maintenance of competitiveness, growth of profit and capitalization.	Adequate standard of living (compared to the regional average value and above) and minimization of social differentiation.

An integral part of sustainable development is the sustainability assessment. According to Kates et al., the assessment of sustainability focused on providing decision makers the results of the analysis of global and local systems "nature-society" in the short and long term to help them determine what actions should or should not be taken in trying to make society more sustainable [43]. At the corporate level, sustainability indicators are also used in decision-making in addition to non-financial reporting. C. Searcy clarifies this application of assessment separately in board-level decision-making, corporate governance and supply chain management [44]. The latter is especially relevant for the LNG industry with a long value chain (gas production, gas liquefaction, LNG transportation, LNG regasification), since the high total value of a product requires sufficient attention at every stage of its creation [45].

According to Wu & Wu, quantitative indicators clarify the meaning of sustainable development and allow for increasing the understanding of the complex interrelationships between the components of sustainability in practical terms, and thereby contribute to the development of science and practice of sustainable development [46]. When developing a list of sustainability indicators, it is necessary to specify which aspects of sustainability in the existing concept should be measured, which of the previously aspects not yet considered should be added and how these properties should be related to each other.

Indicators arise from the content of value illustrated by the phrase "we measure what we are concerned about" and at the same time form that value, which is concurrently illustrated by the phrase "we are concerned about what we are measuring" [47]. Each developed indicator allows for the qualitative or quantitative evaluation of a specific characteristic of a system striving for sustainability. Grouped into independent lists, they reflect the totality of stakeholders' interests and make it possible to assess progress in realizing their expectations.

"Sustainability indicators do not guarantee results, but results are impossible without the use of indicators" [47]. This statement reveals the third important function of sustainability indicators which are indicative planning and monitoring (Figure 1).

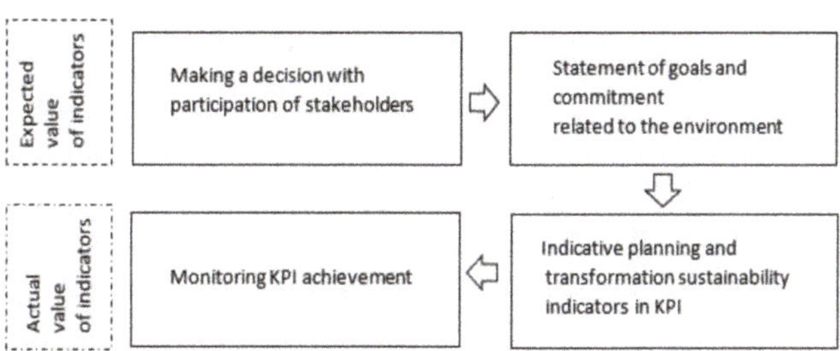

Figure 1. The importance of sustainability indicators in managing stakeholder expectations.

As shown in the diagram above, the stated measured results of sustainable development that meet the interests of stakeholders become a commitment to implement them for the environment. And the owner of the process (in the case of a project, the operator of the project) must integrate the work of creating these results into the content of the initiative. Sustainability indicators are already becoming key performance indicators (KPIs), reflecting the degree of commitment to compliance and as a tool for monitoring. Thus, the implementation of the KPI stage in the Stage-gate project management approach (the most common, since this approach allows to make investment decisions consistently, reducing risks) is a prerequisite for the transition between stages. For this, the KPI must be synchronized in accordance with the capabilities of the stage; the same KPI can be encountered at each stage of the project if the work, the quality of which it characterizes, is performed throughout the project. Regular monitoring of the KPI achievement status is carried out in the monitoring process, which is a key tool in ensuring the compliance of planned indicators with actual ones. Deviations identified during monitoring require an analysis of the causes of their occurrence, after which a cycle of corrective actions is launched.

The number of sustainability assessment tools developed and used today globally and locally is determined by hundreds of indicator and index lists [48]. Some of them are aimed at assessing specific areas of sustainable development, some suggest an integrated assessment. Sustainability assessment systems are based on the concept of sustainability, which does not have a single generally accepted interpretation. For this reason, the estimated parameters in the methodologies are different, and often such texts first explain what the concept is based on, and then disclose the content of the assessment [49].

The most famous and frequently mentioned lists of indicators of sustainability are lists proposed by international organizations. Among them are the United Nations Educational, Scientific and Cultural Organization (UNESCO), International Institute for Sustainable Development (IISD), Organization for Economic Co-operation and Development (OECD), United Nations Commission on Sustainable Development (CSD), Institute for European Environmental Policy, World Bank, European Environmental Agency, which are aimed primarily at assessing sustainable development at the global and national levels. Organizations such as S&P Global, Global 100, Global Reporting Initiative (GRI), and the Russian Union of Industrialists and Entrepreneurs (RSPP) are involved in the assessment of sustainable development at the micro level. At the moment, there is no single, generally accepted approach to assessing the sustainability.

3. Materials and Methods

To address the research questions, open materials of analytical centers and specialized international organizations, the works of Russian and foreign scientists in the field of project management theory, sustainable development, strategic management, as well as regulatory and methodological documentation on research issues were used. A complex approach to the development of a list of sustainability indicators was provided by the use of methods of synthesis, analogy, grouping, comparison, as well as tools for strategic analysis, investment assessment and socio-economic forecasting, using the method of forward and backward linkages. The content of the indicator list is based on the key principles of sustainability assessment noted in the works of a researchers mentioned above.

The Arctic zone of the Russian Federation, which includes the territories of nine constituent entities of the Russian Federation, has been allocated to a separate object of state administration [50,51]. According to the state program titled "Socio-economic development of the Arctic zone of the Russian Federation", it is planned to provide a comprehensive solution of strategic tasks through the implementation of three Sub-Programs. These include the "formation of support development zones and ensuring their functioning, creation of conditions for accelerated social and economic development of the Arctic zone of the Russian Federation", "development of the Northern Sea Route and provision of navigation in the Arctic", and the "creation of equipment and technologies of oil and gas and industrial engineering necessary for the development of mineral and raw materials resources of the Arctic zone of the Russian Federation". The names of subprograms reveal the priority areas of development in the region, and the indicators presented in the document clarify the content of the target results.

It is assumed that the dynamics of a number of indicators may be influenced by industrial complexes operating in the region. Thus, the intensification of production activities largely determines the improvement of macroeconomic indicators, as well as the growth of cargo turnover of the Northern Sea Route. Construction and modernization of industrial systems contributes to the growth of indicators characterizing innovative activity. As a result, it is possible to assess the contribution of industrial complexes to the socio-economic development of the region.

At the same time, each indicator list is aimed at a comprehensive assessment of the object to which it relates. In view of this, the unification of indicators is incorrect and an industrial project, even if it is focused on achieving regional goals, cannot be fully evaluated by the list of regional development indicators. This makes it necessary to develop separate list of indicators, taking into account the capabilities of projects and the interests of their stakeholders.

It should be noted that the assessment of indicators can be done in different ways (Table 3) [52–55].

Based on the analysis, it can be concluded that the assessment tool depends on the purpose of the analysis. The purpose of the analysis, in turn, can influence the list of indicators used, i.e., different indicator systems can be used for different analysis purposes. Their consistency in this case is determined by order, integrity in terms of analyzed characteristics coverage, connection with the final goal of the assessment and the presence of general principles in terms of the approach to the assessment.

In this study, the use of an assessment approach based on the calculation of an integral indicator is proposed in order to compare Arctic projects with each other. This approach is applicable in portfolio analysis and can be used for ranking projects. It allows for the formation of a conclusion about the priority of the project such as when making a decision to launch in conditions of limited resources. The construction of an integral indicator includes such stages as normalization, aggregation and weighting. These are aimed to bring indicators to a dimensionless form, to generalize indicators within individual groups and to differentiate the importance of each indicator in the overall set [56]. The project sustainability assessment algorithm used in this research is shown in the Figure 2.

Table 3. Sustainability indicators assessment tools.

Tool	Description	Benefits	Disadvantages
Qualitative indicators assessment	Description of the indicators content without using quantitative characteristics.	Not all the indicators can be evaluated; not every indicator fully discloses its content in numerical terms.	Lack of objective indicators comparability.
Benchmarking	Evaluation of indicators in dimension relative to the best analogues (values from the maximum/minimum).	Unified dimension of values and their relativity, calculations in terms of industry limits.	Need for the best practices continuous monitoring; limited access to information on each indicator.
Scoring	Experts put points on various indicators within a regulated scale.	Simplicity of the method and uniform dimension of values.	High subjectivity of the assessment.
The integral indicator calculation	All the values are reduced to a single indicator. In most cases, this is done using a system of weights and normalization of values.	Allows to compare evaluation objects with each other comprehensively.	The subjectivity when setting the weights.
Indicators quantification without integral indicator calculation	Evaluation of each indicator based on quantitative values.	The objectivity of the calculations.	Lack of consistency and conclusion about the most sustainable project or an alternative.

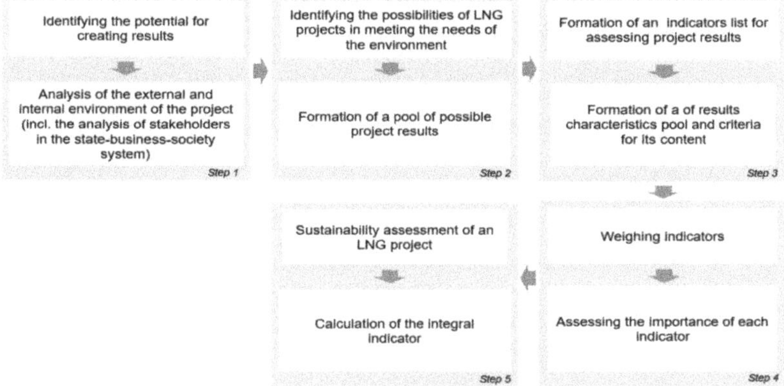

Figure 2. The project sustainability assessment algorithm.

The project sustainability assessment process involves an analytical stage, during which the potential for creating results is analyzed. Understanding current needs will allow to create maximum value for the environment, but it should be noted that the term "current needs" is dynamic, which means that the needs of the environment may change. This indicates a need for sustainable development management processes to function throughout the project, where sustainability indicators are reviewed and refined.

On this stage, special attention should be paid to approved strategies and programs that determine the current needs in the state-business-society system, as well as approved indicators for monitoring the implementation of these strategies and programs. Sources of information about needs in the external environment can be information from the media, as well as the use of various communication methods such as communication sessions, meetings, forums, conferences, etc.

On the next stage, it is necessary to analyze possibilities of LNG projects in solving urgent problems on regional, sectorial and national levels and compare them with the three-pillar conception of sustainability, structuring potential results in the fields of economy, social sphere and ecology.

The following stage involves forming a list of indicators. To do this, it is necessary to offer a quantitative characteristic for each potentially created result that corresponds to a certain interest (with units of measurement) and have previously formed requirements for its content. The requirements for the content of indicators determine the consistency of the list that they form. Furthermore, all the characteristics are grouped to reflect the completeness of the overall result assessment for each direction.

The next stage is weighting by spheres and groups. The individual areas of economy, ecology and social sphere may have an unequal number of indicators, and if not using weights, the assessment of each area may be distorted. Weighting within groups has the same goal. The philosophy of sustainable development is based on the principle of balance and equal coverage of the results in the aforementioned three areas. The same logic should be transposed to interest groups, which involves creating an alternative that provides maximum coverage in each interest group and will be more sustainable. In addition, even in strategic documents, goals may overlap. Indicators for these goals will give a higher value in total than for other goals. To avoid this, these indicators need to be aligned within groups. The weighting is based on the expert scores assigned for each indicator.

The final stage is to evaluate the project based on the developed list of indicators. Considering that each indicator has different scale and unit of measurement, it is necessary to use a normalization method aimed at reducing the indicators to a dimensionless form [56]. Since the purpose of the evaluation is to compare projects based on the principle of best matching their results to the needs of the environment, it is proposed to use the normalization method. In this case, each indicator correlates with a standard among alternatives. The benchmark in this case is the best value of the indicator among the projects under consideration. It should be noted that the benchmark can be both the largest and the lowest value of the indicator, for which it is necessary to have previously determined the orientation of each indicator.

4. Results and Discussion

Despite the fact that projects are initiated by companies, the assessment of their sustainability based on indicators of sustainable development at the corporate level is incorrect. The potential for creating results in the company's operational and project activities is different, because stakeholders and their interests in relation to the company and the project may differ. In addition, the project may involve external participants who create the uniqueness of the asset (Figure 3).

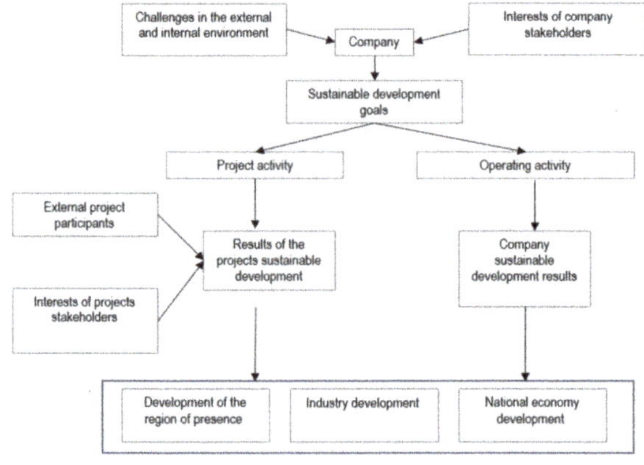

Figure 3. Conceptual model of the results arising in the company's project and operational activities.

In addition to the different potential for creating results, the approach to evaluating project activities will also be different:

(1) The object of project evaluation is its unique result, material or non-material. When evaluating operating activities, the management system is mainly evaluated.
(2) The project is evaluated through effects on the level of influence on a certain object. When analyzing operating activities, usually volatile, constantly changing indicators are used, measured in dynamics.
(3) Project effects can be evaluated before the project is launched (with a certain probability of achievement).
(4) Project effects are focused on the long term and are calculated for the entire period of its implementation, which is strictly limited, with the possibility of clarification during the project lifecycle. Operating indicators are calculated primarily for reporting periods.
(5) Project effects are estimated, as a rule, according to the forecast principle, operating activities according to the actual one.
(6) Approaches to project impact assessment are less standardized than approaches to operational performance assessment. It is generally accepted that the investment performance of a project should be evaluated solely based on discounted cash flow modeling.

The external environment of project implementation is similar to the perimeter of the company that implements it. Accordingly, the external results of projects will affect the same areas as the company's activities, and will mainly affect the development of the region of presence, the industry and the national economy of the country as a whole, while simultaneously affecting the company's activities and satisfying the interests of project investors. The assessment of such an impact is made in the process of project sustainability analysis, aimed at a comprehensive assessment of the project results from the point of view of interests in the state-business-society system.

We propose to evaluate the sustainability of an LNG project as the project's ability to generate economic, social and environmental results that meet the expectations of stakeholders (Figure 4). With regard to LNG projects, such results should be aimed at the following targets:

(1) Minimizing the negative impact on the environment at the site of construction and operation of assets.
(2) The reproduction and efficient use of natural resources.
(3) The support of the local population and promoting the preservation of the cultural heritage of indigenous peoples.
(4) Implementation in the economic interests of project participants, ensuring a long-term contribution to the presence region economy.
(5) The development of its infrastructure framework.
(6) Innovative and technological development in the industry.
(7) Strengthening the position of Russian LNG in the world market.

Sustainable development of the project, i.e., the process of generating these results, must be not less than the period of its life cycle determined by the charter of the project, including the operational phase. The designated targets for sustainable development can be formulated in the form of the following goals:

1. Efficient subsurface using, energy efficiency and transition promotion to environmentally friendly energy sources. This involves a sound approach to resource exploitation that minimizes environmental damage, as well as creating conditions for large-scale LNG using, including regions with vulnerable ecosystems.
2. Creating economic value of the asset. This involves obtaining economic benefits for the main (owners of the LNG asset) and indirect (the state, suppliers and contractors, entities for which the implementation of this project has become a driver for devel-

opment) project participants. It also includes the results that will become sources of additional growth in the future.

3. Participation in solving socially significant problems in the region of presence. This involves the implementation of CSR programs focused on the needs of local population.

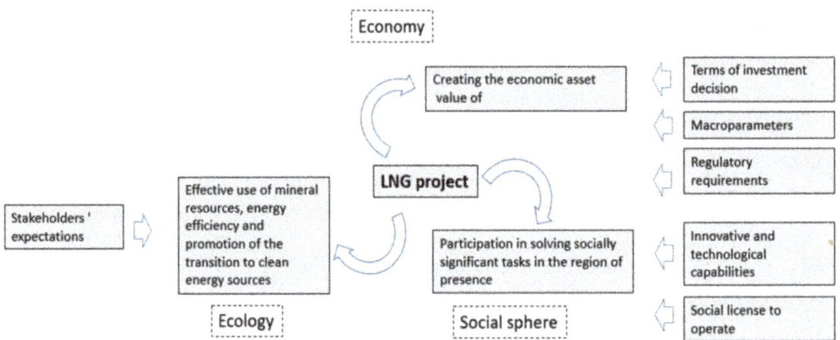

Figure 4. The concept of sustainability LNG project.

Sustainable development of the system is related to its ability to create economic, social and environmental results that meet the interests of stakeholders under existing restrictions. Based on the analysis of LNG projects implementing experience in Russia and in the world, we identified the main groups of stakeholders, their interests and potential contribution to the implementation of projects (Appendix A, Table A1).

For large-scale LNG projects, the list of expectations of stakeholders is quite high, which determines the number of possible effects that reflect its investment attractiveness. Moreover, the consequences of the project's implementation in the external environment, which determine its significance for the region and the industry (Appendix A, Table 2). It is important to note that the geographical and climatic features of the Arctic region in conditions of high dependence on foreign technologies and equipment require the activation of the innovative component at all stages of a project. This includes the organization and conduct of geological exploration, logistic support to industrial systems, organization and conduct of construction and installation work, the operation of production assets, and sales of finished products [57].

It is necessary to involve specialized universities and research institutes, scientific and technical centers and other participants offering advanced technological solutions for safe and efficient work in the extreme conditions of the North. For this reason, the effects of projects related to the development of innovations must also be taken into account.

Based on the analysis of the potential of Arctic LNG projects, the following groups of indicators can be identified for a comprehensive assessment of the results of such projects—macroeconomic, international integration, investment, infrastructure, sectorial, inter-project cooperation, innovative, human development, development of regional culture, improving living standards, environmental pollution, transition to clean energy and energy efficiency (Appendix A, Table 3).

Indicators for assessing the sustainable development of industrial systems, in our opinion, should meet the following requirements:

1. Be universal in relation to other projects in the region of the same sector with an identical set of links in the production chain (for example, production-processing-sales).
2. Meet the interests of stakeholders of the project.
3. Be applicable for evaluation at any stage of the project, including pre-project development.

In accordance with the specified requirements, the following list of indicators for assessing the sustainability of Arctic LNG projects is proposed (Appendix A, Table 4). This list was tested in assessing the sustainability of the Yamal LNG project, as well as on the Arctic LNG-2 project planned for implementation with an investment decision already having been made. Lastly, it will be used on the Ob LNG project, on which an investment decision has not been made yet.

The indicators were evaluated using data from open sources, as well as the authors' own calculations. To set the weight coefficients, we used expert assessments obtained during a survey of a group of five experts in the field of project management and sustainable development of Russian oil and gas company. The principle of setting weights is described in the Methodology section. Bringing the indicators to a single view was also performed according to the principle described earlier—through the ratio of the indicator value to the best among the projects.

Calculations are given in Appendix A in Table 5 and based on data from open sources. The results of the calculations show that the Yamal LNG project is the most sustainable in terms of the characteristics of results that correspond to the interests of stakeholders (Figure 5). However, it is necessary to take into account the high estimation error associated with the limited amount of data on planned projects in open access, which is typical for the Ob LNG project. Even with limited data on the Arctic LNG-2 project, it can be concluded that there is sufficient potential to create results in the external environment comparable to the Yamal LNG project.

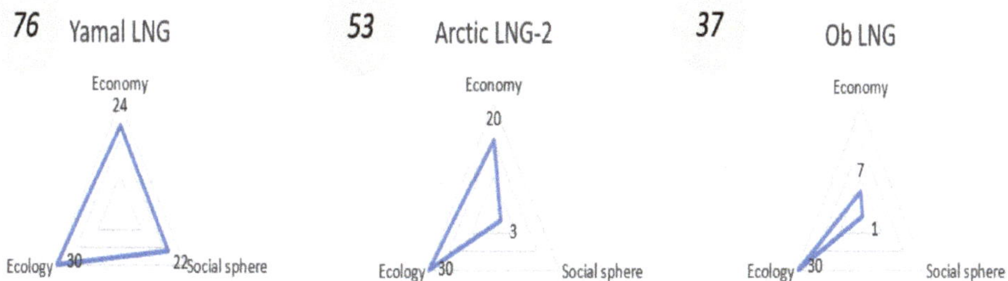

Figure 5. Integrated sustainability indicators for Arctic LNG projects and their distribution by sustainable development areas.

A rating system based on weights, the sum of which gives the final value "100", and normalized according to the principle of bringing to the best result, assumes that the maximum value that can be obtained is 100. This allows us to consider the integral stability indicator not only as an absolute value, but also as a relative one. However, it should be noted that sustainability indicators have different sensitivity to certain project parameters and not every indicator in the project can be increased/decreased to the reference one. For example, a number of indicators depend on the production capacity of a plant, which means that a less productive plant has a certain limit in generating economic, environmental and social results compared to a larger asset. This indicates that sustainability cannot have criteria and the conclusion about it is made in comparison with another object. A more sustainable option is one that provides the greatest coverage of the project results for the key interests of stakeholders in each area of sustainable development.

The proposed assessment approach is applicable in portfolio analysis, since it provides a summary of the status of each project in a specific area, which in this case, is in the field of sustainable development. However, it should be noted that with such an assessment system, the company should not seek to equalize the indicators since this can be achieved by reducing the benchmark indicators, but rather to search for additional potential improvements in lagging areas. Therefore, the sustainability portfolio should be monitored comprehensively for all projects with dynamic changes analysis in the values of each indicator.

A promising area of application of the proposed indicators is mentioned early management of socio-economic development of the region. Large-scale LNG projects are system-forming in regional design and form the core for further development, inter-industry and inter-territorial interaction. The unique results of the projects created in the context of regional expectations and requirements should be evaluated on an early stage, included in comprehensive profile programs and regional development plans, and become specific objects of management, thereby creating maximum value for sustainable development of business, government and society in the long term.

5. Conclusions

Turning a strategy into real action requires quantifying the expected results, and thus allowing for evaluation of the effectiveness of the processes being launched. The existing system of strategic management of the socio-economic development of the Arctic region is in the process of formation. It has complex goals to ensure a functional approach to the creation of modern industrial complexes in remote Northern territories, linking production opportunities with the solution of social and environmental problems. There is an obvious need to synchronize government goals with the production programs of existing industrial systems as well as those that are planned or under development.

Large investment projects initiated on the basis of promising fields are system-forming, participate in solving the problems of integrated spatial development of the Arctic region, generate multiplicative effects for the industry to which they belong, as well as related industries, and stimulate the development of other promising local projects. The specifics of project management in relation to standard processes of production and economic activity determine a specific approach to evaluating the effectiveness of decisions made.

Indicators of sustainable socio-economic development of the region will differ from indicators of industrial systems development. This is due primarily to the fact that each project has its own industry affiliation. The corresponding composition of participants who determine the target vector of development and the profile of interaction with the external environment also contribute to this. Secondly, the indicators of regional development characterize, among other things, socio-demographic processes and the state of the ecosystem, which are determined by general regional trends and depend on the pace of development of the Arctic sectors of the economy as a whole. Subsequently, when conducting comprehensive monitoring of the solution of strategic tasks, an indicator assessment of both the socio-economic development of the region and industrial systems is necessary.

The proposed indicator list takes into account the needs of the Arctic region in the context of LNG projects capabilities. At the same time, the variable nature of the indicators should be noted, such as the list of indicators formed in response to the challenges in the state-business society system. These can be adjusted and clarified as the trends in the external and internal environment of the project change.

Author Contributions: Conceptualization, A.C.; Methodology O.E. and A.C., Formal Analysis O.E.; List of indicators O.E. and A.C., Writing—Original Draft Preparation O.E.; Supervision A.C. All authors have read and agreed to the published version of the manuscript.

Funding: The research was carried out with the financial support of a grant by the President of the Russian Federation for the state support of leading scientific schools of the Russian Federation, the number of the project NSh-2692.2020.5 "Modelling of ecological-balanced and economically sustainable development of hydrocarbon resources of the Arctic".

Data Availability Statement: The research is based on open available data. When calculating the projects integrated sustainability indicator (Appendix A, Table 5), data based on expert survey for indicator weighing were used. The calculated values of the projects integrated sustainability indicators (Appendix A, Table 5) and of projects budget efficiency at the regional level (Table 1) are original.

Conflicts of Interest: The authors declare no conflict of interest.

Appendix A

Table A1. The main groups of LNG project's stakeholders, their interests and potential contribution to project implementation.

Category of Stakeholders	Main Areas of Interests	Potential Contribution to Project Development
Public authorities	Diversification of gas export supplies, development of the raw material base, innovative, technological and socio-economic development, improvement of the country's image on the world stage, budget revenues from projects.	State support for projects in the form of participation in financing and providing tax incentives, opportunities for lobbying, LNG projects integration in industry and regional strategies and promoting their implementation in the context of socio-economic development of regions and the country as a whole.
Investors and credit organizations	Return on investment, sustainable development and socially responsible investment, creating and strengthening partnerships with companies participating in projects, diversifying the project portfolio, and gaining experience in participating in LNG projects.	Provision of financial and other resources for project implementation.
Industrial companies (potential operators)	Achievement of project goals, their implementation in accordance with deadlines and budgets, technological development of companies, increasing the investment attractiveness of business.	Full responsibility for the projects implementation, promoting the development of Russian LNG industry.
Local communities	Safety of used technologies, possibility of employment in the jobs created within the projects, participation of the operator company in socio-economic development of the region, preservation of traditional way of life.	Human resources, an ability to purchase local goods and services, "social license to operate".
Non-governmental environmental organizations	Safety and scientific feasibility of technologies, compliance with all environmental standards and requirements in projects implementation, minimizing negative impact on ecosystems.	Opportunities for lobbying due to credibility of a number of NGOs among the public.
Media	Transparency and accessibility of project information, open dialogue with operator companies and project participants.	A communication tool that contributes to positive reputation formation of operating companies.
Control, supervision and regulation bodies	Reliability and regularity of project data provided, implementation of projects within the framework of current legislation.	Favorable institutional conditions for conducting project work.
Project teams	Social responsibility of operator companies, high wages, decent working conditions, opportunities for professional development.	The main impact on the achievement of project goals and performance indicators.
Suppliers and Contractors	Long-term contracts and the stability of the interaction.	The main impact on project performance in terms of cost, timing, and quality.
Consumers	Stable access to energy that is sold at market prices and meets quality requirements.	Stable demand for LNG.

Table 2. Opportunities for LNG projects in the context of regional and industrial interests.

Region and Industry Strategic Challenges	LNG Project Opportunities
Comprehensive socio-economic development of the Arctic zone of the Russian Federation	Construction and modernization of transport, logistics, energy and social infrastructure facilities, gasification of settlements, provision of employment, implementation of programs for interaction with indigenous peoples, intensification of navigation along the Northern Sea Route (NSR), development of Russian fleet, revenues to the regional budget, growth of capitalization of territories, increased investment attractiveness of the region.
Development of science and technology	Development and implementation of equipment and technologies adapted for use in Arctic conditions, production of innovative knowledge-based products, stimulating the results of intellectual activity, intensifying interaction with research institutes, universities and scientific centers.
Establishment of modern information and communication infrastructure	Construction of information and telecommunication infrastructure facilities enabling the provision of communication services to the population and economic entities.
Ensuring environmental safety	Minimizing the negative anthropogenic impact on the environment by creating conditions for the use of LNG as a fuel source and using of safe and energy-efficient technologies in the construction and operation of a production asset.
International cooperation in the Arctic	Promoting the attraction of foreign capital, creating conditions for the development of mutually beneficial partnerships, improving the efficiency of foreign economic activity.
Developing the natural gas market and meeting domestic energy demand	Increase of natural gas production, gasification of settlements of the region of presence.
Flexible response to global gas market dynamics	Diversification of supplies and expansion of Russian natural gas sales.
Development of LNG production and consumption, ensuring the country's leadership position in the global LNG market	The launch of new process lines has a direct impact on the increase in LNG production and allows to increase Russia's share in the global LNG market.
Increase in production and consumption of GMT (including using LNG)	Part of domestically produced LNG using as a clean energy medium.

Table 3. Indicator groups for LNG project sustainability assessment.

Indicator Group	Indicator Group Description
Macroeconomic	Reflect the project's contribution to macroeconomic indicators. They correspond to the interests of increasing the capitalization of territories, increasing investment activity, increasing trade volumes, developing the budget system and the national economy as a whole.
International integration	They reflect the project's contribution to increasing the investment attractiveness of Russian projects for foreign partners. They correspond to interests of developing international Arctic cooperation.
Investment	Reflect return and return on investment. They correspond to interests on investment efficiency of the project.
Infrastructure	They reflect the number of new, as well as reconstructed and modernized regional infrastructure facilities. They correspond to the interests of developing the infrastructure framework of the region.
Sectorial	They reflect the efficiency of using assets, as well as the project's contribution to increasing the production potential of LNG, increasing production, and intensifying gas exports. It is consistent with the interests of developing the resource base and diversifying the export supply of Russian gas.
Inter-project cooperation	Reflect the impact on the development of related projects. They correspond to the interests of increasing the investment attractiveness of the region, the development of Arctic shipbuilding and the development of the national economy as a whole.
Innovative	They reflect the impact of the project on the increase of domestic developments in specialized areas. They correspond to interests in the field of innovative development and import substitution.
Human development	They reflect the impact of the project on increasing the level of employment of the population, the competitiveness of the labor force, and the expansion of choice opportunities. They correspond to the interests of providing opportunities for professional development and increasing demand in the labor market.
Development of regional culture	They reflect the impact of the project on the preservation of the cultural heritage of the region, including the life of the indigenous peoples of the North. They correspond to the interests of preserving the traditional way of life of local population.
Improving living standards	They reflect the impact of the project on extending the perimeter of access to social benefits. They correspond to the interests of reducing social differentiation of society.
Environmental pollution	They reflect negative impact of the project on the state of the environment and characterize the volume of emissions of harmful substances into the atmosphere. They correspond to the interest to reduce anthropogenic impact on the environment.
Transition to clean energy	Reflect the impact of the project on increasing the share of clean energy in transport and energy sectors. They are in line with the interests of reducing greenhouse gas emissions in the national economy.
Energy efficiency	They characterize the quality of energy resources used in production process. They are consistent with the interests of effective energy using.

Table 4. LNG project sustainability indicators list.

Indicator	Unit of Measurement	The Contents of the Indicator	Group of Indicators	The Field of Sustainable Development
State revenue	$ bln.	An indicator of the project contribution to budget revenues. It meets the interests of the budget system development.	Macroeconomic	Economy
Value of investments	$ bln.	An indicator of the project contribution to the total investment in the fixed capital of the region. It meets the interests to increase investment activity.	Macroeconomic	Economy
Volume of marketable products	$ bln.	An indicator that reflects the project contribution to production growth. It meets the interests of the national economy development.	Macroeconomic	Economy
Export Volume	$ bln.	An indicator that reflects the project contribution to increasing exports of products. It meets the interests of the national economy development.	Macroeconomic	Economy
Number of foreign shareholders of the project	units	Indicator characterizing the project contribution to attract foreign partners. It is in the interests of developing Arctic international cooperation.	International integration	Economy
Volume of foreign capital raised	$ bln.	An indicator characterizing the project contribution to attract foreign capital. It meets interests of developing Arctic international cooperation.	International integration	Economy
Road construction	km.	An indicator of the project contribution to regional road construction. It meets the interests of developing the infrastructure framework of the region.	Infrastructure	Economy
Creation of regional energy infrastructure	units	Indicator of the energy infrastructure facilities number in the region. It meets the interests of developing the infrastructure framework of the region.	Infrastructure	Economy
Creation of regional transport infrastructure	units	Indicator of the transport infrastructure facilities number in the region. It meets the interests of developing the infrastructure framework of the region.	Infrastructure	Economy
Creation of regional information infrastructure	units	Indicator of the information infrastructure facilities number in the region. It meets the interests of developing the infrastructure framework of the region.	Infrastructure	Economy
Reconstruction and modernization of existing transport infrastructure	units	An indicator of the project contribution to the regional infrastructure modernization. It meets the interests of developing the infrastructure framework of the region.	Infrastructure	Economy
Reconstruction and modernization of existing energy infrastructure	units	An indicator of the project contribution to the regional infrastructure modernization. It meets the interests of developing the infrastructure framework of the region.	Infrastructure	Economy
Volume of transportation by NSR	one million tons/year	Indicator characterizing the increase in cargo flow along the routes of the Northern Sea Route (average for the period of the project implementation). It meets the interests to increase the NSR turnover.	Infrastructure	Economy

Table 4. *Cont.*

Indicator	Unit of Measurement	The Contents of the Indicator	Group of Indicators	The Field of Sustainable Development
Payback Period	years	An indicator that reflects the period of return of investments. It meets the interests of the investment efficiency of the project.	Investment	Economy
NPV	$ bln.	Indicator reflecting the income of the project for the investor. It meets the interests of the investment efficiency of the project.	Investment	Economy
PI	-	Indicator reflecting the effectiveness of investment for the investor. It meets the interests of the investment efficiency of the project.	Investment	Economy
Increased gas production	billion m^3	Indicator characterizing the increase in Russian gas production. It meets the interests of developing the resource base.	Sectorial	Economy
Production capacity	one million tons/year	Indicator reflecting the amount of LNG produced per year (average for the project period). It meets the interests of building the productive capacity of the industry.	Sectorial	Economy
Capacity utilization	%/year	Indicator reflecting the annual load level of the asset (average for the period of the project implementation). It meets the interests of efficient production capabilities using.	Sectorial	Economy
Entering new markets	units	An indicator characterizing the number of countries that are new destinations of Russian exports. It meets the interests to diversify the distribution of Russian gas.	Sectorial	Economy
Creation of domestic fundamentally new technologies in natural gas production	units	Indicator characterizing the number of new Russian developments in the field of production. It meets the interests of innovative development and import substitution industrialization.	Innovative	Economy
Creation of domestic fundamentally new technologies in gas liquefaction	units	Indicator of the new Russian developments number in the field of natural gas liquefaction. It meets the interests of innovative development and import substitution industrialization.	Innovative	Economy
Creation of domestic fundamentally new technologies in offshore gas transportation and icebreaking operations	units	Indicator characterizing the new Russian developments number in the field of Arctic shipping and icebreaking operations. It meets the interests of innovative development and import substitution industrialization.	Innovative	Economy
Share of Russian equipment and technologies in project assets	%	An indicator characterizing the volume of domestic products using in the implementation of the project. It meets the interests of innovative development and import substitution industrialization.	Innovative	Economy

Table 4. *Cont.*

Indicator	Unit of Measurement	The Contents of the Indicator	Group of Indicators	The Field of Sustainable Development
Use of project assets in other projects	units	An indicator that reflects the impact of the project on other projects. It meets the interests to increase the investment attractiveness of the region.	Inter-project cooperation	Economy
Creation of specialized complexes to meet the production and technological needs of the project	units	An indicator characterizing the project contribution to stimulate the development of new industries. It meets the interests of the national economy development.	Inter-project cooperation	Economy
Number of domestic vessels built for the needs of the project	units	Indicator reflecting the impact of the project on the shipbuilding industry. It meets the interests for the Arctic shipbuilding development.	Inter-project cooperation	Economy
Job creation	thousand units	An indicator characterizing the impact of the project on the new jobs creation. It meets the interests of increasing the employment of the population and increasing labor productivity in the economy.	Human development	Social sphere
Employment of the local population in the project	%	An indicator characterizing the participation of the local population in the project. It meets the interests of increasing employment in the region.	Human development	Social sphere
Contribution to the preservation of the traditional way of life and the distinctive culture of indigenous peoples	units	Indicator of the number of programs for interaction with indigenous peoples. It is in the interests of protecting the rights of the indigenous peoples of the North.	Development of regional culture	Social sphere
Targeted training of workers	units	Indicator characterizing the number of specialized educational programs developed for the purpose of the project. It meets the interests of human development.	Human development	Economy
Building social infrastructure in the region	units	Indicator of the social infrastructure facilities number in the region. It meets the interests of developing the infrastructure framework of the region.	Improving living standards	Social sphere
Reconstruction and modernization of social infrastructure in the region	units	An indicator of the project contribution to the regional infrastructure modernization. It meets the interests of developing the infrastructure framework of the region.	Improving living standards	Social sphere
Gasification of Russian regions	units	Indicator characterizing the number of gasified settlements. It meets the interests to improve the quality of life.	Improving living standards	Social sphere
Greenhouse gas emissions	million tons/year	Indicator of greenhouse gas emissions (average for the project period). It meets the interests to reduce the anthropogenic impact on the ecosystem.	Environmental pollution	Ecology

Table 4. *Cont.*

Indicator	Unit of Measurement	The Contents of the Indicator	Group of Indicators	The Field of Sustainable Development
Energy intensity of production	LNG KW/ton	Indicator of production energy efficiency. It meets the interests of rational use of resources.	Energy efficiency	Ecology
Gas flaring	thousand m^3	Indicator characterizing the amount of gas burned. It meets the interests to reduce the anthropogenic impact on the ecosystem.	Environmental pollution	Ecology
LNG utilization volumes for domestic vessel refueling	thousand m^3	An indicator of the increase in the use of LNG in marine refueling. It meets the interests to reduce the anthropogenic impact on the ecosystem.	Transition to clean energy	Ecology
Volume of LNG used as gas engine fuel for motor vehicles and large machinery	thousand m^3	An indicator of the increase in the use of LNG as motor fuel. It meets the interests to reduce the anthropogenic impact on the ecosystem.	Transition to clean energy	Ecology
LNG usage volumes for power generation	thousand m^3	An indicator of the increase in the use of LNG as a source for electricity generation. It meets the interests to reduce the anthropogenic impact on the ecosystem.	Transition to clean energy	Ecology

Table 5. Calculation of integrated sustainability indicators for Arctic LNG projects.

Indicator, Vector, Unit	Indicator Value			Normalized Value			Weight Normalized Value			Justification
	Yamal LNG	Arctic LNG-2	Ob LNG	Yamal LNG	Arctic LNG-2	Ob LNG	Yamal LNG	Arctic LNG-2	Ob LNG	
Volume of marketable products, max, $ bln.	224.2	239.8	60.1	0.9	1.0	0.3	1.0	1.1	0.3	Authors' calculation based on LNG production volume and price.
State revenue, max, $ bln.	5.7	5.1	1.3	1.0	0.9	0.2	1.2	1.1	0.3	Authors' calculation based on modeling and discounting of cash flows in terms of budget efficiency.
Value of investments, max, $ bln.	28	21	6	1.0	0.8	0.2	1.3	1.0	0.3	Open source project costs data.
Export volume, max, $ bln.	224.2	239.8	60.1	0.9	1.0	0.3	1.1	1.2	0.3	Authors' calculation based on LNG production volume and price (authors' assumption that the project volume of production will correspond to the volume of exports).
Number of foreign shareholders of the project, max, units	3	4	0	0.8	1.0	0.0	1.8	2.3	0.0	Data on project participants from open sources. Information is shown in the Table 1 of this article.
Volume of foreign capital raised, max, $ bln.	13	9	0	1.0	0.6	0.0	2.4	1.5	0.0	Data on the structure of participation in project capital from open sources.
Road construction, max, km.	0	0	0	0	0	0	0	0	0	Data on road construction volumes from open sources (not available currently).
Creation of regional energy infrastructure, max, units	1	0	0	1.0	0.0	0.0	0.7	0.0	0.0	Data on construction of regional energy infrastructure from open sources. For the Yamal LNG project was created the Yamal LNG TPP. There is no information about other projects.
Creation of regional transport infrastructure, max, units	2	1	0	1.0	0.5	0.0	0.7	0.3	0.0	Data on construction of regional transport infrastructure facilities from open sources. For the Yamal LNG project were created the seaport and the airport. For the Arctic LNG-2 project is planned the marine terminal. There is no data for Ob LNG.
Creation of regional information infrastructure, max, units	1	0	0	1.0	0.0	0.0	0.6	0.0	0.0	Data on construction of regional information infrastructure facilities from open sources. For the Yamal LNG was built a network of mobile communication towers. There is no data for other projects.

Table 5. *Cont.*

Indicator, Vector, Unit	Indicator Value			Normalized Value			Weight Normalized Value			Justification
	Yamal LNG	Arctic LNG-2	Ob LNG	Yamal LNG	Arctic LNG-2	Ob LNG	Yamal LNG	Arctic LNG-2	Ob LNG	
Reconstruction and modernization of existing transport infrastructure, max, units	0	0	0	0	0	0	0	0	0	Data on reconstruction and modernization of facilities from open sources (not available currently).
Reconstruction and modernization of existing energy infrastructure	0	0	0	0	0	0	0	0	0	Data on reconstruction and modernization of facilities from open sources (not available currently).
Volume of transportation by NSR, max, one million tons/year	17.5	19.8	4.8	0.9	1.0	0.2	0.6	0.7	0.2	On the basis of data on production capacity.
Payback Period, min, years	20	16	10	0.5	0.6	1.0	0.8	1.0	1.6	Authors' calculation based on modeling and discounting the cash flows in terms of commercial efficiency.
NPV, max, $ bln.	4.9	5.2	1.2	0.9	1.0	0.2	1.5	1.6	0.4	Authors' calculation based on modeling and discounting the cash flows in terms of commercial efficiency. For the Yamal LNG project, tax benefits are included in the cash flows. For other projects, there is no data on tax benefits yet. The calculations are based on the project capacity of the plant, taking into account current prices and authors' assumption of 100% export sales (data on domestic use are not available) and 40-year operational phase period. We also assumed that domestic consumption can be achieved by exceeding the project capacity due to low temperatures. Excess of project plant capacity was not taken into account in the cash flow calculations.
PI, max	1.26	1.57	1.54	0.8	1.0	1.0	1.3	1.6	1.6	Authors' calculation based on modeling and discounting the cash flows in terms of commercial efficiency.
Increased gas production, max, billion m^3	540	560	140	1.0	1.0	0.3	1.2	1.3	0.3	Authors' calculation based on the production demand in the raw material base of the liquefaction plant.
Production capacity, max, one million tons/year	17.5	19.8	4.8	0.9	1.0	0.2	1.0	1.1	0.3	Open source data. Information is shown in the Table 1 of this article.

Table 5. *Cont.*

Indicator, Vector, Unit	Indicator Value			Normalized Value			Weight Normalized Value			Justification
	Yamal LNG	Arctic LNG-2	Ob LNG	Yamal LNG	Arctic LNG-2	Ob LNG	Yamal LNG	Arctic LNG-2	Ob LNG	
Capacity utilization, max, %/year	111	111	111	1.0	1.0	1.0	1.1	1.1	1.1	Based on the authors' assumptions, taking into account current experience with exceeding production volumes over production capacity about 11%. There is no reason to believe that excess production capacity will differ between others Arctic LNG projects.
Entering new markets, max, units	6	0	0	1.0	0.0	0.0	1.2	0.0	0.0	Data from open sources. For the Yamal LNG project, these are the USA, Canada, Malta, Spain, Brazil, Panama. There is no data for the Ob LNG and Arctic LNG-2 projects.
Creation of domestic fundamentally new technologies in natural gas production, max, units	0	0	0	0	0	0	0	0	0	Data on the number of new technologies from open sources (not available currently).
Creation of domestic fundamentally new technologies in gas liquefaction, max, units	1	0	0	1.0	0.0	0.0	1.4	0.0	0.0	Data on the number of new technologies from open sources. For the Yamal LNG project, the first Russian gas liquefaction technology «Arctic cascade» was created. There is no data for other projects.
Creation of domestic fundamentally new technologies in offshore gas transportation and icebreaking operations, max, units	0	0	0	0	0	0	0	0	0	Data on the number of technologies from open sources (not available currently).
Share of Russian equipment and technologies in project assets, max, %	15	0	0	1.0	0.0	0.0	1.2	0.0	0.0	Open source data (not available currently for the Arctic LNG-2 and the Ob LNG projects).

Table 5. *Cont.*

Indicator, Vector, Unit	Indicator Value			Normalized Value			Weight Normalized Value			Justification
	Yamal LNG	Arctic LNG-2	Ob LNG	Yamal LNG	Arctic LNG-2	Ob LNG	Yamal LNG	Arctic LNG-2	Ob LNG	
Use of project assets in other projects, max, units	8	0	0	1.0	0.0	0.0	1.6	0.0	0.0	According to data from open sources, the Yamal LNG project infrastructure will be used to implement the Arctic LNG-2, Ob LNG, Arctic LNG-1, Arctic LNG-3 projects, and project of production coal, gold, non-ferrous and rare metal ores (the number and projects titles are not reported, the calculation takes into account the value 4). There is no publicly available data for other projects.
Creation of specialized complexes to meet the production and technological needs of the project max, units	0	1	0	0.0	1.0	0.0	0.0	1.4	0.0	For the Arctic LNG-2 project, it is a Center for the construction of large-capacity offshore structures. There is no data for other projects.
Number of domestic vessels built for the needs of the project max, units	0	15	0	0.0	1.0	0.0	0.0	1.8	0.0	Open source data. No data available for the Ob LNG project.
Job creation max, thousand units	32	24.7	6.7	1.0	0.8	0.2	3.9	3.0	0.8	Open source data for the Yamal LNG project and the authors' assumption that for the other projects the approximate number of jobs will depend on the amount of investment (specific value of the jobs number per investment unit).
Employment of the local population in the project max, %	0	0	0	0	0	0	0	0	0	Open source data (not available currently).
Targeted training for workers max, units	1	0	0	1.0	0.0	0.0	4.0	0.0	0.0	Open source data. For the Yamal LNG project, it is a specialized training program based on the «Innopolis» center. For other projects there is no information.
Contribution to the preservation of the traditional way of life and the distinctive culture of indigenous peoples max, units	1	0	0	1.0	0.0	0.0	11.1	0.0	0.0	Open source data. For the Yamal LNG project transitions for deer were built. For other projects there is no information.

Table 5. *Cont.*

Indicator, Vector, Unit	Indicator Value			Normalized Value			Weight Normalized Value			Justification
	Yamal LNG	Arctic LNG-2	Ob LNG	Yamal LNG	Arctic LNG-2	Ob LNG	Yamal LNG	Arctic LNG-2	Ob LNG	
Creation of social infrastructure in the region max, units	3	0	0	1.0	0.0	0.0	3.3	0.0	0.0	Open source data. For the Yamal LNG project, it is a shift settlement, a hospital, and a sports complex. For other projects there is no information.
Reconstruction and modernization of social infrastructure in the region max, units	0	0	0	0	0	0	0	0	0	Open source data (not available currently).
Gasification of Russian regions max, units	0	0	0	0	0	0	0	0	0	Open source data (not available currently).
Greenhouse gas emissions, min, million tons/year	1	1	1	1.0	1.0	1.0	5.8	5.8	5.8	Based on the assumption of similar liquefaction technologies, there is no reason to believe that greenhouse gas emissions will be different.
Gas flaring, min, thousand m^3	1	1	1	1.0	1.0	1.0	5.3	5.3	5.3	Based on the assumption of liquefaction technologies similar in technological characteristics, there is no reason to believe that the volumes of gas flaring will be different.
Energy intensity of production, min, LNG KW/ton	1	1	1	1.0	1.0	1.0	11.1	11.1	11.1	Based on the assumption of liquefaction technologies similar in technological characteristics, there is no reason to believe that the energy intensity of production will be different.
LNG utilization volumes for domestic vessel refueling, max, thousand m^3	0	0	0	0	0	0	0	0	0	Open source data (not available currently).
Volume of LNG used as gas engine fuel for motor vehicles and large machinery, max, thousand m^3	1	1	1	1.0	1.0	1.0	3.7	3.7	3.7	Based on the information on the plans for the integrated use of the Arctic LNG in Chukotka for electric generation and transport, there is no reason to believe that the consumption volumes will vary between projects.
LNG usage volumes for power generation, max, thousand m^3	1	1	1	1.0	1.0	1.0	3.6	3.6	3.6	Based on the information on the plans for the integrated use of the Arctic LNG in Chukotka for electric generation and transport, there is no reason to believe that the consumption volumes will vary between projects.
Total							75.5	52.6	36.9	

References

1. Moe, A. The dynamics of arctic development. In *Asia in the Arctic. Narratives, Perspectives and Policies*; Sakhuja, V., Narula, K., Eds.; Springer: Singapore, 2016; pp. 3–15.
2. Glomsrod, S.; Mäenpää, I.; Lindholt, L.; McDonald, H.; Goldsmith, S. Arctic economies within the Arctic nations. In *The Economy of the North*; Glomsrod, S., Aslaksen, I., Eds.; Statistics Norway: Oslo, Norway, 2008; pp. 37–68.
3. Russian International Affairs Council. *Arctic Region: Problems of International Cooperation. Volume 1*; Ivanov, I.S., Ed.; Aspect Press: Moscow, Russia, 2013. (In Russian)
4. Center for Strategic Research. Arctic 18-24-35: The View of the Young. 2018. Available online: https://www.csr.ru/upload/iblock/077/077bf5da0f50eeea20d3459409938f6a.pdf (accessed on 15 May 2020). (In Russian).
5. Nikulin, A.A. Mineral resources of Russia: Opportunities and prospects for development. *Natl. Strategy Probl.* **2017**, *40*, 163–187. (In Russian)
6. Dodin, D.A.; Kaminskiy, V.D.; Suprunenko, O.I.; Pavlenko, V.I. Nodal problems of ensuring the economic development of the Russian Arctic. *Arct. Ecol. Econ.* **2011**, *4*, 64–79. (In Russian)
7. Asheim, B.T.; Smith, H.L.; Oughton, C. Regional innovation systems: Theory, empirics and policy. *Reg. Stud.* **2011**, *45*, 875–891. [CrossRef]
8. Skufyina, T.P. Alternatives for the development of the Russian North. *Reg. Econ. Theory Pract.* **2011**, *187*, 2–9. (In Russian)
9. Krivovichev, S.V. Editorial for special issue "Arctic mineral resources: Science and technology". *Minerals* **2019**, *9*, 192. [CrossRef]
10. Gaynanov, D.A.; Kirillova, S.A.; Kuznetsova, Y.A. Russian Arctic in the context of sustainable development. *Econ. Soc. Chang. Facts Trends Forecast* **2013**, *30*, 79–88. (In Russian)
11. Smirnova, O.O.; Lipina, S.A.; Sokolov, M.S. Modern prospects and challenges for sustainable development of the Arctic zone of the Russian Federation. *Trends Manag.* **2017**, *1*, 1–15. (In Russian) [CrossRef]
12. Ermida, G. Strategic decisions of international oil companies: Arctic versus other regions. *Energy Strategy Rev.* **2014**, *2*, 265–272. [CrossRef]
13. Petrov, A.N.; BurnSilver, S.; Chapin, F.S., III; Fondahl, G.; Graybill, J.K.; Keil, K.; Nilsson, A.E.; Riedlsperger, R.; Schweitzer, P. *Arctic Sustainability Research: Past, Present and Future*; Taylor & Francis Group Ltd.: Oxford, UK, 2017.
14. Carson, M.; Sommerkorn, M.; Behe, C.; Cornell, S.; Gamble, T.; Mustonen, T.; Peterson, G.; Vlasova, T.; Chapin, F.S., III. An Arctic resilience assessment. In *Arctic Resilience Report*; Carson, M., Peterson, G., Eds.; Arctic Council: Stockholm, Sweden, 2016; pp. 2–26.
15. Elhuni, M.R.; Ahmad, M.M. Key performance indicators for sustainable production evaluation in oil and gas sector. *Procedia Manuf.* **2017**, *11*, 718–724. [CrossRef]
16. Dale, B.; Veland, S.; Hansen, A. Petroleum as a challenge to arctic societies: Ontological security and the oil-driven 'push to the north'. *Extr. Ind. Soc.* **2019**, *6*, 367–377. [CrossRef]
17. Bogachev, V.F.; Veretennikov, N.P.; Sokolov, P.V. Regional interests of Russia in the concept of development of the Arctic. *Vestnik MGTU* **2015**, *18*, 373–376. (In Russian)
18. O'Garra, T. Economic value of ecosystem services, minerals and oil in a melting Arctic: A preliminary assessment. *Ecosyst. Serv.* **2017**, *24*, 180–186. [CrossRef]
19. Ilinova, A.; Chanysheva, A. Algorithm for assessing the prospects of offshore oil and gas projects in the Arctic. *Energy Rep.* **2020**, *6*, 504. [CrossRef]
20. Nilsson, A.E.; Larsen, J.N. Making regional sense of global sustainable development indicators for the Arctic. *Sustainability* **2020**, *12*, 1027. [CrossRef]
21. Litvinenko, V. The role of hydrocarbons in the global energy agenda: The focus on liquefied natural gas. *Resources* **2020**, *9*, 59. [CrossRef]
22. Hönig, V.; Prochazka, P.; Obergruber, M.; Smutka, L.; Kučerová, V. Economic and technological analysis of commercial LNG production in the EU. *Energies* **2019**, *12*, 1565. [CrossRef]
23. Moe, A. A new Russian policy for the Northern sea route? State interests, key stakeholders and economic opportunities in changing times. *Polar J.* **2020**. [CrossRef]
24. Mitrova, T.; Kapitonov, S.; Klimentiev, A. Volume 2. Russian Small and Medium-Sized LNG. Regional Series. Arctic. 2019. Available online: https://energy.skolkovo.ru/downloads/documents/SEneC/Research/SKOLKOVO_EneC_RU_Arc_Vol2.pdf (accessed on 15 May 2020).
25. Holden, E.; Linnerud, K.; Banister, D. Sustainable development: Our common future revisited. *Glob. Environ. Chang.* **2014**, *26*, 130–139. [CrossRef]
26. Report of the World Commission on Environment and Development: Our Common Future. 1987. Available online: https://sustainabledevelopment.un.org/content/documents/5987our-common-future.pdf (accessed on 25 November 2020).
27. Giddings, B.; Hopwood, B.; O'Brien, G. Environment, economy and society: Fitting them together into sustainable development. *Sustain. Dev.* **2002**, *10*, 187–196. [CrossRef]
28. Jeronen, E. Sustainability and sustainable development. In *Encyclopedia of Corporate Social Responsibility*; Idowu, S.O., Capaldi, N., Zu, L., Gupta, A.D., Eds.; Springer: Berlin, Germany, 2013.

29. Purvis, B.; Mao, Y.; Robinson, D. Three pillars of sustainability: In search of conceptual origins. *Sustain. Sci.* **2019**, *14*, 681–695. [CrossRef]
30. Barbier, E.B. The concept of sustainable economic development. *Environ. Conserv.* **1987**, *14*, 101–110. [CrossRef]
31. Elkington, J. *Cannibals with Forks: The Triple Bottom Line of 21st Century Business*; Capstone Publishing: Oxford, UK, 1997.
32. Coşkun Arslan, M.; Kisacik, H. The corporate sustainability solution: Triple bottom line. *J. Acc. Fin.* **2017**, 18–34.
33. Shchukina, L.V. Theoretical aspects of sustainable development of regional socio-economic systems. *Pskov Reg. J.* **2015**, *21*, 38–50. (In Russian)
34. UN. Transforming Our World: The 2030 Agenda for Sustainable Development. Available online: https://sustainabledevelopment.un.org/post2015/transformingourworld (accessed on 11 November 2020).
35. Stepanova, N.; Gritsenko, D.; Gavrilyeva, T.; Belokur, A. Sustainable development in sparsely populated territories: Case of the Russian Arctic and far east. *Sustainability* **2020**, *12*, 2367. [CrossRef]
36. Azapagic, A. Developing a framework for sustainable development indicators for the mining and minerals industry. *J. Clean. Prod.* **2004**, *12*, 639–662. [CrossRef]
37. Ville, M.; Yang, L.; Tzong-Ru, L.; Jurgen, P. Extracting key factors for sustainable development of enterprises: Case study of SMEs in Taiwan. *J. Clean. Prod.* **2018**, *209*. [CrossRef]
38. Dvořáková, L.; Zborkova, J. Integration of sustainable development at enterprise level. *Procedia Eng.* **2014**, *69*, 686–695. [CrossRef]
39. Padash, A.; Ghatari, A. Toward an innovative green strategic formulation methodology: Empowerment of corporate social, health, safety and environment. *J. Clean. Prod.* **2020**, *261*, 121075. [CrossRef]
40. Filho, W.; Brandli, L. *Engaging Stakeholders in Education for Sustainable Development at University Level*; World Sustainability Series; Springer: Cham, Switzerland, 2016. [CrossRef]
41. Donaldson, T.; Preston, L. The stakeholder theory of the corporation: Concepts, evidence, and implications. *Acad. Manag. Rev.* **1995**, *4*, 65–91. [CrossRef]
42. Cherepovitsyn, A.E.; Ilinova, A.A.; Evseeva, O.O. Stakeholders management of carbon sequestration project in the state–business–society system. *J. Min. Inst.* **2019**, *240*, 731–742. [CrossRef]
43. Kates, R.W.; Clark, W.C.; Corell, R.; Hall, M.J.; Jaeger, C.C.; Lowe, I.; McCarthy, J.J.; Schellnhuber, H.J.; Bolin, B.; Dickson, N.M. Sustainability science. *Science* **2001**, *292*, 641–642. [CrossRef] [PubMed]
44. Searcy, C. *The Role of Sustainable Development Indicators in Corporate Decision-Making*; International Institute for Sustainable Development: Winnipeg, MB, Canada, 2009.
45. Wang, J.; Chi, H.-L.; Shou, W.; Chong, H.-Y.; Wang, X.A. Coordinated approach for supply-chain tracking in the liquefied Natural gas industry. *Sustainability* **2018**, *10*, 4822. [CrossRef]
46. Wu, J.; Wu, T. Sustainability indicators and indices: An overview. In *Handbook of Sustainability Management*; Madu, C.N., Kuei, C.-H., Eds.; World Scientific Publishing: Singapore, 2012; pp. 65–86. [CrossRef]
47. Meadows, D. Indicators and Information Systems for Sustainable Development. Available online: http://donellameadows.org/wp-content/userfiles/IndicatorsInformation.pdf (accessed on 15 May 2020).
48. Poveda, C.A. *Sustainability Assessment: A Rating System Framework for Best Practices*; Emerald Publishing Limited: Bingley, UK, 2017.
49. Gunnarsdottir, I.; Davidsdottir, B.; Worrell, E.; Sigurgeirsdottir, S. Review of indicators for sustainable energy development. *Renew. Sustain. Energy Rev.* **2020**, *133*, 110294. [CrossRef]
50. Official Website of the Ministry of Economic Development of the Russian Federation. Available online: https://economy.gov.ru/ (accessed on 22 April 2020). (In Russian)
51. Official Website of the Ministry of the Russian Federation for the Development of the Far East and the Arctic. Available online: https://eng.minvr.ru/ (accessed on 22 April 2020).
52. Velasquez, M.; Hester, P. An analysis of multi-criteria decision making methods. *Int. J. Oper. Res.* **2013**, *10*, 56–66.
53. Abdelli, A.; Mokdad, L.; Hammal, Y. Dealing with value constraints in decision making using MCDM methods. *J. Comput. Sci.* **2020**, *44*. [CrossRef]
54. Lamichhane, S.; Eğilmez, G.; Bhutta, M.K.; Gedik, R.; Erenay, B. Benchmarking OECD countries' sustainable development performance: A goal-specific principal component analysis approach. *J. Clean. Prod.* **2020**, 125040. [CrossRef]
55. Tilla, I.; Blumberga, D. Qualitative indicator analysis of a sustainable remediation. *Energy Procedia* **2018**, *147*, 588–593. [CrossRef]
56. Zhou, L.; Tokos, H.; Krajnc, D.; Yang, Y. Sustainability performance evaluation in industry by composite sustainability index. *Clean Technol. Environ. Policy* **2012**, *14*. [CrossRef]
57. Litvinenko, V.S.; Sergeev, I.B. Innovations as a factor in the development of the natural resources sector. *Stud. Russ. Econ. Dev.* **2019**, *30*, 637–645. [CrossRef]

resources

Review

New Concepts of Hydrogen Production and Storage in Arctic Region

Mikhail Dvoynikov [1], George Buslaev [1,*], Andrey Kunshin [1], Dmitry Sidorov [1], Andrzej Kraslawski [2,3] and Margarita Budovskaya [1]

[1]   Arctic Competence Center, Saint Petersburg Mining University, 199106 Saint Petersburg, Russia; Dvoynikov_MV@pers.spmi.ru (M.D.); kunshin.a.a@gmail.com (A.K.); dmitrysidorov95@gmail.com (D.S.); Budovskaya_ME@pers.spmi.ru (M.B.)
[2]   School of Engineering Science, Lappeenranta-Lahti University of Technology, FI-53851 Lappeenranta, Finland; Andrzej.Kraslawski@lut.fi
[3]   Faculty of Process and Environmental Engineering, Lodz University of Technology, 90-924 Lodz, Poland
*   Correspondence: Buslaev_GV@pers.spmi.ru

Abstract: The development of markets for low-carbon energy sources requires reconsideration of issues related to extraction and use of oil and gas. Significant reserves of hydrocarbons are concentrated in Arctic territories, e.g., 30% of the world's undiscovered natural gas reserves and 13% of oil. Associated petroleum gas, natural gas and gas condensate could be able to expand the scope of their applications. Natural gas is the main raw material for the production of hydrogen and ammonia, which are considered promising primary energy resources of the future, the oxidation of which does not release CO_2. Complex components contained in associated petroleum gas and gas condensate are valuable chemical raw materials to be used in a wide range of applications. This article presents conceptual Gas-To-Chem solutions for the development of Arctic oil and gas condensate fields, taking into account the current trends to reduce the carbon footprint of products, the formation of commodity exchanges for gas chemistry products, as well as the course towards the creation of hydrogen energy. The concept is based on modern gas chemical technologies with an emphasis on the production of products with high added value and low carbon footprint.

Keywords: hydrogen production; methanol production; ammonia production; storage of hydrogen; hydrogen transport; development of the Arctic; sustainable development

Citation: Dvoynikov, M.; Buslaev, G.; Kunshin, A.; Sidorov, D.; Kraslawski, A.; Budovskaya, M. New Concepts of Hydrogen Production and Storage in Arctic Region. *Resources* **2021**, *10*, 3. https://doi.org/10.3390/resources10010003

Received: 23 November 2020
Accepted: 4 January 2021
Published: 7 January 2021

Publisher's Note: MDPI stays neutral with regard to jurisdictional claims in published maps and institutional affiliations.

1. Introduction

Despite the existing trends of transition to renewable energy, a complete elimination of hydrocarbon raw materials is not possible in a near perspective because of constant growth of energy consumption. Simultaneously, the problem of climate change, especially in Arctic and other ecologically fragile regions, starts to be more and more acute [1,2].

Therefore, use of hydrogen starts to attract attention as an alternative way to reduce greenhouse gas emissions [3]. The European Union (EU) prognosis to create a hydrogen economy recognizes the limits in satisfying the EU's demand for low-carbon hydrogen on the basis of its own renewable energy resources [4]. Therefore, supply from external sources is considered to be a viable option. On the other hand, according to Energy Strategy of the Russian Federation for the period up to 2035, it is planned to export 0.2 million tons by 2024, and by 2035, 2 million tons of hydrogen [5].

The growing interest in the use of hydrogen has attracted attention to a need of classification of its origin and production methods. It is motivated by the fact that different methods of hydrogen production have their own specific carbon and toxic footprint. One of the attempts to build a taxonomy is the color classification [6]. The following colors of hydrogen are identified:

Green hydrogen—produced by electrolysis of water, while electrolysis uses only electricity from renewable energy sources. Regardless of the electrolysis technology chosen, the production of hydrogen has no carbon footprint, as 100% of the energy must come from renewable sources.

Gray hydrogen—produced from fossil fuels, typically natural gas. It produces waste gases containing CO_2 and increases the global greenhouse effect. The production of one ton of hydrogen in this way produces about 10 tons of CO_2.

Blue hydrogen—like gray hydrogen, but during the production process CO_2 must be separated and buried (Carbon Capture and Storage, CCS) or used for the production of fuels or chemical raw materials (Carbon Capture and Utilization, CCU) [7,8].

Turquoise hydrogen—produced by the thermal decomposition of natural gas (methane pyrolysis). Instead of CO_2, the process releases solid carbon (soot). At the same time, the use of renewable energy for the operation of a high-temperature reactor, as well as the long-term binding or storage of solid carbon, is a necessary condition for the CO_2-neutrality of the process.

In the future, gray hydrogen will not be used at all due to its highest carbon footprint. The potential for obtaining green hydrogen is limited due to high energy consumption in the process of water electrolysis.

The cost of turquoise hydrogen obtained by pyrolysis of methane, according to experts in the field of simulation and computer modeling, is much lower than that of hydrogen produced by electrolysis of water (green hydrogen). This is due to the fact that part of the energy is lost due to the efficiency of the electrolyzer 70–80%. Whereas pyrolysis of methane requires several times less energy than electrolysis. In addition, water used in electrolysis has no free energy, whereas methane has the potential to release energy during oxidation. When separating a methane molecule, hydrogen has a greater energy potential for further oxidation than that the energy spent on pyrolysis. In this case, the cost of hydrogen will depend on the cost of gas, which is minimal in the places of its production. According to our preliminary estimates, the cost of turquoise hydrogen may be lower than the cost of hydrogen produced by traditional steam reforming of natural gas with the capture of greenhouse gases (blue hydrogen). This is due to the fact that there is no need to build and maintain a CSS and CCU infrastructure for turquoise hydrogen.

The growing demand of European countries for hydrogen could create an opportunity for hydrogen production in the Russian Arctic region and its transport using new trunk pipelines. However, it should be noted that building trunk pipelines for transportation of pure hydrogen is economically unprofitable due to the high capital (use of special types of steels and coatings) and operating costs of installing and maintaining the infrastructure. The use of the existing gas transportation infrastructure is possible only when natural gas is transported with the addition of maximum 20% of hydrogen. These circumstances require the rebuilding of existing gas transportation infrastructure or the use of other means of transport [9].

The above-mentioned limitations related to production and transport of hydrogen lead to the idea of situating hydrogen production in an area with a nearby well-established transport system, and not to build transport infrastructure of remote green field. The Northern Sea Route for transportation of the Russian Federation production to the countries of Western Europe and the Asia-Pacific region, Figure 1, is considered as a perspective due to the following factors [10]:

- The presence of the Northern Sea route (product-distribution routes, both to the countries of the EU and to the countries of the Asia-Pacific region);
- possibility of bound hydrogen transport in a solid form using dry cargo vessels and in liquid form by tankers.

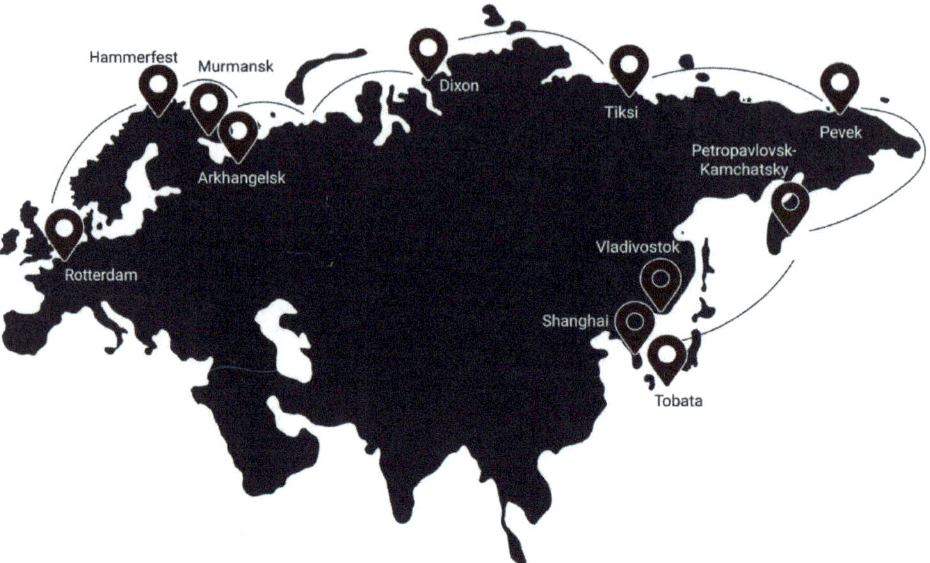

Figure 1. Product transportation routes of the Arctic zone of the Russian Federation to the countries of the European Union and the Asia-Pacific region.

In consideration of the foregoing, the task of considering new concepts for the production, storage and transport of hydrogen at the facilities of the Arctic region corresponds to actual global needs.

For this, the article attempts to revise the existing approaches to the development of Arctic oil, gas and gas condensate fields in order to focus on the emerging markets for low-carbon products.

First of all, we will consider promising methods of hydrogen production that can be integrated into Arctic industrial production.

2. Promising Methods of Hydrogen Production

More than 100 existing and proposed technologies for the production of hydrogen are described [11,12]. More than 80% of the described technologies are based on the steam conversion of fossil fuels and over 70% among them on the conversion of natural gas. However, to reduce carbon dioxide emissions from remote Arctic oil and gas facilities, a broader range of hydrogen production technologies needs to be considered, including methane pyrolysis and seawater electrolysis using renewable energy sources.

2.1. Blue Hydrogen

The main industrial technology for hydrogen production is steam methane reforming (SMR) [13,14]. The autothermal reforming (ATR) is the most common production method [14–19]. Also popular is dry reforming of methane (DRM), partial oxidation (POX) and their combinations.

In recent years, the technology of production and purification of synthesis gas has attracted much attention from both researchers and industry [19]. Since its development, it has become one of the main directions for more efficient, sustainable and environmentally safe use of hydrocarbon resources.

The reference [17] presents the topic of hydrogen-based fuel cells and energy, and addresses important trends in the contemporary energy industry, in particular, how to integrate fuel reprocessing into modern systems.

According to [16], it is important to remember that the route of hydrogen production and the choice of a specific technology depends on the type of energy and available raw materials, as well as on the required purity of the final product.

POX technologies were originally proposed by Texaco and Shell [17,18]. They are widely used in industry, and at present there are over 300 operating installations [11].

In Russia, many research centers have carried out theoretical and experimental investigations of the process of hydrocarbon partial oxidation [18,20]. However, there are no industrial POX installations. The source [20] describes in detail a method for producing synthesis gas for low-tonnage methanol production based on partial oxidation of natural gas in original three-component (hydrocarbon feedstock—oxidizer—water) synthesis gas generators (SGG) (hydrogen, carbon monoxide and carbon dioxide, water steam and ballast gases).

To carry out POX, it is possible to use chemical reactors based on power plants with high productivity and relatively low energy consumption for conversion and small weight and size characteristics. It favorably distinguishes them from any other partial oxidation devices [18,20,21]. The source [21] considers the physical model and the design of the SGG. A method for calculating the nominal geometric dimensions of the SGG is proposed, which makes it possible to assess the mass and size characteristics of the SGG already at the stage of implementation of the basic project.

The recent research [22] found that in the case of synthesis gas production by POX of natural or associated petroleum gas, the estimated cost of hydrogen production is 1.33 euros/kg H_2. The cost of large-scale production of H_2 varies from 1 to 1.5 euro/kg H_2. It is worth remembering that the decision on the economic feasibility of using natural or associated petroleum gases for the production of H_2 should be made in the context of transport infrastructure or the use of hydrogen directly at the site of oil or gas production.

2.2. Turquoise Hydrogen

Thermal or thermocatalytic decomposition (cracking or pyrolysis) is one of the methods to produce hydrogen with a low carbon footprint. The reaction of splitting methane molecules into hydrogen and carbon is carried out in a pyrolysis reactor at temperature from 600 to 1200–1400 °C [23]. Many studies are devoted to the search of effective catalysts in pyrolysis. However, the problem of catalyst deactivation, highlighted in these studies, cannot be solved by burning carbon from the catalyst surface, since in this case CO_2 emissions become comparable to the emissions in steam reforming of methane [24]. The problem of coking is less acute when using carbon catalysts—activated and black carbon, graphite, nanostructured carbon, etc. [25–27]. In this case, the reactor contains a fluidized bed with carbon catalyst particles [24]. However, all catalysts are subject to deactivation, and the formation of solid carbon as a product can lead to reactor fouling [28]. Another method of methane pyrolysis is the removal of these disadvantages thanks to the use of molten metal as a heat carrier and a bubble column for the extraction of solid carbon and gaseous hydrogen. In addition, the produced carbon, as well as the liquid metal, can act as a catalyst for the process [29]. As a rule, Sn [15,29] and Pb [17] are used as molten metal; however, there are studies aimed at finding new materials, e.g., Ni-Bi melt is proposed in [16]. Ni-Bi melt possesses catalytic properties capable of increasing methane conversion, in contrast to non-catalytic Sn and Pb melt [13,30–39]. The effect of temperature on the direct cracking of methane was studied in [30]. This research confirmed that the accumulation of carbon black in the reaction tube is the main technological obstacle in the implementation of the process of direct thermal cracking of methane. There were also studies on the types of catalysts used for methane cracking in order to reduce the amount of carbon dioxide generated in hydrogen production from methane [32,35]. The thermochemical model for the assessment of factors influencing the process of obtaining hydrogen by pyrolysis of methane was presented in [34]. It was discovered that the temperature and residence time of the gas have the greatest impact on the yield of methane conversion. It was also confirmed in [37,38]. In [37,39] there was analyzed the process of thermocatalytic decomposition of methane in a fluidized bed reactor aimed at reducing the amount of harmful emissions.

According to preliminary modeling, hydrogen with a negative carbon footprint can be produced by pyrolysis of biogas, in which carbon has been captured naturally from the air during biomass formation. In addition, it is assumed that solid carbon produced by pyrolysis of methane can be used for the production of nanostructured materials such as carbon nanotubes and fullerenes [40]. At present, the high cost of such materials limits their broad use, e.g., in the construction of highways or cement production.

2.3. Green Hydrogen

The electrolysis of water is also a common method of hydrogen production. The electrochemical dissociation of water into oxygen and hydrogen was studied in [41]. Based on recent research, low-temperature electrolysis is used to produce hydrogen. It allows to store electricity from renewable sources in chemical bonds in the form of high-purity H_2 [42]. Low-temperature electrolysis of water uses a concentrated potassium hydroxide solution KOH, a proton exchange membrane (PEM), or an alkaline anion exchange membrane (AEM) as an electrolyte. The main advantage of AEM electrolysis over other options is its cost as in this case there is no need to use platinum group metals (PGM) as catalysts. However, the major difficulty is the instability of the alkaline method, due to the sensitivity to pressure drop and the low rate of hydrogen production [43]. A traditional electrolysis device consists of two metal electrodes, an anode and a cathode, placed in an electrolyte solution and separated by a membrane [44]. When a current passes through the solution, oxygen bubbles rise above the anode and hydrogen bubbles above the cathode. To minimize the amount of energy required to liberate hydrogen from water, both electrodes are usually coated with a catalyst. However, to produce hydrogen, significant volumes of fresh water will be required, the reserves of which are already limited. The solution of this problem would be the conversion of seawater.

The seawater is mainly a mixture of Cl and Na ions. Unfortunately, Cl ions cause corrosion of the anode metal and prevent the production of hydrogen. To overcome this problem, the anode is designed as a porous nickel foam pan collector coated with an active and cheap nickel and iron catalyst [45]. Moreover, there is added negatively charged sulfate and carbonate molecules to the catalyst bed [46].

Currently, almost all the hydrogen is used in the immediate vicinity of the place of its production [47]. One of the main challenges in the construction of sustainable hydrogen energy systems is the problem of its transport. Therefore, it is necessary to consider promising technologies for the transport of hydrogen in a bound state.

3. Concept of Transport and Storage of Hydrogen in a Bound Form

The major issues related to the creation of the global hydrogen energy industry and the use of hydrogen as a fuel are its storage and transport.

To ensure sustainable development of hydrogen energy, it is proposed to consider the possibility of using sea transport, for example, bulk carriers and tankers. Their production and operation are more economical and safer than LNG ships. In addition, at the moment there is no experience in operating tankers for the transport of liquefied hydrogen.

When considering the role of Arctic regions in the development of hydrogen economy it is worth mentioning that actually the Russian ship owners control 356 tankers of various size [48].

In Saint Petersburg Mining University, the possibilities of pipeline transport of gas and methane through existing pipelines has been extensively studied [49].

In a situation where the construction of a gas pipeline or an LNG plant is not economically feasible, it is proposed to consider an option of producing hydrogen from methane. Hydrogen can be converted into various chemical compounds. The production, storage and transportation methods of these products must be well understood, the regulatory framework well structured, and the international markets for such substances well developed. It looks that ammonia and methanol meet these requirements.

3.1. Ammonia

Ammonia can be dissociated into nitrogen and hydrogen. It is a very interesting option when considering the development of hydrogen economy, as ammonia transport over long distances is much easier than that of hydrogen. At the same time, unlike hydrogen, storage and transportation of ammonia does not require the use of special expensive cryogenic containers. Ammonia can be stored and transported in standard liquid hydrocarbon storage tanks [50].

A mixture of nitrogen and hydrogen of the required composition can be obtained by reforming natural gas using atmospheric air and water, followed by the separation of CO_2 from the synthesis gas. Then, in the column for ammonia synthesis on a catalyst (promoted iron) at temperature 390–530 °C, an exothermic reaction of ammonia formation from a nitrogen-air mixture takes place:

$$3H_2 + N_2 \leftrightarrow 2NH_3 + 111.5 \text{ kJ/mol,} \tag{1}$$

The circulation gas after leaving the synthesis column, depending on the technology used, comes out with a temperature of up to 350 °C and an ammonia content of up to 19.9%. The heat of reaction of ammonia synthesis is used to heat the feed water supplied to the waste heat boiler to produce high pressure steam.

The ammonia condensation is carried out in two stages. The gas is cooled with water, air, and evaporating ammonia. The condensed ammonia is separated in a separator and the gas is directed to a circulation compressor.

Ammonia refrigeration is provided to the units by ammonia compressor or absorption refrigeration units. The produced ammonia can be supplied to consumers both in liquid and gaseous form [51]. In liquid form, ammonia can be transported over long distances in a special tanker or containers on cargo ship.

3.2. Methanol

To achieve a low carbon footprint of the products obtained from the gas and oil fields, the use of carbon capture technologies is needed. CO_2 obtained in the process of hydrogen production can be used in the synthesis of chemical raw materials, during the processing of which carbon will be tightly bound in high molecular weight compounds. This will prevent carbon from escaping into the atmosphere for several decades. Such compounds can include resins, plastics, synthetic fabrics and other components that can be obtained from such a simple monohydric alcohol as methanol [52].

In addition, ammonia plants could be successfully integrated with methanol production, and as a result the energy consumption in the production of both products is reduced.

A mixture of carbon dioxide and hydrogen can be used to produce methanol. Thus, CO_2 will be bound in the form of raw materials used in the chemical industry for the production of a wide range of products, such as formaldehyde, formalin, acetic acid, methyl tert-butyl ether, dimethyl ether, isoprene, etc. In addition, methanol can be used for the inhibition of gas hydrate formation in wells and pipelines of gas treatment units in remote Arctic fields.

The modern industrial production method is a synthesis from carbon monoxide (II) and hydrogen on a copper-zinc oxide catalyst at the temperature—250 °C and pressure—7 MPa.

The scheme of the mechanism for the catalytic production of methanol is complex and can be summarized as a reaction [53]:

$$CO + 2H_2 \rightarrow CH_3OH, \ \Delta H = -128.93 \text{ kJ/mol.} \tag{2}$$

Methanol can be also produced from CO_2 and H_2, where carbon dioxide is captured from the waste gases. In this case, the precious metals Ir and Rh or Cu, Ni, Fe, Co are usually proposed as catalysts [33,54].

3.3. Cyclohexane

In order to store and transport hydrogen, it is necessary to consider the technology of hydrogenation of aromatic hydrocarbons, such as benzene-toluene-xylene, which can be obtained at the field with the release of C_1–C_2 gas and hydrogen.

So, in the process of hydrogenation, benzene-toluene-xylene will be converted into cyclohexane by the reaction, for example, of benzene:

$$C_6H_6 + 3H_2 \overset{t,Pt}{\leftrightarrow} C_6H_{12}, \Delta H = -239.13 \ kJ/mol. \tag{3}$$

This reaction occurs at a pressure of 5 MPa and a temperature above 300 °C, after cooling to 25 °C (standard conditions), cyclohexane becomes a liquid, and in the liquid state, 1 L of cyclohexane will contain 673 L of hydrogen.

The processes of catalytic conversion of unsaturated and saturated light hydrocarbons C_2–C_5 into benzene, toluene and xylene have recently attracted the attention of many researchers [16,23,26,27,55–57].

3.4. Hydrides

Another promising method for hydrogen storage is the use of metal hydrides [58]. This method enables keeping of hydrogen in a solid-state using hydride-forming metals and alloys, hydrogen-absorbing materials with a high specific surface area, metal-hydride-carbon and amide-imide composites, hydrolysable metals and hydrides, reversibly hydrogenated organic compounds [59]. The technology under consideration, according to [60], facilitates the transportation of hydrogen fuel and reduces the volume of the system by almost three times, without requiring high economic costs for the conversion and liquefaction of hydrogen, as well as the manufacture of special vessels. It is mainly used for stationary and portable devices, as well as transport systems [61].

Hydrogen production is carried out in two reactions—hydrolysis and dissociation. As noted in [62], hydrogen production doubles during hydrolysis; however, it is an irreversible process. Hydrogen accumulators are created by thermal dissociation of hydride. In this case, minor changes of temperature and pressure cause a significant change in the equilibrium of the reaction of hydride formation.

The chemical reaction of the formation of metal hydrides is the interaction of a hydride-forming metal with H_2 in a gaseous phase [63]:

$$M \ (solid) + H_2 \ (gas) \leftrightarrow MH_2 \ (solid) + Q \tag{4}$$

where M—metal used.

A rapid increase in the pressure of hydrogen in the gaseous phase and, at the same time, a decrease in temperature, leads to a direct shift of equilibrium towards hydrides, and the reverse process causes their decomposition [64].

The whole process of hydrogen absorption consists in the mobility of H_2 molecules to the surface of the material, dissociation of adsorbed hydrogen molecules, as well as the transition of its atoms into the bulk of the material with the formation of an interstitial solid solution—α-phase, then the hydride—β-phase [56].

Kinetics of absorption/release of hydrogen under mild conditions in intermetallic hydrides is suitable for creating hydrogen storage systems [65].

As a rule, in the literature, metal hydrides are divided into high- and low-temperature ones. When considering low-temperature metal hydrides, it should be noted that stable hydrogen pressure above atmospheric pressure is observed in the temperature range up to 400 K. Low-temperature metal hydrides are intermetallic compounds: AB5, AB2, AB, as well as hydrides of body-centered crystal lattice alloys based on vanadium (V) and the

Ti-Cr system. The reaction of the formation of hydrogen hydrides with these compounds takes place with a heat of formation (<45 kJ/mol H_2), as well as high rates of sorption and desorption at medium temperature and pressure [66]. It should be noted that it is advisable to use such materials precisely for the formation of stable hydrogen fuel storage systems with an emphasis on simplifying operation and reducing energy costs. However, they have a low hydrogen content (1.5–3 wt.%). In addition, it should be noted that high-temperature metal hydrides include intermetallic compounds based on magnesium and alloy hydrides [56], which are characterized by elevated temperatures of hydrogen sorption and desorption in a temperature range of about 600 K and a high hydrogen capacity [67].

Metals-N-H systems, consisting of metal amides and hydrides are promising materials for storing hydrogen on the board of vehicles [59].

For instance:

$$2Li_2NH_2 + MgH_2 \rightarrow LiMg(NH)_2 + 2H_2MgH_2 \leftrightarrow Mg[NH_2]_2 + 2LiH \text{ (5.6 mass\% hydrogen).} \quad (5)$$

It should be noted that the methods to optimize the magnesium-hydrogen system are gaining popularity [68,69]. The kinetic properties of Mg/MgH_2 are improved with the help of nanostructured materials and the addition of catalysts, e.g., transition metals, their oxides or rare earth metals. As mentioned in [25], an increase in the rate of absorption and the release of hydrogen is achieved by grinding magnesium into a powder with particles of 50–75 μm and alloying with Ni, La, Ce, Cd, Fe, Lu, Sn, Er, Ti, Mn. The addition of destabilizing agents, for example Si, helps to reduce the dehydrogenation temperature. In the reaction of Si with MgH_2, stable $MgSi_2$ is formed, rather than Mg, and the enthalpy of the process decreases by almost 40 kJ/mol [66,70].

According to [70], in the near future, the combined hydrogen storage composed of several systems of storage and processing of hydrogen fuel, will be a serious competitor to the current technologies. The new systems can be composed of small to medium sized metal hydride hydrogen storage units as well as thermo-sorption compressors. It is worth noting that the buffer installations for storage, purification and controlled supply of hydrogen to the consumer may also be in demand.

In addition to the reaction of formation of metal hydrides, it is important to determine how the reverse process of obtaining hydrogen from them is carried out. The hydrogen from metal hydride is produced either by its heating to above 400 °C or by its reaction with water. In heating process, the metal hydrides are extruded as rods, and are decomposed by heating using electrical heaters or flue gas [56].

It can be concluded that the application of the metal hydrides is promising, as it will allow for compact, environmentally friendly and cost-effective storage and transportation of hydrogen fuel in a chemically bound state.

4. Concepts of Development of Oil and Gas Fields in the Arctic Shelf

The development of gas condensate fields is composed of many operations during which a wide range of gaseous and liquid products is obtained, e.g., methane, ethane, propane-butane mixture, unstable gas condensate, etc. As a result, in order to transport those products, the large investments are needed like gas or condensate pipelines. In the case of sea routes, it is additionally needed to build plants for the liquefaction of natural gas [71].

When arranging an Arctic oil field, similar problems arise related to logistics and transportation of products. For the export of oil, pipeline or tanker transport can be used, but the situation is complicated by the presence of significant volumes of associated petroleum gas. For its transport, it is not economically feasible to build LNG production units and long gas pipelines [10]. In this case, the possibility of its utilization on site for the production of heat and electricity should be considered. In the case when heat and electric energy is not required in volumes comparable to the amount of assisted associated petroleum gas (APG), surplus flaring is used. However, there are fields where there is so much APG that the field should be considered as oil and gas, and APG as a valuable raw material that should be converted into final products with high added value [72].

In these cases, it would be recommended to apply a Gas-To-Chem approach of creating installations, a concept developed in Saint-Petersburg Mining University. The proposed concepts for the development of remote Arctic oil and gas and gas condensate fields with the possibility of producing and transporting low-carbon products by tankers are presented in Figure 2a,b, respectively.

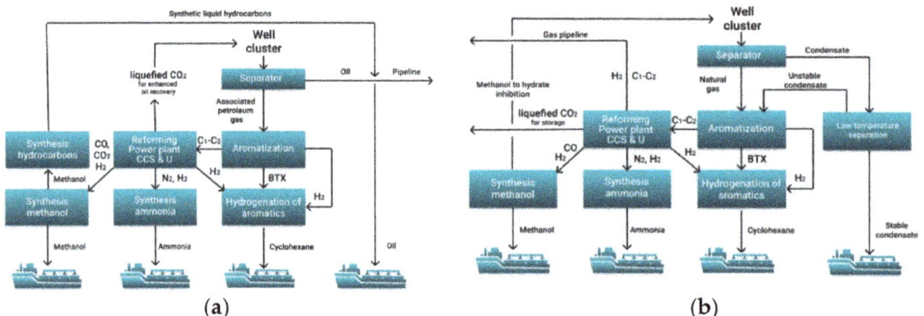

(a)   (b)

Figure 2. Concept for the development of a remote Arctic oil and gas field (**a**), gas condensate (**b**) with the possibility of producing and transporting blue hydrogen bound to methanol, ammonia and cyclohexane.

In this article there is introduced the concept of development of oil and gas and gas condensate fields, taking into account the current trends to reduce the carbon footprint of products, creation of chemical commodities, as well as hydrogen-based power generation.

4.1. Ammonia, Methanol and Cyclohexane as Hydrogen Carriers

As a promising technology for the preparation of fat gases, it is proposed to use the process of aromatization of the produced gas, which will allow converting the fat gas with C_1–C_5 components and unstable gas condensate into the lean C_1–C_2 gas (methane and ethane), liquid aromatic hydrocarbons benzene-toluene-xylene, as well as hydrogen. In this case, the lean gas can be used as a gas-chemical feedstock for the production of hydrogen, ammonia and methanol, or it will be pumped into the pipeline mixed with hydrogen.

Benzene-toluene-xylene is a valuable chemical raw material. It can be hydrogenated to cycloalkanes and next used as an intermediate product for transport and storage of hydrogen as an energy source. Additional hydrogen for the hydrogeneration of the benzene-toluene-xylene can be obtained by steam reforming and partial oxidation of natural gas, and the greenhouse gases formed in the process, together with exhaust gases from turbine power generators. It could be injected into the underground reservoir carbon capture and storage or processed into methanol for subsequent implementation carbon capture and utilization.

The main elements of the proposed schemes are the unit for reforming C_1–C_2 hydrocarbon gases, a power generation unit and waste gas utilization. Electricity generation block will include a steam turbine with a binary cycle to utilize excess heat from other gas chemical processes [73].

A schematic diagram of a methane steam reforming unit with water gas reforming reactors and amino purification is shown in Figure 3. Natural gas, over 90% methane, is mixed with the required amount of water vapor. The mixture is heated in heat exchangers H-1-5 and fed to the mixer M-1, where the prepared O_2 mixture with an equal volume of water vapor is supplied. The conversion of methane with steam takes place in the R-1 reactor. Next, the products of conversion are sent to equilibrium reactors R-2-3 for the conversion of CO with water vapor and obtaining a mixture of carbon dioxide with hydrogen. In the absorber A-1, the mixture is separated into components by an aqueous solution of MDEA. After amine purification, the hydrogen-containing mixture is sent to a pressure swing adsorption unit (PSA) to obtain hydrogen with a purity of up to 99.999%.

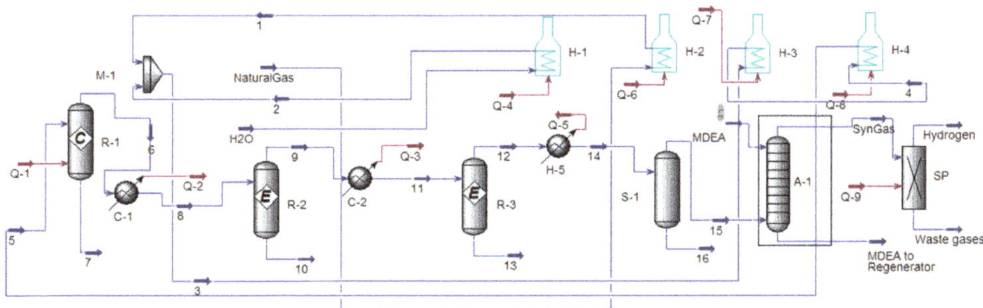

Figure 3. Steam methane reforming unit with water gas reforming reactors and amine purification. R-1—conversion reactor; R-2-3—equilibrium reactor for additional oxidation of CO and CO_2; M-1—mixer; H-1-5—heat exchanger; C-1-2—refrigerator-condenser; S-1—separator for separating liquid products; A-1—absorber; SP—adsorber.

In the reforming unit, depending on the selected configuration of the equipment, it is proposed to use the technology of steam reforming or partial oxidation of C_1–C_2 components. It is important to note that in the production of hydrogen, it is possible to implement one of several methods of utilization of CO_2 released from synthesis gas and exhaust gases of a power plant (except for injection into wells to maintain intra-reservoir pressure and increase oil recovery). The method consists in mixing in the presence of a catalyst hydrogen and carbon dioxide with the formation of products—water and carbon monoxide, which, when hydrogen is added, forms synthesis gas suitable for processing to produce methanol and motor fuels [74,75]. The produced synthesis gas will be next processed into methanol, Figure 4.

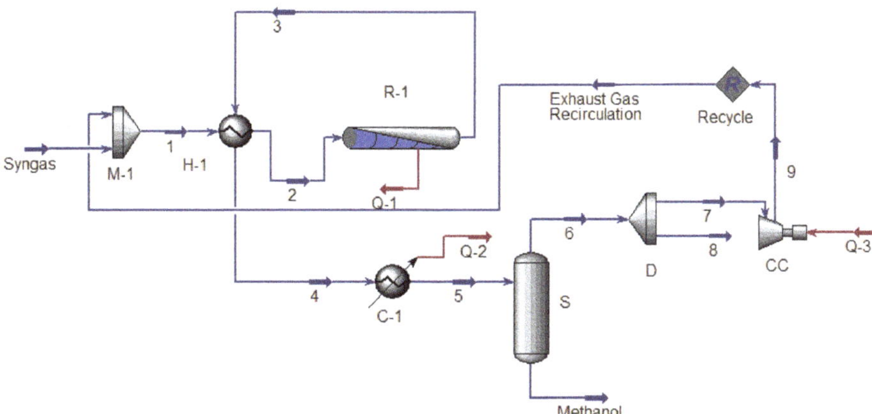

Figure 4. Preparation of synthesis gas with its subsequent processing into methanol. R-1—methanol synthesis reactor; M-1—mixer; H-1—heat exchanger; C-1—refrigerator-condenser; S—separator for separating liquid products; D—flow divider; CC—centrifugal compressor.

A mixture of carbon monoxide and hydrogen is purified from impurities, compressed in a multistage compressor CC, then mixed with unreacted gases in a mixer M-1. Next the gas enters the tubular heat exchanger H-1, heating up to 320 °C with hot reaction products. The heated gas enters the contact synthesis reactor R-1, where methanol is formed in the catalyst bed. The main stream of the gas mixture, heated in the heat exchanger H-1, is introduced into the upper part of the column R-1 and enters the catalyst bed. The reaction products are cooled in the tubes of the heat exchanger and removed through the bottom of the column. The synthesis products leaving the column are cooled to 100 °C in the heat exchanger H-1, and then in the cooler-condenser C-1 to 25–30 °C. The resulting liquid methanol is separated from the unreacted substances in the separator S and collected in the collector.

In the absence of the possibility of building a gas pipeline, hydrogen could be directed to the hydrogenation of benzene-toluene-xylene with the production of cyclohexane, methanol and ammonia from a nitrogen-hydrogen mixture associated petroleum gas, Figure 5.

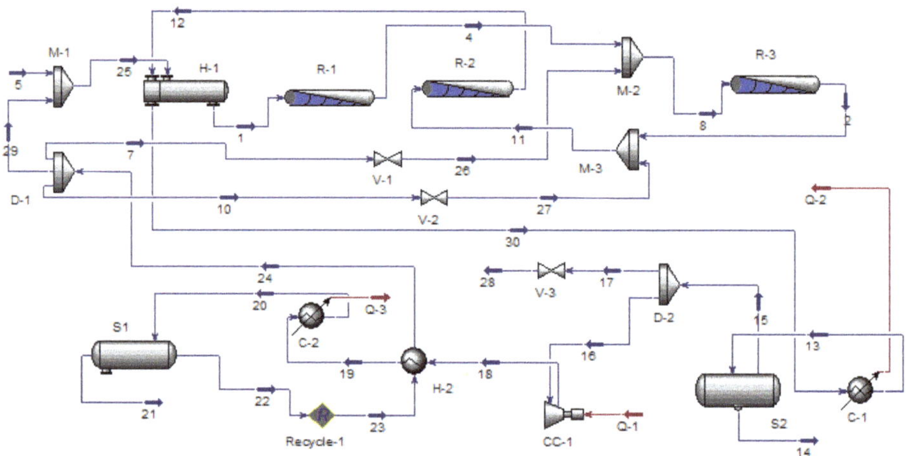

Figure 5. Scheme of ammonia production from a nitrogen-hydrogen mixture. R-1—ammonia synthesis column; R-2-3—condensation column; M-1-3—mixer; H-1—remote heat exchanger; C-1-2—refrigerator-condenser; S-1-2—separator for separating liquid products; D-1-2 filter—flow divider; CC-1—turbocharger; V-1-3—valves.

Raw gas and unreacted circulating gases enter M-1, then flow through the tubes of the heat exchanger H-1 to the synthesis column R-1. Next, liquid ammonia, passing through M-2, mixes with the circulation gas, entering the top of the condensation column R-3, where, due to the evaporation of liquid ammonia in the annular space, the gas mixture is further cooled down to −10–15 °C. A mixture of gas and condensed ammonia enters the separation section of the R-2 column to separate liquid ammonia from gases. The mixture from the separation column R-3 is discharged into the external heat exchanger H-1 and, the flow enters the coolers C-1-2 The condensed ammonia is separated in the separator S1, and the gas mixture enters the suction of the circulation stage of the CC-1 compressor, where it is compressed to pressure not higher than 24 MPa, compensating for pressure losses in the system.

From the circulation stage of the CC-1 compressor, the circulation gas is fed to the secondary condensation system, which consists of a condensing column R-1 and liquid ammonia evaporators R-2-3.

In this case, associated petroleum gas for the synthesis of ammonia is obtained from the waste gases of the methane reforming unit. There is used the method of partial oxidation and conversion of steam with subsequent removal of CO_2 by amine purification. In addition, according to the presented scheme, the other gases, e.g., exhaust gases from turbine electric generators operating on methane, can be purified by the removal of CO_2.

At the same time, benzene-toluene-xylene is a valuable chemical raw material; in addition, aromatic hydrocarbons can be used as an intermediate product for the transport and storage of hydrogen by hydrogenation. Additional hydrogen for benzene-toluene-xylene hydrogenation can be obtained by reforming natural gas.

This arrangement will ensure a low carbon footprint of the products, and the produced oil, condensate and produced hydrogen, ammonia, cyclohexane and methanol will be classified as blue. In addition, hydrogen can be released in close proximity to the consumer in the process of dehydrogenation of ammonia and cyclohexane, and reduced nitrogen and aromatic hydrocarbons could be sold on local markets as an independent product.

It is important to note that it is possible to implement one of the several methods of utilizing CO_2 released from synthesis gas and exhaust gases of a power plant, e.g., CO_2 injection into the well to increase the oil recovery factor by 5–40% [76].

The proposed schemes for the development of remote gas condensate and oil and gas fields will ensure a low carbon footprint of the products. In addition, hydrogen can be released in close proximity to the consumer in the process of dehydrogenation of ammonia and cyclohexane, and the reduced nitrogen and aromatic hydrocarbons could be sold at the local markets. The produced oil, condensate and hydrogen, ammonia, cyclohexane and methanol will have a low carbon footprint.

4.2. Low Carbon Hydrogen Bound in Hydrates

Another option for converting hydrocarbon gas into pure hydrogen can be based on combining pyrolysis of hydrocarbon gases and formation of metal hydrides. In this case, in the process of pyrolysis, solid carbon is formed, which could be collected in still bags or briquetted for transportation. The produced hydrogen could be converted into the solid hydrides. At the same time, the products could be transported from a remote field by dry cargo ship, Figure 6. In this case, solid carbon could be used locally and the hydride would be dissociated with the release of hydrogen in the immediate vicinity of the consumer. The hydride-forming materials are returned to the remote Arctic port for reuse. The produced hydrogen can be classified as turquoise when it uses a renewable source [77].

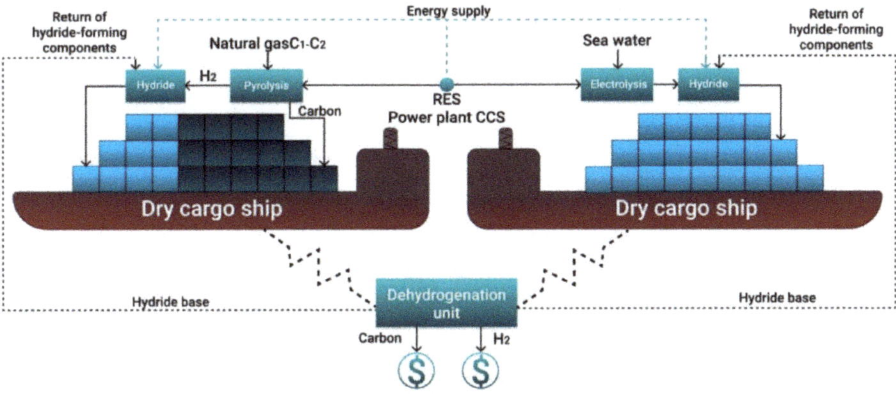

Figure 6. Concept of arranging a remote Arctic gas field with the possibility of producing and transporting solid carbon and turquoise hydrogen as a hydride using a dry cargo ship.

In the absence of a source of hydrocarbon gases, it is suggested to consider the process of electrolysis of seawater for hydrogen production. The disadvantage of this option is its high energy consumption. Therefore, there is an argument to implement the process of electrolysis from seawater and the binding of hydrogen into hydrides, to use low-carbon high-potential green energy sources, e.g., hydro energetics or geothermal systems as well as nuclear power plants.

5. Extraction of Hydrogen from Ammonia, Methanol and Cyclohexane

To facilitate the use of hydrogen by consumers, it is important to consider ways of its recovery from substances in which hydrogen has been bound to facilitate its storage and transport.

5.1. Extraction of Hydrogen from Ammonia

A nitrogen-hydrogen mixture is created in the decomposition reaction of ammonia (6) at a temperature of 900 °C:

$$2NH_3 \rightarrow N_3 + 2H_2, \; \Delta H = +46.22 \; kJ/mol. \tag{6}$$

Ammonia in liquid form enters a high-pressure liquid evaporator heated by electric heaters, where it is transformed into a gaseous state. From the evaporator, gaseous ammonia, NH_3, enters into a cracker (dissociator), in which it is decomposed into nitrogen and hydrogen. The use of a catalyst (Fe_2O_3, NiO) lowers the dissociation temperature by 100 °C or more. Next, the mixture of hydrogen and nitrogen is returned to the evaporator, where heat is recovered, and the liquid ammonia evaporates. Furthermore, the dissociated ammonia is dried to a dew point of −45 to −60 °C. The obtained mixture contains up to 75% hydrogen and 25% nitrogen [78].

Next, the reaction mixture is sent to a membrane nitrogen recovery unit. After membrane separation, the hydrogen-containing gas is fed to the adsorbers, where the remaining nitrogen is removed. The commercial hydrogen will be sent to the consumer.

5.2. Extraction of Hydrogen from Methanol

Dehydrogenation of methanol can be carried out with the formation of formaldehyde in a heterogeneous process that occurs in the gas phase on a solid catalyst according to reaction (7):

$$CH_3OH \rightarrow HCOH + H_2, \; \Delta H = +84.0 \; kJ/mol. \tag{7}$$

The process takes place in a reactor with a fixed bed of catalyst. The degree of conversion increase is controlled by a gas flow distributor. Reactor for the oxidative dehydrogenation of methanol is proposed in [79]. Formaldehyde formed during the reaction is an intermediate product of the organic industry. It is widely used in the production of synthetic resins and plastics, and in the synthesis of drugs and dyes. Thus, in the process of methanol dehydrogenation, hydrogen will be obtained, and the carbon-containing component, formaldehyde, will be bound in the target product.

5.3. Extraction of Hydrogen from Cyclohexane

Cyclohexane is a product for obtaining a wide range of chemical raw materials, such as caprolactam, adipic acid and cyclohexanone. It is also used as a solvent. The release of hydrogen from cyclohexane occurs in the dehydrogenation reaction (8), with the formation of an additional intermediate benzene, which is widely used in industry and is a feedstock for the production of drugs, synthetic rubber and dyes. Based on the kinetics of the process, a simplified reaction mechanism is given as:

$$C_6H_{12} \rightarrow C_6H_6 + 3H_2, \; \Delta H = +239.13 \; kJ/mol. \tag{8}$$

The substances considered are raw materials used in the production of many compounds used in the production of fertilizers, plastic, resins, etc. If needed, they could be decomposed with the release of hydrogen as an energy resource. Carbonaceous substances can be used for obtaining the products binding CO_2 for a long time and preventing it from entering the atmosphere.

6. Discussion and Conclusions

The role of hydrogen as an energy source is constantly growing amid trends to reduce greenhouse gas emissions. In this context, the exploitation of oil and gas deposits in Arctic regions starts to be confronted with new tasks. They are related to development of new methods of hydrogen production and transport under sever climatic conditions, strict environmental rules and lack of appropriate infrastructure. The described above factors were considered in the development of the Gas-To-Chem concept for the development of oil and gas fields in the Arctic region with the production and transport of bound hydrogen. Preparation of natural or associated petroleum gas, reduction of carbon footprint of produced products, as well as production and transport of bound hydrogen, would require implementation and the improvement of several key processes:

1. The most important are aromatization of C_{3+} hydrocarbon gases and unstable gas condensate to obtain benzene-toluene-xylene, as well as hydrogenation of benzene-toluene-xylene to produce cyclohexane.
2. The conversion of C_1–C_2 hydrocarbons for the further production of hydrogen, ammonia and methanol must combine processes with the addition of methane, water steam, oxygen and even carbon dioxide.
3. All production chains must be harmoniously linked into one production process, achieving a synergistic effect. This effect can be achieved for the combined hydrogen-methanol-ammonia production.
4. Due to the lack of fleet for shipping of compressed or liquefied hydrogen, it is suggested to use tankers and dry cargo vessels for the transport of bound hydrogen in liquid and solid form. For the existing gas transportation systems and port infrastructure, it is proposed to use C_1–C_2 pyrolysis to obtain turquoise hydrogen and solid carbon from renewable energy sources and to ship the compressed carbon and hydrogen bound into hydride.

Some of the processes described above are well known. However, they are practically not used in oil and gas fields, despite the fact that the development of gas chemical plants directly at the place of hydrocarbon production will reduce the cost of production. Such an oil and gas facility may be more efficient and more environmentally friendly than existing ones since it will have high-tech equipment and systems for the utilization of industrial wastewater and gases into absorption wells.

This article introduces conceptual Gas-To-Chem solutions for oil, gas and gas condensate fields taking into account the current trends to reduce carbon footprint of energy and chemical raw materials, as well as efforts towards the creation of low-carbon economics based on hydrogen. A low-carbon economy creates several opportunities for the use of products originating in the discussed regions: in the energy sector (hydrogen and ammonia burnt without CO_2 emissions), chemical industry (cyclohexane and methanol binding carbon in chemical compounds), production of food and biomass (nitrogen from ammonia as the basis for fertilizers). The concept is based on modern gas chemical technologies with an emphasis on the production of substances with high added value and low carbon footprint. Ammonia, methanol and cyclohexane are considered as sources of a clean energy resource of hydrogen. In the article, the methods to increase production efficiency due to the complementary nature of each other gas chemical process and to ensure selling flexibility through diversification of production are presented. Such diversification will reduce investment risks and maintain profitability amid the transformation of international markets.

The concepts presented in the article could be the starting point for the transformation of the oil and gas sector in Arctic regions, considering the actual global environmental and climate agenda.

Author Contributions: Conceptualization, M.D. and G.B.; methodology, A.K. (Andrzej Kraslawski); validation, M.D.; formal analysis, A.K. (Andrzej Kraslawski); investigation, A.K. (Andrey Kunshin), D.S. and M.B.; resources, M.D.; data curation, G.B.; writing—original draft preparation, A.K. (Andrey

Kunshin); writing—review and editing, G.B.; visualization, D.S.; supervision, M.D.; project administration, G.B. All authors have read and agreed to the published version of the manuscript.

Funding: This research received no external funding.

Institutional Review Board Statement: Not applicable.

Informed Consent Statement: Not applicable.

Data Availability Statement: The data presented in this study are available on request from the corresponding author. The data are not publicly available due to Federal Law of the Russian Federation "On Export Control".

Conflicts of Interest: The authors declare no conflict of interest.

References

1. Alekseeva, M.B.; Bogachev, V.F.; Gorenburgov, M.A. Systemic Diagnostics of the Arctic Industry Development Strategy. *J. Min. Inst.* **2019**, *238*, 450–458. [CrossRef]
2. Vasiltsov, V.S.; Vasiltsova, V.M. Strategic Planning of Arctic Shelf Development Using Fractal Theory Tools. *J. Min. Inst.* **2018**, *234*, 663–672. [CrossRef]
3. Evangelopoulou, S.; De Vita, A.; Zazias, G.; Capros, P. Energy System Modelling of Carbon-Neutral Hydrogen as an Enabler of Sectoral Integration within a Decarbonization Pathway. *Energies* **2019**, *12*, 2551. [CrossRef]
4. A Hydrogen Strategy for a Climate Neutral Europe. Available online: https://ec.europa.eu/commission/presscorner/detail/en/FS_20_1296 (accessed on 20 July 2020).
5. Energy Strategy of the Russian Federation for the Period up to 2035. Available online: https://minenergo.gov.ru/node/1026 (accessed on 10 November 2020).
6. Eckpunktepapier der Ostdeutschen Kohleländer zur Entwicklung Einer Regionalen Wasserstoffwirtschaft. Available online: https://www.medienservice.sachsen.de/medien/medienobjekte/130485/download (accessed on 13 November 2020).
7. Ilinova, A.A.; Romasheva, N.V.; Stroykov, G.A. Prospects and social effects of carbon dioxide sequestration and utilization projects. *J. Min. Inst.* **2020**, *244*, 493–502. [CrossRef]
8. Tcvetkov, P.S.; Cherepovitsyn, A.E.; Fedoseev, S.V. The Changing Role of CO_2 in the Transition to a Circular Economy: Review of Carbon Sequestration Projects. *Sustainability* **2019**, *11*, 5834. [CrossRef]
9. Melaina, M.W.; Antonia, O.; Penev, M. Blending Hydrogen into Natural Gas Pipeline. Networks: A Review of Key Issues. 2013. Available online: https://www.nrel.gov/docs/fy13osti/51995.pdf (accessed on 5 October 2020).
10. Litvinenko, V. The Role of Hydrocarbons in the Global Energy Agenda: The Focus on Liquefied Natural Gas. *Resources* **2020**, *9*, 59. [CrossRef]
11. Saphin, A.K. *Production of Hydrogen Plant Sand Construction of Hydrogen Infrastructure in Industrialized Countries. Technical and Investment Indicators of Installation Sand Hydrogen Stations*; Issue 2; LLC Prima-Khimmash: St. Petersburg, Russia, 2015; p. 226.
12. Salikhov, K.M.; Stoyanov, N.D.; Stoyanova, T.V. Using Optical Activation to Create Hydrogen and Hydrogen-Containing Gas Sensors. *Key Eng. Mater.* **2020**, *854*, 87–93. [CrossRef]
13. Parkinson, B.; Tabatabaei, M.; Upham, D.C.; Ballinger, B.; Greig, C.; Smart, S.; McFarland, E. Hydrogen production using methane: Techno economics of decarbonizing fuels and chemicals. *Int. J. Hydrogen Energy* **2018**, *43*, 2540–2555. [CrossRef]
14. Susmozas, A.; Iribarren, D.; Dufour, J. Assessing the Life-Cycle Performance of Hydrogen Production via Biofuel Reforming in Europe. *Resources* **2015**, *4*, 398–411. [CrossRef]
15. Dahl, P.Y.; Christensen, T.S.; Winter-Madsen, S. Autothermal Reforming Technology for modern large-capacity methanol plants. In Proceedings of the International Conference Nitrogen and Syngas, Paris, France, 24–27 February 2014; p. 14.
16. Dawood, F.; Anda, M.; Shafiullah, G.M. Hydrogen production for energy: An overview. *Int. J. Hydrog. Energy* **2019**, *7*, 107–154. [CrossRef]
17. Liu, K.; Song, C.; Subraman, V. *Hydrogen and Syngas Production and Purification Technologies*; John Wiley & Sons: Hoboken, NJ, USA, 2010; p. 564. [CrossRef]
18. Machlin, V.A.; Cetaruk, J.R. Modern technologies of producing synthesis gas from natural and associated gas. *Sci. Tech. J. Chem. Ind. Today* **2010**, *3*, 6–17.
19. Indarto, A.; Palguandi, J. *Syngas Production, Application and Environmental Impact*; Nova Science Publishers: New York, NY, USA, 2013; p. 365.
20. Zagashvili, U.V.; Levain, A.A.; Kuzmin, A.M.; Aniskevich, Y.V.; Vasilieva, O.V. Technology of hydrogen production using small transportable units based on high-temperature syngas generators. *Sci. Tech. J. Vopr. Mater. Edeniya* **2017**, *2*, 92–109.
21. Zagashvili, U.V.; Levain, A.A.; Kuzmin, A.M. Principles of design of three-component gas synthesis gas. Oil and gas. *LLC Obrakademnauka* **2017**, *4*, 9–16.
22. Buslaev, G.V. Technology of associated petroleum gas processing at remote Arctic oil and gas facilities to produce synthetic liquid hydrocarbons. In Proceedings of the Plenary Report at the XIX International Youth Scientific Conference SeverGeoEcoTech-2018, Ukhta, Russia, 21–23 March 2018.

23. Abalaev, A.V.; Del Toro Fonseca, D.A.; Lewiner, I.I.; Vyatkin, Y.L. Optimization of the operating mode of shelf reactors with a fixed layer in the process of aromatization of hydrocarbons C_5. Scientific service on the Internet: Search for new solutions. In Proceedings of the International Supercomputing Conference, Hamburg, Germany, 17–21 June 2012; pp. 273–277.
24. Azhazha, V.M.; Tikhonovskiy, M.A.; Shepelev, A.G.; Kurilo, Y.P.; Ponomarenko, T.A.; Vinogradov, D.V. *Materials for Hydrogen Storage: Analysis of Development Trends Based on Data on Information Flows. Questions of Nuclear Science and Technology. Series: Vacuum, Pure Materials, Superconductors*; National Scientific Center, Kharkiv Institute of Physics and Technology: Kharkov, Ukraine, 2006; pp. 145–153. Available online: https://vant.kipt.kharkov.ua/ARTICLE/VANT_2006_1/article_2006_1_145.pdf (accessed on 10 July 2020).
25. Baraban, A.P.; Gabis, I.E.; Dmitriev, V.A.; Dobrotvorskii, M.A.; Kuznetsov, V.G.; Matveeva, O.P.; Titov, S.A.; Voyt, A.P.; Yelets, D.I. Luminescent properties of aluminum hydride. *J. Lumin.* **2015**, *166*, 162–166. [CrossRef]
26. Bazhin, V.Y.; Trushnikov, V.E.; Suslov, A.P. Simulation of partial oxidation of natural gas in a resource-saving reactor mixer. *IOP Conf. Ser. Mater. Sci. Eng.* **2020**, *862*, 862. [CrossRef]
27. Belousova, O.Y.; Kutepov, B.I. *Textbook of Aromatization of Hydrocarbons on Pentasyl-Containing Catalysts*; Khimiya: Moscow, Russia, 2000; p. 95.
28. Bilera, I.V. Education of Fine Soot When Receiving the Synthesis Gas under Conditions of Combustion of Methane. In *The Gas Chemistry*; Billera, I.V., Borisov, A.A., Borunova, A.B., Kolbanoskiy, Y.A., Korolev, Y.M., Rossokhin, I.V., Troshin, I.V., Eds.; Metaprocess Ltd.: Saint-Brice-Courcelles, France, 2010; Volume 3, pp. 72–78.
29. Cooper, H.; Donnis, B.B.L. Aromatic saturation of distillates: An overview. *Appl. Catal. A Gen.* **1996**, *137*, 203–223. [CrossRef]
30. Abanades, A.; Ruiz, E.; Ferruelo, E.M.; Hernández, F.; Cabanillas, A.; Martínez-Val, J.M.; Rubio, J.A.; López, C.; Gavela, R.; Barrera, G.; et al. Experimental analysis of direct thermal methane cracking. *Int. J. Hydrog. Energy* **2011**, *36*, 12877–12886. [CrossRef]
31. Abbas, H.F.; Wan Daud, W.M.A. Hydrogen production by methane decomposition: A review. *Int. J. Hydrog. Energy* **2010**, *35*, 1160–1190. [CrossRef]
32. Amin, A.M.; Croiset, E.; Epling, W. Review of methane catalytic cracking for hydrogen production. *Int. J. Hydrogen Energy* **2011**, *36*, 2904–2935. [CrossRef]
33. Bode, A.; Anderlohr, C.; Bernnat, J.; Flick, F.; Glenk, F.; Klingler, D.; Kolios, G.; Scheiff, F.; Wechsung, A.; Hensmann, M.; et al. *Feste und Fluide Produkte ausGas—FfPaG, Schlussbericht*; BMBF: Bonn, Germany, 2018.
34. Geißler, T.; Plevan, M.; Abánades, A.; Heinzel, A.; Mehravaran, K.; Rathnam, R.; Rubbia, C.; Salmieri, D.; Stoppel, L.; Stuckrad, S.; et al. Experimental investigation and thermo-chemical modeling of methane pyrolysis in a liquid metal bubble column reactor with a packed bed. *Int. J. Hydrog. Energy* **2015**, *40*, 14134–14146. [CrossRef]
35. Moliner, R.; Suelves, I.; Lazaro, M.; Moreno, O. Thermocatalytic decomposition of methane over activated carbons: Influence of textural properties and surface chemistry. *Int. J. Hydrog. Energy* **2005**, *30*, 293–300. [CrossRef]
36. Muradov, N.; Chen, Z.; Smith, F. Fossil hydrogen with reduced CO_2 emission: Modeling thermocatalytic decomposition of methane in a fluidized bed of carbon particles. *Int. J. Hydrog. Energy* **2005**, *30*, 1149–1158. [CrossRef]
37. Muradov, N.; Smith, F.; Huang, C.; T-Raissi, A. Autothermal catalytic pyrolysis of methane as a new route to hydrogen production with reduced CO_2 emissions. *Catal. Today* **2006**, *116*, 281–288. [CrossRef]
38. Plevan, M.; Geißler, T.; Abanades, A.; Mehravaran, K.; Rathnam, R.; Rubbia, C.; Salmieri, D.; Stoppel, L.; Stückrad, S.; Wetzel, T. Thermal cracking of methane in a liquid metal bubble column reactor: Experiments and kinetic analysis. *Int. J. Hydrog. Energy* **2015**, *40*, 8020–8033. [CrossRef]
39. Schultz, I.; Agar, D.W. Decarbonization of fossil energy via methane pyrolysis using two reactor concepts: Fluid wall flow reactor and molten metal capillary reactor. *Int. J. Hydrog. Energy* **2015**, *40*, 11422–11427. [CrossRef]
40. Osman, A.I.; Farrell, C.; Al-Muhtaseb, A.H.; Harrison, J.; Rooney, D.W. The production and application of carbon nanomaterials from high alkali silicate herbaceous biomass. *Sci. Rep.* **2020**, *10*, 1–13. [CrossRef]
41. Ostadi, M.; Paso, K.G.; Rodriguez-Fabia, S.; Oi, L.E.; Manenti, F.; Hillestad, M. Process Integration of Green Hydrogen: Decarbonization of Chemical Industries. *Energies* **2020**, *13*, 4859. [CrossRef]
42. Olivier, P.; Bourasseau, C.; Bouamama, P. Low-temperature electrolysis system modelling: A review. *Renew. Sustain. Energy Rev.* **2017**, *78*, 280–300. [CrossRef]
43. Yu, L.; Zhu, Q.; Song, S.; McElhenny, B.; Wang, D.; Wu, C.; Qin, Z.; Bao, J.; Yu, Y.; Chen, S.; et al. Non-noble metal-nitride based electrocatalysts for high-performance alkaline seawater electrolysis. *Nat. Commun.* **2019**, *10*, 1–10. [CrossRef]
44. Zhu, C.; Liu, C.; Fu, Y.; Gao, J.; Huang, H.; Liu, Y.; Kang, Z. Construction of CDs/CdS photocatalysts for stable and efficient hydrogen production in water and seawater. *Appl. Catal. B Environ.* **2019**, *242*, 178–185. [CrossRef]
45. Hung, W.-H.; Xue, B.-Y.; Lin, T.-M.; Lu, S.-Y.; Tsao, I.-Y. A highly active selenized nickel-iron electrode with layered double hydroxide for electrocatalytic water splitting in saline electrolyte. *Mater. Today Energy* **2021**, *19*, 100575. [CrossRef]
46. Jamesh, M.I.; Harb, M. Recent advances on hydrogen production through seawater electrolysis. *Mater. Sci. Energy Technol.* **2020**, *3*, 780–807. [CrossRef]
47. Materials of SKOLKOVO Energy Centre, Moscow School of Management SKOLKOVO. Available online: https://energy.skolkovo.ru/downloads/documents/SEneC/Research/SKOLKOVO_EneC_Hydrogen-ecomy_Eng.pdf (accessed on 20 June 2019).
48. Buyanov, S. Prospects for the construction of ships for Russian shipowners. In Proceedings of the Current State and Prospects for Development of Russian Bunker Services Market, Saint Petersburg, Russia, 27–28 June 2019; Available online: http://cniimf.ru/press-tsentr/news/870/ (accessed on 2 July 2019).

49. Litvinenko, V.S.; Tsvetkov, P.S.; Dvoynikov, M.V.; Buslaev, G.V. Barriers to implementation of hydrogen initiatives in the context of global energy sustainable development. *J. Min. Inst.* **2020**, *244*, 421–431. [CrossRef]

50. *Anhydrous Liquefied Ammonia. Specifications*; Russian Government Standard GOST 6221-90; Standartinform: Moscow, Russia, 2011; Available online: http://www.gostrf.com/normadata/1/4294823/4294823189.pdf (accessed on 15 September 2020).

51. Production of Ammonia, Mineral Fertilizers and Inorganic Acids. Information and Technical Directory (ITD) 2019. Available online: http://docs.cntd.ru/document/564068887 (accessed on 13 November 2020).

52. Samuel, S.A.; Vincenzo, L.; Xiaoti, C.; Na, L.; Jimin, Z.; Simon, L.S.; Søren, H.J.; Mads, P.N.; Søren, K.K. A Review of The Methanol Economy: The Fuel Cell Route. *Energies* **2020**, *13*, 1–32.

53. Yurieva, T.M.; Plyasova, L.; Makarova, O.; Krieger, T. Mechanisms for hydrogenation of acetone to isopropanol and of carbon oxides to methanol over copper-containing oxide catalysts. *J. Mol. Catal. A Chem.* **1996**, *113*, 455–468. [CrossRef]

54. Chen, W.H.; Chen, C.Y. Water gas shift reaction for hydrogen production and carbon dioxide capture: A review. *Appl. Energy* **2020**, *258*, 114078. [CrossRef]

55. Angell, V.W.; Graham, H.J.; Post, G.J. Case History: Ice Island Drilling Application and Well Considerations in Alaskan Beaufort Sea. *Soc. Pet. Eng.* **1990**. [CrossRef]

56. Bolotov, V.A.; Cheremisina, O.V.; Ponomareva, M.A.; Alferova, D.A. Prospects for the use of the sorbent for purification of gases from sulfur-containing components on the basis of manganese ore. *Key Eng. Mater.* **2020**, *836*, 13–18. [CrossRef]

57. Vyatkin, Y.L.; Lishchiner, I.I.; Sinitsyn, S.A.; Kuz'min, A.M. Perspective directions of chemical processing of hydrocarbon raw materials. *Neftegaz* **2020**, *4*, 114–118.

58. Rusman, N.A.A.; Dahari, M.A. Review on the current progress of metal hydrides material for solid-state hydrogen storage applications. *Int. J. Hydrog. Energy* **2016**, 12108–12126. [CrossRef]

59. Tarasov, B.P. Problems and prospects of creating materials for hydrogen storage in a bound state. *Int. Sci. J. Altern. Energy Ecol.* **2006**, *2*, 11–17.

60. Tarasov, B.P.; Burnasheva, V.V.; Lototsky, M.V.; Yartys, V.A. Methods of hydrogen storage and the possibility of using metallohydrides. *Int. Sci. J. Altern. Energy Ecol.* **2005**, *12*, 14–37.

61. Tarasov, B.P.; Lototsky, M.V.; Yartys, V.A. The Problem of hydrogen storage and prospects for using hydrides for hydrogen storage. *Russ. Chem. J.* **2006**, *6*, 34–48.

62. McKay, M. *Hydrogen Compounds of Metals*; Mir: Moscow, Russia, 1968; p. 244.

63. York, A.P.E.; Xiao, T.; Green, M.L.H. Brief overview of the partial oxidation of methane to synthesis gas. *Top. Catal.* **2003**, *22*, 345–358. [CrossRef]

64. Schlapbach, L. Hydrogen as a fuel and its storage for mobility and transport. *MRS Bull.* **2002**, 675–679. [CrossRef]

65. Kalashnikov, J.A. *Physical Chemistry of Substances at High Pressures*; Higher School: Moscow, Russia, 1987; p. 237.

66. Hydride Hydrogen Storage System. Available online: https://lektsia.com/2x8693.html (accessed on 13 November 2020).

67. Landrum, L.; Holland, M.M. Extremes become routine in an emerging new Arctic. *Nat. Clim. Chang.* **2020**, *10*, 1108–1115. [CrossRef]

68. Fateev, V.N.; Alekseeva, O.K.; Korobtsev, S.V.; Seregina, E.A.; Fateeva, T.V.; Grigoriev, A.S.; Aliev, A.S. Problems of hydrogen accumulation and hydrogen storage. *KimyaProblemleri Baku Aliyev Akif Shihanoglu* **2018**, *4*, 453–483. [CrossRef]

69. Liu, C.; Li, F.; Ma, L.-P.; Cheng, H.-M. Advanced materials for energy storage. *Adv. Mater.* **2010**, *22*, E28–E62. [CrossRef]

70. Kornev, A.V.; Barkan, M.S. Prospects for the use of associated gas of oil development as energy product. *Int. J. Energy Econ. Policy* **2017**, *7*, 374–383.

71. Koz'menko, S.Y.; Masloboev, V.A.; Matviishin, D.A. Justification of Economic Benefits of Arctic LNG Transportation by Sea. *J. Min. Inst.* **2018**, *233*, 554–560. [CrossRef]

72. Buslaev, G.; Morenov, V.; Konyaev, Y.; Kraslawski, A. Reduction of carbon footprint of the production and field transport of high-viscosity oils in the Arctic region. *Chem. Eng. Process. Process Intensif.* **2020**, 108189. [CrossRef]

73. Morenov, V.A.; Leusheva, E.L.; Buslaev, G.V.; Gudmestad, O.T. System of comprehensive energy-efficient utilization of associated petroleum gas with reduced carbon footprint in the field conditions. *Energies* **2020**, *13*, 4921. [CrossRef]

74. Bode, A.; Agar, D.W.; Buker, K.; Göke, V.; Hensmann, M.; Janhsen, U.; Klingler, D.; Schlichting, J.; Schunk, S.A. Research cooperation develops innovative technology for environmentally sustainable syngas production from carbon dioxide and hydrogen. In Proceedings of the 20th World Hydrogen Energy Conference, Gwangju, Korea, 15–20 June 2014.

75. Church, J.A.; Clark, P.U.; Cazenave, A.; Gregory, J.M.; Jevrejeva, S.; Levermann, A.; Merrifield, M.A.; Milne, G.A.; Nerem, R.S.; Nunn, P.D.; et al. Sea Level Change. Contribution of Working Group I to the Fifth Assessment Report of the Intergovernmental Panel on Climate Change. In *Climate Change 2013: The Physical Science Basis*; Stocker, T.F.D., Qin, G.-K., Plattner, M., Tignor, S.K., Allen, J., Boschung, A., Nauels, Y., Xia, V.B., Midgley, P.M., Eds.; Cambridge University Press: Cambridge, UK; New York, NY, USA, 2013.

76. *Accelerating the Uptake of CCS: Industrial Use of Captured Carbon Dioxide*; Global CCS Institute: Melbourne, Australia, 2011.

77. Kirsanova, N.Y.; Lenkovets, O.M.; Nikulina, A.Y. The Role and Future Outlook for Renewable Energy in the Arctic Zone of Russian Federation. *Eur. Res. Stud. J.* **2018**, *21*, 356–368.

78. Wang, M.; Li, J.; Chen, L.; Lu, Y. Miniature NH_3 cracker based on microfibrous entrapped Ni-CeO2/Al2O3 catalyst monolith for portable fuel cell power supplies. *Int. J. Hydrog. Energy* **2009**, *34*, 1710–1716. [CrossRef]
79. Mukhovikova, N.K.; Nepomnyakschikh, Y.V. Preparation of Formaldehyde by Oxidative Dehydration of Methanol. In Proceedings of the XI All-Russian Scientific-Practical Conference of Young Scientists «YOUNG RUSSIA», Moscow, Russia, 2–4 October 2019; Available online: http://science.kuzstu.ru/wp-content/Events/Conference/RM/2019/RM19/pages/Articles/70207.pdf (accessed on 22 October 2020).

 *resources*

MDPI

Article

Risk-Based Methodology for Determining Priority Directions for Improving Occupational Safety in the Mining Industry of the Arctic Zone

Semyon Gendler and Elizaveta Prokhorova *

Department of Industrial Safety, Saint Petersburg Mining University, 2, 21st Line, 199106 St. Petersburg, Russia; Gendler_SG@pers.spmi.ru
* Correspondence: Prokhorovaea96@gmail.com

Citation: Gendler, S.; Prokhorova, E. Risk-Based Methodology for Determining Priority Directions for Improving Occupational Safety in the Mining Industry of the Arctic Zone. *Resources* **2021**, *10*, 20. https://doi.org/10.3390/resources10030020

Academic Editors: Pavel Tcvetkov, Nikolay Didenko and Eleni Iacovidou

Received: 18 December 2020
Accepted: 24 February 2021
Published: 1 March 2021

Publisher's Note: MDPI stays neutral with regard to jurisdictional claims in published maps and institutional affiliations.

Abstract: Over the past 10 years, the mining industry of Russia has seen a greater than three-fold decrease in injury rates, thanks to the successful implementation of innovative labor safety technologies. Despite this, injury levels remain unacceptably high compared to the leading mining countries, which results in increased mining costs. For the mining areas of the Arctic Zone—unlike other regions located in areas with a more favorable climate—the injury rates are influenced not only by the underground labor conditions, but also by the adverse environmental factors. For the Russian Arctic zone, the overall injury risk is proposed to be calculated as the combined impact of occupational and background risk. In this article, we have performed correlation analysis of the overall injury risks in regions of the Arctic zone and regions with favorable climate conditions. Using the Kirov branch of "Apatit", Joint-stock company (JSC) as an example, we have calculated the risks related specifically to occupational injury rates. We have constructed the relative injury risks and their changes over time and have developed a "basic injury rate matrix" that makes it possible to visualize the results of the comparative analysis of the injury rates on the company's production sites and to determine priority avenues for improving the occupational safety and lowering the injury rates.

Keywords: arctic zone; mining industry; mines; labor safety; occupational injury rate; risk-based methodology; risk of injuries; injury risk diagram; correlation analysis

1. Introduction

Occupational safety remains a priority for a great number of countries. Successful solutions to issues of labor safety and measures aimed at reducing the occupational injury rates influence the production efficiency and the product's cost effectiveness. Neglect or inattention towards labor safety can lead to not only technical and economic losses, but also to social ones [1].

Occupational safety issues are most relevant for the mining industry, in which the majority of production facilities are classified as hazardous. If we consider the occupational injury rate to be an indicator of occupational safety, calculated as the total number of injuries divided by the total number of workers, then the occupational risks for the mining industry would be 4–5 times higher than their average levels in other Russian industries.

Occupational injuries are even higher for the Arctic zone industries, where the miners are subject to polar stress syndrome and unfavorable environmental living conditions that result from the habitation zones being located next to the mining areas. Such conditions include low air temperatures, high precipitation, strong winds, polar nights, lack of ultraviolet radiation, comparatively high air and water pollution levels, and also excessive noise and vibration levels [2]. The polls of the Arctic zone inhabitants show interesting results. Two age groups of the Arctic zone inhabitants participated in the sociological survey, "youth"—whose age was between 25 and 44—as well as "the elderly"—those older than 60 years old. Regardless of age, during polar nights, 87% of the young people and 80% of the elderly reported feeling drowsy and depressed. Furthermore, 54% of the young people and 60% of the elderly participants reported high levels of stress, and 47% and 50% of them (respectively) reported feeling unwell. During the polar nights, 87% of young people and 80% of the elderly reportedly suffer from excessive agitation, which leads to insomnia. The low temperatures of the Arctic zone were difficult to tolerate for 60% of the young people and 80% of the elderly. Additionally, 20% of the young people and 50% of the elderly reportedly feel the changes in atmospheric pressure. Taken together, these negative environmental factors cause irreversible changes to people's mental and physical conditions, which leads to increased injury risks compared to other regions of Russia.

These high injury risks inhibit exploration of the Arctic zone's mineral deposits, which are strewn across an area of 3.1 million sq. km—18% of Russia's total area [3].

As per the annex to the Decree of the President of Russian Federation No. 296 dated 2 May 2014, the land area of the Arctic zone includes: Murmansk region, Nenets Autonomous Area, Chukotka Autonomous District, Yamalo-Nenets Autonomous District, Komi Republic, Republic of Karelia, Sakha Republic (Yakutia), Krasnoyarsk Territory, Arkhangelsk Region, and the lands and islands located in the Arctic ocean as specified by the Decree of the Presidium of the Central Executive Committee of the USSR (dated 15 April 1926) "On declaring the lands and islands of the Arctic oceans the territories of the Union of Soviet Socialist Republics" and other USSR acts [4] (Figure 1).

Figure 1. The Arctic zone of the Russian Federation. 1—Murmansk region; 2—Republic of Karelia; 3—Arkhangelsk region; 4—Nenets Autonomous Area; 5—Yamalo-Nenets Autonomous District; 6—Krasnoyarsk Territory; 7—Sakha Republic (Yakutia); 8—Chukotka Autonomous District; 9—Komi Republic (as part of Vorkuta City District).

The Arctic zone accounts for 90% of Russia's nickel and cobalt production, 60% of copper, over 96% of platinum metals, and around 80% of natural gas and 60% of oil production (Figure 2). When it comes to reserves of hydrocarbons, the zone's share becomes even larger. The Arctic shelf could be considered a strategic reserve for strengthening Russia's resource security [4–6].

Efficient development of natural resources is impossible without proper labor safety standards having been implemented in mining enterprises, some of which incorporate multiple companies (i.e., vertically integrated companies).

Injury prevention and occupational disease and illness prevention have been the topic of many studies by Russian and foreign scientists [7–10]. Importantly, when it comes to risk assessment, Russian methodology involves the use of qualitative indices that characterize various risk types (personal risk, collective risk, economic risk, expected value of damages) [11].

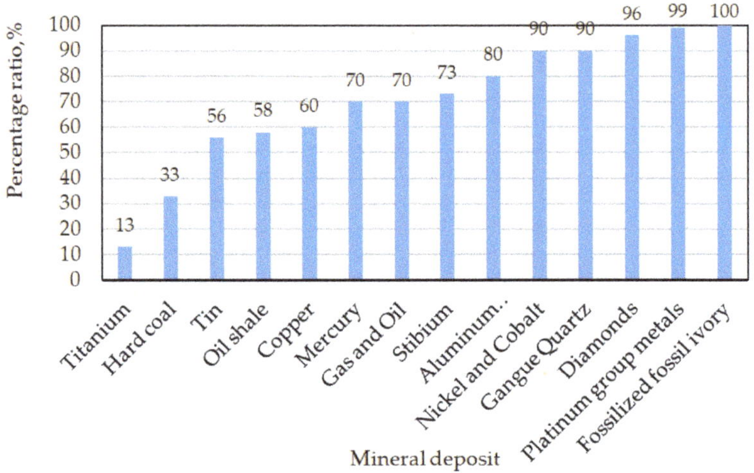

Figure 2. Share of the Arctic zone's mineral deposits in the total of Russia's mineral reserves.

In Finland, for instance, the Elmeri method of occupational safety assessment [12] has become widely used. This method allows for the determination of the likelihood of occurrence of conditions leading to injury or occupational diseases. It is based on observations on labor safety aspects, such as how orderly the workplaces are, how safe the machinery exploitation is, what personal protection equipment the workers use, how ergonomic the work processes are, and what the labor hygiene and sanitation practices are. All of these components are categorized into seven item groups: safety behavior, order and tidiness, machine safety, industrial hygiene, ergonomics, walkways, and first aid and fire safety. Each group is rated from "bad" to "good". "Good" means the group meets the minimal legal requirements and is in line with positive safety practices of the company in question. The Elmeri index is calculated as a ratio of the "good" scores to the total number of item groups, and it ranges from 0 to 100. As such, a score of 60% indicates that a potential injury risk from non-compliance with the occupational safety standards equals 40%.

The Canadian Centre for Occupational Health and Safety (CCOHS) permits use of different methods for occupational safety assessment. The priority in each case is to select the most suitable method. The organization provides an approximate step-by-step guide for risk assessment [13,14]. Based on it, they have developed forms that make it possible to properly document procedures and decision-making processes. However, it must be noted that, despite the existence of a sizeable body of work (by both national and foreign scientists) related to the matter of occupational safety assessment, the issue of injury risk assessment with the subsequent selection of priority avenues for lowering the risk levels remains understudied.

The methods of risk assessment such as FMEA (Failure Mode and Effects Analysis), HAZOP (Hazard and operability studies), and FTA (Fault Tree Analysis) are used in the USA. The HAZOR method, for example, is a risk assessment procedure consisting of the process of detailing and identifying operational disturbances and malfunctions of the equipment as well as a process, a production unit or a system resulting in some undesirable consequences [15]. Similar risk assessment procedures are applied in countries such as Sweden and Norway [16].

A similar standard for determining the risk of occupational injuries was used in the research by Russian scientists [17]. These studies are based on correlation analysis of the assessment of occupational injury risks. However, they do not take the features of the Arctic zone of Russia into full consideration. The methodology proposed differs from the known ones as it considers the impact of the adverse environmental factors on the risk of injuries and also includes an additional indicator—the average risk of injuries for the period under consideration, allowing for a more complete description of the comparative dynamics of occupational injury risk rates.

The versatility of the proposed methodology for assessing the risk of occupational injuries and determining priority directions for its reduction makes it possible to extend this methodology for use in other mining enterprises located in different climates and characterized by various mining and engineering conditions.

However, it should be noted that despite a significant number of studies (by both domestic and foreign scientists) related to the issue of assessing labor safety for various industries, the problem of assessing the risk of injuries in the Arctic zone with the subsequent selection of priority areas for its reduction has not been fully studied.

The correlation analysis of published statistical data characterising occupational safety, both in the regions of the Arctic zone and in the specific company, as an example, the mining areas of the Kirov branch of "Apatit", JSC is considered to be an advantage of the risk-based methodology proposed in the paper.

The fatal injury risk indicator has been used to conduct a comparative analysis of accidents in the industries of the countries located in areas with a climate similar to that of the Arctic zone of Russia (Figure 3) [18].

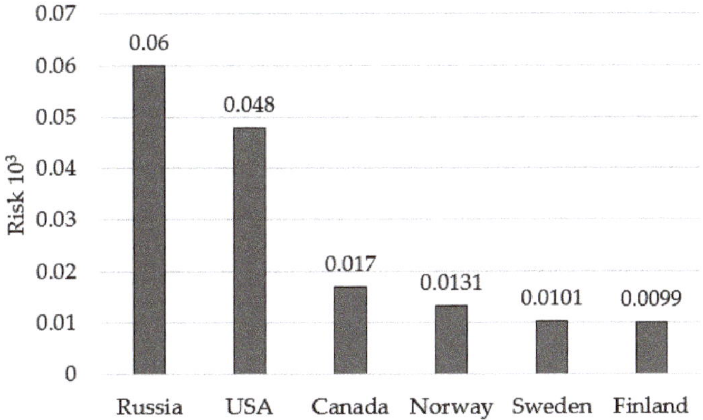

Figure 3. Fatal injury risk for different countries.

The data presented in the figure indicate that the highest fatal injury risk is evident for the Arctic zone of Russia, which stresses the importance of studying this issue, specifically for the Russian Federation.

However, the risk also remains high for other countries. As a result, the risk-based methodology for determining priority directions for reducing the occupational injury rate will be useful for other countries with a similar climate.

The present study owes its relevance to the existing need to decrease the occupational injury rates in the mining industry of Russia's Arctic zone, which helps optimize investments in occupational safety.

The goal of the present study is to provide a risk-based approach for selecting priority directions for occupational injury risk prevention in the mining industry of the Arctic zone.

The practical significance of the study lies in the development of an occupational risk assessment methodology that takes into account the background risk levels in calculations of injury rates caused by labor conditions. The proposed methodology also determines the relative and absolute shifts in occupational injury risk levels. The "basic injury rate matrix", built based on these parameters, allows for the visualisation of the correlation analysis of injury rates in vertically integrated companies and facilitates selection of priority avenues for lowering the injury rates and improving occupational safety.

We use the injury risk level—calculated as the ratio of the injury rate to the total region population number—as a primary indicator of labor safety.

To exemplify the use of the developed methodology, we use it to assess the situation at the mines of the Kirov branch of "Apatit", JSC (located in Murmansk), which belongs to the Arctic zone as per the Decree of the President of Russian Federation No. 296 dated 2 May 2014 [19].

2. Materials and Methods

Methods for selecting priority avenues for lowering the injury rates at the facilities of the Arctic circle are an important basis for occupational safety. The development of mineral resources of the Arctic zone has been a priority for Russia and the global community. The human factor becomes one of the main factors of efficiency in the extreme environmental conditions of the Arctic. When it comes to Russia, the problem is exacerbated by the fact that the vast and resource-filled territory of the Arctic zone is only populated by slightly over 2 million people, which necessitates attracting rotational workers from other regions.

The adverse environmental conditions of the Arctic zone, paired with the difficult labor conditions for the miners (such as polar nights, excessive noise and vibration levels and other), lead to the injury risks being higher compared to regions with more favorable conditions. In order to reduce the injury risks, the occupational health and safety systems at mining facilities need to be constantly improved upon, and the first step of this process is the assessment of labor safety conditions [20–22].

The results of the occupational safety assessment allow for the selection of priority directions for reducing the occupational injuries (Figure 4).

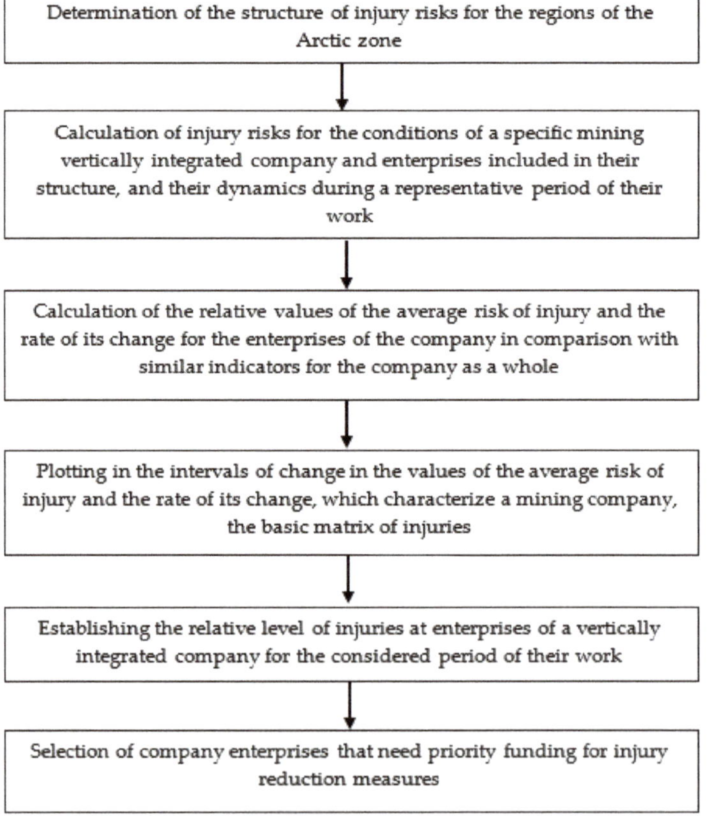

Figure 4. Algorithm for selecting priority avenues for reducing the injury rate.

The first step in the selection is to determine the risk structure (Figure 5). The overall risk—aside from its occupational component, which is determined by labor conditions—includes background risk, which is determined by the unfavorable ecological situation and the adverse environmental effects such as harsh climate conditions, high precipitation, low air temperatures, strong winds, polar nights, polar days, and lack of ultraviolet radiation.

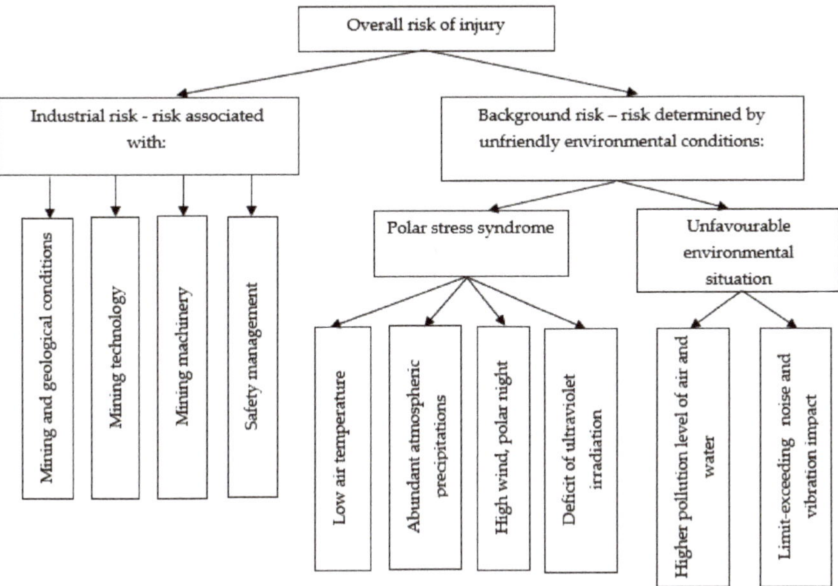

Figure 5. Injury risk diagram for the Arctic region.

To calculate the overall risk of injuries in the Arctic regions of Russia, we used official statistical data [23]. Figure 6 illustrates the calculation results.

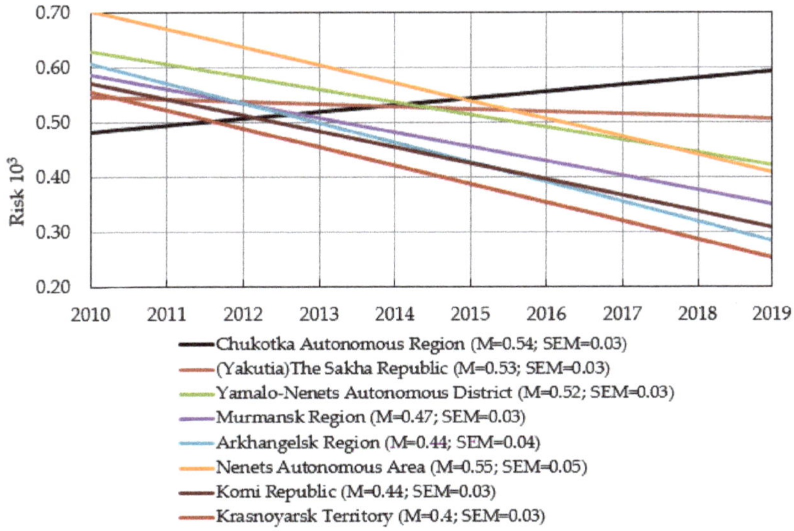

Figure 6. Overall injury risk for the Arctic zone.

In general, the injury rate in the Arctic zone displays a decade-long downward trend. The only outlier in this case is the Chukotka Autonomous District, where we see an increase of 22%.

Figure 6 shows that the linear correlation dependencies of injury rates on time are characterized by the correlation coefficient of over 0.75 and have individual regression coefficients that determine the trends in injury rates over the 10-year period. For each linear correlation dependence of the injury risk for the regions of the Arctic zone, the mean value (M) was calculated, which describes the central trend, and the standard error (SEM) indicating the accuracy of the average value calculation was computed.

Figure 7 shows the main causes of occupational accidents for the Arctic zone of Russia.

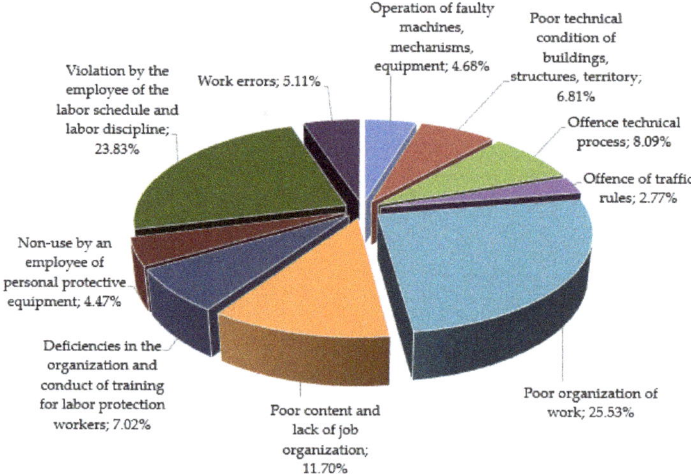

Figure 7. Causes of accidents for the Arctic zone.

As shown in Figure 7, the highest risk of accidents relates to poor management, violations of labor procedures as well as employees' misconduct and unsatisfactory maintenance of workplaces. The listed causes of accidents are organizational due to the insufficient professional training of employees, as well as their low motivation to follow the safety regulations at work.

At the core of the procedure for determining the risk structure lies the assumption that the overall injury risk is the result of the combined effects of two types of risk: the overall risk for the territories of the Arctic zone ($R_{F.N.t.}$), which is calculated based on the average risk for Russia ($R_{RF.av.}$), and the risk determined by the external factors, stemming from the territories themselves ($R_{EXT.C.}$).

In turn, the overall injury risk for the mining enterprises ($R_{M.I.t.}$) is calculated based on the value of $R_{EXT.C.}$ and the risk stemming from the individual company's operations ($R_{I.A.}$). Risk analysis of the occupational injury rates at the mining enterprises pre-supposes that the risk structure for them is identical to the one for the corresponding territory.

The equations linking these risks are based on the addition law of probability and can be written as follows [24,25]:

$$R_{F.N.t.} = R_{RF.av.} + R_{EXT.C.} - R_{RF.av.} \cdot R_{EXT.C.} \tag{1}$$

$$R_{M.I.t.=} = R_{EXT.C.} + R_{IA.} - R_{EXT.C.} \cdot R_{IA.} \tag{2}$$

Knowing the overall injury risks for a given territory ($R_{F.N.t.}$), a given enterprise ($R_{M.I.t.}$), and knowing the average risk for the Russian Federation ($R_{RF.av.}$) makes it trivial to calculate the external risks for each territory ($R_{EXT.C.}$), as well as the injury risks ($R_{IA.}$).

Equation (1) makes it possible to represent the background injury risks for the separate territories as follows:

$$R_{EXT.C.} = (R_{F.N.t.} - R_{RF.av.})/(1 - R_{RF.av.}) \tag{3}$$

With the background risk value being known, the occupational risk for a particular enterprise ($R_{I.A.}$) can be written as follows (Equation (4)):

$$R_{I.A.} = (R_{M.I.t.} - R_{EXT.C.})/(1 - R_{EXT.C.}) \tag{4}$$

The $R_{F.N.t.}$, $R_{RF.av.}$, $R_{M.I.t.}$, and $R_{I.A.}$ risk values were determined based on the statistical data for each region for a given period, and then correlation and regression analysis was performed.

3. Results

3.1. Results of Calculating the Occupational and Background Risks for the Arctic Zone

The results of the $R_{F.N.t.}$, $R_{RF.av.}$ $R_{M.I.t.}$, and $R_{I.A.}$ risk values were determined on the basis of statistical data for each region of the Arctic zone over the past 10 years, which were then subjected to correlation and regression analysis—see Table 1.

Table 1. Background risks of occupational injury for the Arctic zone (10^3).

Place \ Year	2010	2011	2012	2013	2014	2015	2016	2017	2018	2019
Chukotka Autonomous Region	0.077	0.022	0.123	0.021	0.325	0.068	0.207	0.514	0.397	0.579
The Sakha Republic (Yakutia)	0.054	0.315	0.235	0.135	0.092	0.180	0.136	0.239	0.338	0.553
Yamalo-Nenets Autonomous District	0.232	0.206	0.324	0.360	0.139	0.070	0.289	0.210	0.199	0.321
Murmansk Region	0.033	0.049	0.306	0.175	0.188	0.232	0.110	0.065	0.025	0.226
Arkhangelsk Region	0.052	0.167	0.265	0.141	0.160	0.123	0.066	0.012	0.037	0.077
Nenets Autonomous Area	0.218	0.183	0.444	0.635	0.370	0.309	0.119	0.181	0.039	0.372
Arkhangelsk Region without JSC	0.241	0.477	0.524	0.254	0.328	0.242	0.177	0.083	0.088	0.123
Komi Republic	0.025	0.056	0.141	0.109	0.124	0.120	0.236	0.081	0.043	0.012
Krasnoyarsk Territory	0.057	0.134	0.044	0.035	0.000	0.059	0.050	0.029	0.013	0.004

Table 1 provides the background risk values for each Arctic zone territory, calculated using Equation (3).

The data from Table 1 show that the background risk levels have remained practically the same in the past decade. It should be noted that the Krasnoyarsk Territory displays the lowest average background risk values, while the Nenets Autonomous Area has the highest ones.

Figure 8 illustrates the relationship between background risk and injury risk for each region, indicating that background risk has an impact on the magnitude of overall injury risk. The results indicate that the background risks can have significant impact on the overall injury rates. As such, the background risk accounts for 35% of the overall risk in the Nenets Autonomous Area and for over 30% in the Chukotka Autonomous Region and the Sakha Republic (Yakutia).

Thus, by determining the risk structure, we can elucidate the relations between its determining factors, and whether they are controllable or uncontrollable [26].

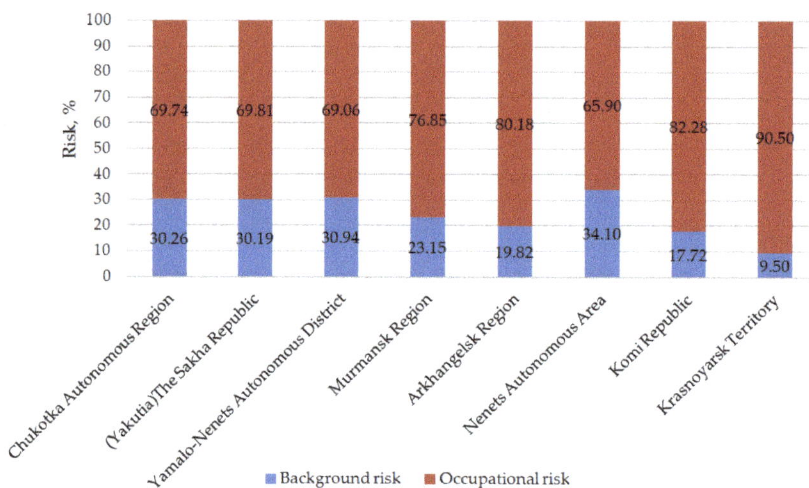

Figure 8. Share of the background risks and occupational risks in the overall injury risks.

The controllable factors in this case are the factors linked to the enterprises' operations, and the uncontrollable ones comprise the external conditions determined by the environment (such as air pollution), placement of workers' habitations, length of the polar day and night, intensity of the UV radiation, location of the facilities, weather conditions, etc. The external conditions affect the mental and physical states of the workers.

3.2. The Specifics of Determining the Priority Directions for Reducing Injury Rates at the Kirov Branch of "Apatit", JSC

We use the vertically integrated Kirov branch of "Apatit" (based in Murmansk) as an example subject for our methodology. The company incorporates the following: the United Kirovsky mine, the Rasvumchorr mine, the East mine, and the Central mine. The original data used in the study were taken from the reports by the State Committee for Supervision of Industrial and Mining Practices (Rosgortekhnadzor), the internal reports by the Kirov branch of "Apatit", and the reports collected by other authors [27]. The calculated occupational risk values are presented as linear functions in Figure 9.

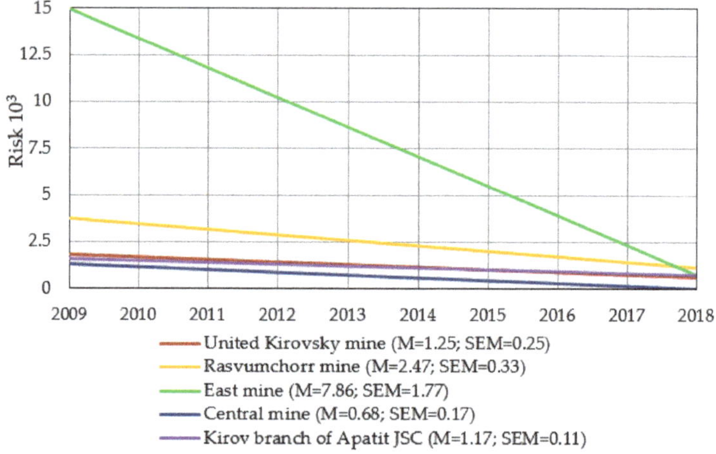

Figure 9. Occupational injury rates at the mines of Apatit.

Figure 9 illustrates that all the mines of the company have seen a decrease in injury rates over the 10-year period studied. The risk trend is described by a linear function with the correlation coefficient exceeding 0.7, with the confidence interval being 0.95.

At the East mine, the occupational injury rate of the first few years is largely different from the other mines' and the company as a whole. However, by 2018, the injury rates have become comparable to other mines'. For each linear correlation dependence of the injury risk for the mines of the Kirov branch of Apatit, the mean value (M) and standard error (SEM) were calculated.

The occupational risk situation at the mines is characterized by the following two indicators: the average risk of injury $(\overline{R})$ and the average rate of change $(\overline{V})$ in the risk of injury. The average rate of change in the injury risks corresponds to the linear correlation's regression coefficient, and the average injury risk represents the mean between the injury risks at the beginning and at the end of the period in question [28,29]. The relative changes in values $(\Delta\overline{R},\ \Delta\overline{V})$ are calculated as the ratio of the average injury risk and the average injury risk of change at a specific mine to the company-wide values.

On the basis of the data on occupational injury rates at the Apatit company from 2008 to 2018, we have calculated the occupational injury risks for the four mines [30,31].

Table 2 shows the relative changes in occupational injury risks and the rate of changes in the injury risks ($\Delta\overline{R}$ and $\Delta\overline{V}$) respectively) for the four company mines.

Table 2. Values of the relative change in injury risks ($\Delta\overline{R}$) and the relative change in their rates $(\Delta\overline{V})$ and their reciprocals $1/\overline{\Delta V}$ for the company mines.

Year \ Minename	$\Delta\overline{R}_{kir}$	$\Delta\overline{R}_{ras}$	$\Delta\overline{R}_{Est}$	$\Delta\overline{R}_{Cent}$	$\Delta\overline{V}_{Kir}$	$\Delta\overline{V}_{ras}$	$\Delta\overline{V}_{est}$	$\Delta\overline{V}_{Cent}$	$\frac{1}{\Delta\overline{V}_{Kir}}$	$\frac{1}{\Delta\overline{V}_{Ras}}$	$\frac{1}{\Delta\overline{V}_{Est}}$	$\frac{1}{\Delta\overline{V}_{Cent}}$
2009	1.998	3.419	12.574	0.812								
2010	1.532	2.706	8.547	1.196								
2011	0	2.776	12.719	0.684								
2012	1.282	2.137	5.983	0.940								
2013	1.538	2.307	9.052	0.812	2	3	16	1	0.5	0.33	0.06	1
2014	1.538	1.709	10.265	0.979								
2015	1.282	2.137	6.229	0.385								
2016	0.932	2.137	0.427	0								
2017	0.085	1.709	1.282	0								
2018	0.502	0.085	0.085	0								

The relative change in the injury risk rate and the relative injury risk value serve as the basis for the "basic injury rate matrix", where the reciprocal of the relative change in the injury risk rate is shown on the X axis, and the relative value of risk, on the Y axis.

For the sake of clarity, the segments of the matrix that correspond to various occupational injury risks are colored differently: green means acceptable labor safety; yellow—satisfactory; red—unsatisfactory; dark red—critical.

The matrix allows for the results of comparative assessment of occupational injury risks at different company enterprises to be visualised and primary measures for their reduction to be determined [32–34].

Figure 10 shows the matrix of relative change in injury risk rates for the mines of the Kirov branch of the "Apatit" company. The results of the analysis and the calculated average values of the changes in injury risk rates over the decade make it possible to assess the occupational safety measures in their entirety.

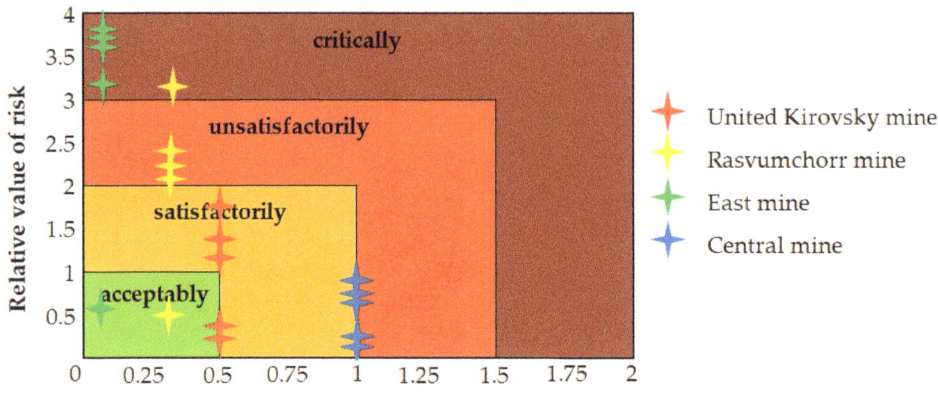

Figure 10. Matrix of the relative change in injury risk rates at the mines of the Kirov branch of "Apatit", JSC.

Figure 11 shows the "Matrix of changes in occupational injury risk levels over a 10-year period at the mines of the Kirov branch of "Apatit", JSC".

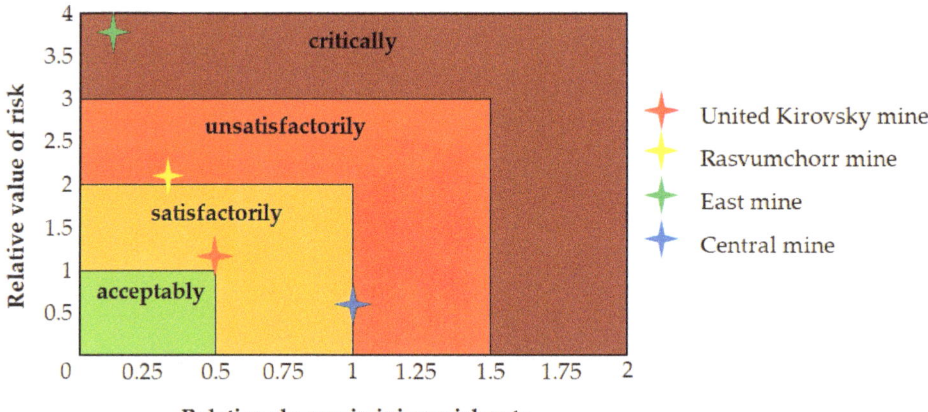

Figure 11. Matrix of changes in occupational injury risk levels over a 10-year period at the mines of the Kirov branch of "Apatit", JSC.

As shown in Figure 11, over the last decade, labor safety conditions pertaining to the occupational injury risks at the Kirov mine were satisfactory; at the Rasvumchorr mine—unsatisfactory; at the East mine—critical; at the Central mine—between satisfactory and unsatisfactory.

The analysis provided allows for the assessment of labor safety conditions pertaining to occupational injury risks both over the last 10 years and during the current period of enterprise operation [35,36].

The assessment makes it possible to determine the priority avenues for reducing injury risk and improving occupational safety.

In conclusion, it should be noted that the article advocates the methodology of identifying the enterprise with higher occupational injury rate in comparison with others within the company as a whole based on the comparative analysis of the occupational injury rates at specific enterprises and companies including these enterprises. The proposed methodology allows companies to identify the priority directions for improving occupational safety, which will be targeted at specific enterprises.

For further research, it is useful to develop software to implement the proposed algorithm for identifying the priority areas for reducing occupational injury rates.

4. Discussion

The proposed methods for assessing the current state of injuries and determining priority areas for reducing occupational injuries, as well as the results obtained, are important for the prevention of injuries in the mining industry. Despite the wide spread of this topic among domestic and foreign scientists, the proposed study most fully describes the situation of industrial injuries in the Arctic zone of Russia, and also demonstrates the methodology for determining priority areas for reducing industrial injuries and improving labor safety using the Kirov branch of Apatit JSC as an example. Based on the studies, the overall injury risk is proposed to be calculated as the combined impact of two types of occupational risks: the occupational risk determined as an average risk for the Arctic zone and the risk caused by the adverse environmental factors. The background risk for the mining areas of the Arctic zone has been calculated for a period of 10 years and used to create the occupational risk-based structure for the mining industry in the Arctic zone of Russia. Despite the fact that the "background risk" has practically not changed during the period under consideration, this risk has an impact on the indicator of the overall injury risk. Thus, the determination of the occupational risk structure allows the relationship between controlled and uncontrolled factors that have an impact on the risk of injuries within the mining industry to be established. This is vital for the subsequent assessment of the impact of industrial factors on the occupational injury rate.

The main advantage of the methodology employed is in providing a comprehensive approach to the analysis of the risk of occupational injuries at mining enterprises located in the Arctic zone of Russia, taking into account the following for the first time

- the risk structure consisting of the occupational risk related to underground labor conditions and a "background risk" determined by the influence of the environment;
- comparative dynamics and average levels of injury risks for mining companies and enterprises within their structure during the period under review;
- possibility of visualisation of the results obtained from the assessment of the occupational injury rates at enterprises within member companies, based on the analysis of risk matrices of occupational safety, which allows for priority directions of the targeted occupational safety and health measures to be determined.

Based on the results obtained, the following important conclusions should be noted:

1. Determining the priority avenues for reducing the occupational injury risks is one of the steps for optimizing financial investments strategies aimed at improving the occupational safety at the mines of the Arctic zone.
2. The overall occupational risk levels are impacted by the background risks, stemming from the influence of the polar stress syndrome and unfavorable environmental conditions.
3. The ranking of vertically integrated companies by occupational injury rates should be done according to two criteria: the average injury risk level and its rate of change at every corporate division compared to the same indicators for the company as a whole.
4. "The matrix of injury risks"—with colored segments signifying acceptable, satisfactory, unsatisfactory, and critical injury rates—can be used to visualize the results of the comparative analysis of occupational injury rates and to determine the priority directions for labor safety improvement.

Author Contributions: S.G.–formulation of the study purpose and goals, elaboration of methodology for determining priority directions for the improvement of occupational safety as well as the development of the assessment and calculation methods for the effects of the background and occupational risks on labor safety. E.P.–assessment of influence of the background and occupational risks on the structure of overall injury rates, development of the matrix showing the relative changes in injury rates for mining areas of the Kirov branch of "Apatit", JSC. All authors have read and agreed to the published version of the manuscript.

Funding: This research received no external funding.

Conflicts of Interest: The author declares no conflict of interest.

References

1. Gridina, E.B.; Pasynkov, A.V.; Andreev, R.E. Comprehensive approach to managing the safety of miners in coal mines. In *Innovation-Based Development of the Mineral Resources Sector: Challenges and Prospects-11th conference of the Russian-German Raw Materials, 7–8 November 2018*; CRC Press: Boca Raton, FL, USA, 2018; pp. 85–94.
2. Arsentiev, E.N. *Efficiency and Human Health in the North*; Kola Research and Development Center "VALEOS": Murmansk, Russia, 1993; pp. 34–35. Available online: https://sharikov.jofo.me/1552261.html (accessed on 28 February 2021).
3. Karnachev, I.P. Analysis of statistical indicators of occupational safety and health used in the study of the dynamics of industrial injuries. *Vestn. MGTU* **2011**, *14*, 751–757.
4. Pavlenko, V.I. The Arctic zone of the Russian Federation in the system of ensuring the national interests of the country. *Arct. Ecol. Econ.* **2013**, *4*, 16–25.
5. Litvinenko, V.S. Digital Economy as a Factor in the Technological Development of the Mineral Sector. *Nat. Resour. Res.* **2019**, *28*, 1–21. [CrossRef]
6. Iakovleva, E.; Belova, M.; Soares, A. Specific features of mapping large discontinuous faults by the method of electromagnetic emission. *Resources* **2020**, *9*, 135. [CrossRef]
7. Dal, N.N. Improving the occupational safety of personnel in coal mines in Vorkuta based on environmental, socio-economic and organizational factors: Avtoref. Dis. cand. tech. Sciences, Saint Petersburg State Mining University, SPb, 2011; 20p. Available online: https://dlib.rsl.ru/viewer/01005003932#?page=1 (accessed on 15 September 2020).
8. Galkin, V.A.; Makarov, A.M.; Kravchuk, I.L. Safety production organization theory and methodology. *Coal* **2016**, *4*, 39–43. [CrossRef]
9. Fainburg, G.Z.; Fedorets, A.G. Current issues of labor protection at the present stage. *Saf. Labor Prot.* **2018**, *3*, 1–22.
10. Rudakov, M.L.; Kolvakh, K.A.; Derkach, I.V. Assessment of environmental and occupational safety in mining industry during underground coal mining. *J. Environ. Manag. Tour.* **2020**, *11*, 579–588. [CrossRef]
11. Samarov, L.Y. Substantiation of the system of indicators for assessing industrial injuries in vertically integrated coal companies: Avtoref. Dis. cand. tech. Sciences, Saint Petersburg Mining University, SPb, 2017; 20p. Available online: https://spmi.ru/sites/default/files/imci_images/sciens/dissertacii/2017/2017-3/avtoreferat_samarov.pdf (accessed on 20 September 2020).
12. Laitinen, H.; Rasa, P.-L.; Lankinen, T.; Lechtel, J.; Leskinen, T. *Manual on Monitoring Working Conditions in the Workplace in Industry. The Elmery System*; Institute of occupational health of Finland: Helsinki, Finland, 2000; pp. 3–5.
13. Canadian Centre for Occupational Health and Safety. Available online: https://www.ccohs.ca/oshanswers/hsprograms/risk_assessment.html (accessed on 3 September 2020).
14. Canadian Centre for Occupational Health and Safety. Available online: https://www.ccohs.ca/oshanswers/hsprograms/sample_risk.html (accessed on 3 September 2020).
15. Head, G.L. *Essentials of Risk Control*; Insurance Institute of America: Malvern, PA, USA, 1989; Volumes 1 and 2.
16. Siddiqui, N.A.; Nandan, A.; Sharma, M.; Srivastava, A. Risk Management Techniques HAZOP & HAZID Study. *Int. J. OHSFE-Allied Sci.* **2014**, *1*, 5–8.
17. Gendler, S.G.; Grishina, A.M.; Samarov, L.Y. *Assessment of the Labour Protection Condition in the Vertically Integrated coil Companies on the Basis of Risk-Based Approach to Analysis of Industrial Injuries*; Saint Petersburg Mining University: Saint Petersburg, Russia, 2018; pp. 507–517.
18. Eurostat Statistics Explained. Available online: https://ec.europa.eu/eurostat/statistics-explained/index.php/Accidents_at_work_statistics (accessed on 20 January 2021).
19. Decree of the President of the Russian Federation No. 296 of 2 May 2014 on the land territories of the Arctic zone of the Russian Federation. Available online: http://kremlin.ru/acts/bank/38377 (accessed on 7 November 2020).
20. Chernova, G.V.; Kudryavtsev, A.A. *Risk Management*; Prospect: Moscow, Russia, 2003; pp. 94–97. Available online: https://www.elibrary.ru/item.asp?id=19744981 (accessed on 28 February 2021).
21. Iphar, M.; Cukurluoz, A.K. Fuzzy Risk Assessment for Mechanized Underground Coal Mines in Turkey. *Int. J. Occup. Saf. Ergon.* **2018**, *3*, 110–158. [CrossRef] [PubMed]
22. Kabanov, E.I. Expert System for Complex Express-Assessment and Forecast of Accidents Risk and Professional Risks on Coal Mines. *Min. Inf. Anal. Bull.* **2019**, *4*, 78–86.

23. Federal State Statistics Service of Russia. Available online: https://rosstat.gov.ru/working_conditions?print=1 (accessed on 15 September 2020).
24. Radosavljevića, S.; Radosavljević, M. Risk assessment in mining industry: Apply management. *Serb. J. Manag.* **2009**, *4*, 91–104.
25. Filimonov, V.A.; Gorina, L.N. Development of an occupational safety management system based on the process approach. *J. Min. Inst.* **2019**, *235*, 113–122. [CrossRef]
26. Artemiev, V.B.; Lisovsky, V.V.; Tcinoshkin, G.M.; Kravchuk, I.L. SUEK Heading to "Zero Injury" Target. *Coal* **2018**, *8*, 71–73. [CrossRef]
27. Gendler, S.G.; Rudakov, M.L.; Falova, E.S. Analysis of the risk structure of injuries and occupational diseases in the mining industry of the Far North of the Russian Federation. *Nauk. Visnyk Natsionalnoho Hirnychoho Universytetu* **2020**, *3*, 81–85. [CrossRef]
28. Botin, J.A.; Guzman, R.R.; Smith, M.L. A methodological model to assist in the optimization and risk management of mining investment decisions. *Dyna* **2011**, *78*, 221–226.
29. Abdrakhimova, I.R.; Zagrieva, G.D.; Mukhametshin, A.K.; Pashkevich, V.C. Development of a risk assessment methodology. *Young Sci. Bull. USPTU* **2016**, *4*, 139–146.
30. Kretschmann, J.; Plien, M.; Nguyen, T.H.N.; Rudakov, M. Effective capacity building by empowerment teaching in the field of occupational safety and health management in mining. *J. Min. Inst.* **2020**, *242*, 248–256. [CrossRef]
31. Bohus Leitner, A. General Model for Railway Systems Risk Assessment with the Use of Railway Accident Scenarios Analysis. *Procedia Eng.* **2017**, *187*, 150–159. [CrossRef]
32. Krause, M. Hazards and occupational risk in hard coal mines—A critical analysis of legal requirements. *Semant. Sch.* **2017**, *268*, 5–6. [CrossRef]
33. Tian, D.H.; Zhao, C.L.; Wang, B.; Zhou, M. Media-in method for assessing security risks in the oil and gas industry based on interval numbers and risk approaches. *Artif. Intell. Eng. Appl.* **2019**, *85*, 269–283. [CrossRef]
34. Wang, W.; Jiang, X.; Xia, S.; Cao, Q. Incident tree model and incident tree analysis method for quantified risk assessment. An in-depth accident study in traffic operation. *Saf. Sci.* **2010**, *48*, 1248–1262. [CrossRef]
35. Shi, X.; Wong, Y.D.; Li, M.Z.F.; Chai, C. Key risk indicators for accident assessment conditioned on pre-crash vehicle trajectory. *Accid. Anal. Prev.* **2018**, *117*, 346–356. [CrossRef] [PubMed]
36. Cherepovitsyn, A.E.; Ilyinova, A.A.; Evseeva, O.O. Stakeholders management of carbon sequestration project in the state-business—Society system. *J. Min. Inst.* **2019**, *240*, 731–742. [CrossRef]

Article

Migration Attractiveness as a Factor in the Development of the Russian Arctic Mineral Resource Potential

Amina Chanysheva [1],*, Pierre Kopp [2], Natalia Romasheva [1] and Anni Nikulina [1]

[1] Economics, Organization and Management Department, Saint-Petersburg Mining University, 199106 Saint-Petersburg, Russia; Smirnova_NV@pers.spmi.ru (N.R.); Nikulina_AYu@pers.spmi.ru (A.N.)

[2] Economic Center of Sorbonne, University Panthéon-Sorbonne (Paris I), 75231 Paris, France; pierreakopp@gmail.com

* Correspondence: aminusha@yandex.ru; Tel.: +7-911-754-07-08

Abstract: The development of mineral resources in the Arctic territories is one of the priorities of the state policy of Russia. This endeavor requires modern technologies, high-quality personnel, and a large number of labor resources. However, the regions of the Arctic are characterized by difficult working and living conditions, which makes them unattractive to the working population. The research objectives were to study the importance of Arctic mineral resources for the Russian economy, the Arctic mineral resource potential, and the migration attractiveness of Arctic regions. The migration processes in these locations were analyzed and modeled using a new econometric tool—complex-valued regression models. The authors assume that the attractiveness of the Arctic regions is determined by the level of their social and economic development and can be assessed using a number of indicators. A comparative analysis of four regions that are entirely in the Arctic zone of the Russian Federation was carried out based on the calculation of integral indicators of the social and economic attractiveness of these territories. Forecasting migration growth using the proposed complex-valued models produced better results than simple trend extrapolation. The authors conclude that complex-valued economic models can be successfully used to forecast migration processes in the Arctic regions of Russia. Understanding and predicting migration processes in the Arctic will make it possible to develop recommendations for attracting labor resources to the region, which will contribute to the successful development of its resource potential. The methodology of this study includes desk studies, a graphical method, arithmetic calculations, correlation analysis, statistical analysis, and the methods of the complex-valued economy.

Keywords: labor resources; mineral resources; Arctic; resource potential; migration attractiveness; econometric models; modeling; migration processes; complex-valued economy

Citation: Chanysheva, A.; Kopp, P.; Romasheva, N.; Nikulina, A. Migration Attractiveness as a Factor in the Development of the Russian Arctic Mineral Resource Potential. *Resources* **2021**, *10*, 65. https://doi.org/10.3390/resources10060065

Academic Editors: Pavel Tcvetkov and Nikolay Didenko

Received: 29 April 2021
Accepted: 15 June 2021
Published: 20 June 2021

Publisher's Note: MDPI stays neutral with regard to jurisdictional claims in published maps and institutional affiliations.

1. Introduction

The development of rich mineral resources in the Arctic has been of global interest for several decades [1–3]. Five countries with coastal access to the Arctic seas (Canada, USA, Russia, Norway, and Denmark) have long been seeking opportunities to explore and extract or expand their exclusive rights to these resources [1,4]. The development of the Arctic is of great interest to both business and science communities around the world [5–7], including Russia, which is primarily due to the depletion of proven natural resources in traditional mining regions.

Oil and gas are the most attractive for exploration and production in the Arctic zone, and the sustainable development of the country is impossible without the constant replenishment of proven oil and gas reserves, since 30–40% of the Russian budget depends on oil and gas revenues [5–10]. In addition to hydrocarbon resources, the Arctic territory contains reserves and resources of platinum metals, nickel and cobalt, copper, oil and gas, and so on [11].

The successful development of technologically complex Arctic fields implies the creation of a technologically modern, competitive industry [12]. With the development of the mining industry, including the oil and gas industry as the largest driver, the Arctic could potentially attract over $100 billion of investments [13].

Nevertheless, the development of the Arctic mineral resource potential is limited by political, climate, environmental, and other problems [14,15], which are characterized by an outflow of human resources. The rapid population growth that occurred during the Soviet period ceased at the end of the 1980s, after which all regions of the Arctic experienced a sharp decline in population, primarily due to migration outflow.

The need to increase industrial production combined with the reduction in the working-age population is one of the most important problems in the development of the Arctic mineral resource potential, namely, a lack of human resources. Therefore, the study of migration attractiveness as a factor of the successful development of the Arctic mineral resource potential is relevant.

The development of Arctic territories and mineral resource potential, especially the functioning of high-tech industries, is impossible without skilled personnel [16,17]. However, residing in the Arctic is not appealing due to difficult living and working conditions, including a severe climate, poor infrastructure (including healthcare), large distances from economic and cultural centers, and other factors [18–23].

The Arctic presents unique challenges for human occupation, with snow cover for up to 10 months a year, up to 24 h of darkness during the winter, a limited variety of resources, and sea-ice-dependent travel and food [24]. People in the Arctic face many interrelated social and economic challenges that add to the many difficulties of daily life in the region [25].

The regions of the Russian Federation located entirely or partially beyond the Arctic Circle are leaders in terms of the number of people involved in territorial movement. These regions are not very attractive as places of permanent residence and work, which is confirmed by an analysis conducted by the authors. We assume that there is a need to create certain social and economic incentives that compensate for the specific working and living conditions in the Arctic to attract labor resources. The combination of such incentives in the region impacts the level of its socio-economic development and reflects its attractiveness.

The research goal is to identify how the attractiveness of an Arctic region affects the migration of the labor resources necessary to ensure its development. This will allow us to effectively regulate and predict the inflow and outflow of the population in Arctic regions, which is essential for the development of its mineral resource potential.

In the research process, extensive practical materials and academic literature were analyzed. Particular attention was paid to the United Nations Development program and reports, the Directorate-General for Regional and Urban Policy of European Commission report, the report of The Reason Public Policy Institute in the USA, and publications by international experts in scientific electronic and printed journals such as *"Energy Research"*, *"Social Science"*, *"Marine Policy"*, *"Polar Science"*, and others.

2. Materials and Methods

In the first stages of the research, desk studies were carried out based on an academic literature review focused on the Arctic resource potential; northern territory development; the definition of the concepts of migration, economic and social attractiveness, and regional competitiveness; and approaches to their assessment. We identified the role of resource potential in the Russian economy and the main challenges and prospects of the Arctic region and analyzed approaches and indicators for assessing regional attractiveness.

The analysis of the literature described above led us to the first research hypothesis:

1. The development of the Arctic mineral resource potential is impossible without qualified personnel, and there is a strong relationship between mining activities and migration processes in the Arctic regions.

To confirm this hypothesis, we investigated key indicators that reflect migration processes in the Russian Arctic regions—the number of arrivals and departures in the region from 2010 to 2019. To this end, we analyzed time series of official state statistics and used a graphical method and arithmetic calculations.

Based on the analysis of migration indicators' dynamics, we formed the second research hypothesis:

2. The migration processes taking place in the Arctic regions depend on the level of attractiveness of the region, which can be determined by a number of social and economic indicators.

In order to choose the number of indicators for assessing the regional economic and social attractiveness, the following process was implemented. In the first stage, the authors reviewed the literature and created a list of the most common quantitative indicators, which were readily found in official Russian statistics. Then, we used an individual expert survey and interviewed representatives from Russian universities and scientific centers (Saint-Petersburg Polytechnic University, Saint-Petersburg Mining University, Saint-Petersburg State University, Kola Scientific Center of the Russian Academy of Sciences), as well as representatives from mining and energy companies (PJSC Gazpromneft, PJSC Gazprom). The experts were chosen on the basis of their ability to quickly communicate with the researchers, specific knowledge in the research field, high level of erudition, and industrial or scientific experience. The authors asked the experts to choose 5 social and 5 economic indicators that, in their opinion, could be used to assess regional attractiveness. The survey was conducted online by sending the form (Appendix A) to the participants through e-mail. The information in the responses was processed, and the most frequently mentioned indicators in the experts' answers were selected. Thus, the experts helped the authors to create a list of 12 social and economic indicators affecting the migration attractiveness of the Arctic regions.

Then, we performed a correlation analysis to identify linear relationships between each of the 12 regional attractiveness and migration indicators. Based on the statistical analysis, we identified social and economic indicators that could potentially be used to model migration processes in the Arctic regions.

The third research hypothesis was Equationted as follows:

3. Modern tools of the complex-valued economy can be successfully used to model migration processes in the Arctic regions.

To test this hypothesis, we applied methods for analyzing the complex-valued economy to create econometric models. Four simple linear regression models were formed with the identified social and economic indicators.

Finally, we aimed to determine which of the four Arctic regions under consideration (Murmansk region, Nenets Autonomous district, Yamalo-Nenets Autonomous district, and Chukotka Autonomous district) is more appealing to working migrants in terms of its social and economic attractiveness. For this purpose, we converted the value of each indicator to a value on a relative scale using the "maximum–minimum" method. The attractiveness of the regions was assessed by calculating integral indicators as weighted averages of specific social and economic indicators.

The structure of the research is presented in Figure 1.

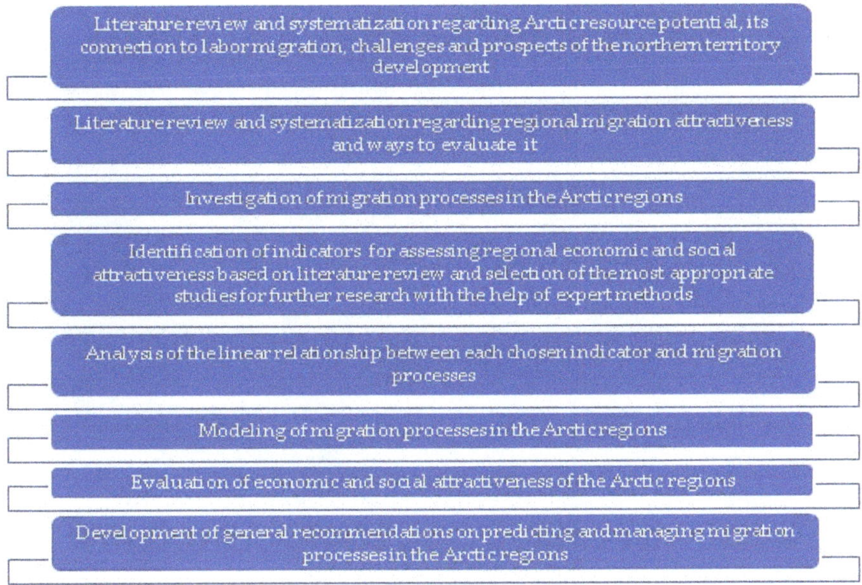

Figure 1. The structure of the research.

3. Results

3.1. Resource Potential of the Russian Arctic and Its Role in the Russian Economy

The Arctic zone of the Russian Federation covers an area of about 9 million km^2; it is home to more than 2.5 million people, which is about 40% of the population of the entire Arctic. The Arctic zone of the Russian Federation includes seven regions: the Republic of Sakha (Yakutia); the Murmansk, Arkhangelsk, and Krasnoyarsk regions; and the Nenets, Yamalo-Nenets, and Chukotka Autonomous districts (Figure 2) [26].

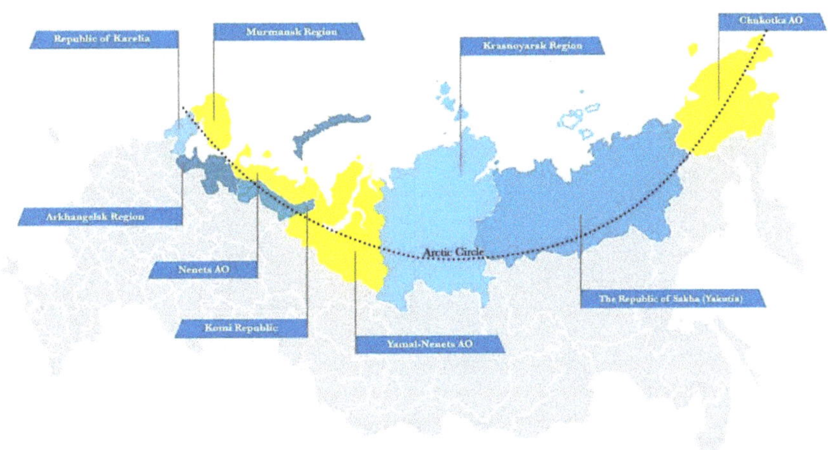

Figure 2. Regions included in the Arctic zone of the Russian Federation [27].

The Arctic is a territory rich in natural resources with strategic geopolitical importance. The Arctic contains more than 97% of Russia's reserves of platinoids, 43% of the tin reserves, and a significant amount of nickel, titanium, and apatite ores and rare earth metals. The proven reserves provide almost 98% of the internal production of platinum, 100% of its titanium, zirconium, rare earth metal, and apatite ores, and more than 97% of its nickel. In the Arctic, about half of the volume of copper and bauxite is extracted, and up to a quarter of the production of diamonds, gold and silver, iron ores, and coal are mined [11] (Figure 3).

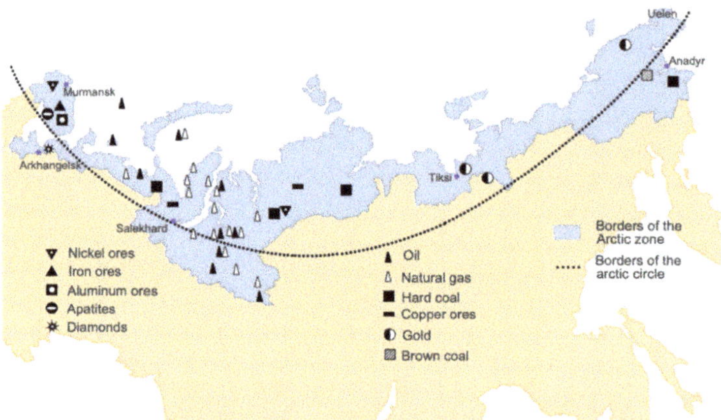

Figure 3. Mineral deposits in the Russian Arctic [28].

According to the Minister of Natural Resources of the Russian Federation (Dmitry Kobylkin), the Russian Arctic zone is estimated to have 7.3 billion tons of oil reserves, 2.7 billion tons of gas condensate, and about 55 trillion cubic meters of natural gas. The Arctic produces 17% of all Russian oil and 83% of its gas. The Yamalo-Nenets Autonomous district has the greatest potential. It accounts for approximately 43.5% of the initial total resources of the Arctic zone. Approximately 41% of the region's oil and gas resources are located on the Arctic shelf [11,29]. The largest fields of the Russian Arctic shelf are presented in Table 1.

Table 1. The largest fields of the Russian Arctic shelf.

Field Name	Location	Year of Discovery	Type of Deposit
Severo-Kamennomysskoye		2000	Gas
Kamennomyskoye	Kara Sea	2000	
Leningradskoye		1990	Gas condensate
Rusanovskoye		1989	
Prirazlomnoye		1989	Oil
Dolginskoye	Pechora sea	1999	
Medynskoye more		1997	
Severo-Gulyayevskoye		1986	Oil and gas condensate
Ludlovskoye		1990	Gas
Murmanskoye	Barencevo sea	1983	
Shtokmanovskoye		1988	Gas condensate
Ledovoye		1982	

Source: Data from oil and gas companies.

Oil and gas resources play a fundamental role in the stability of the Russian economy. The Russian budget is calculated on the basis of three main indicators: the price of oil, the price of gas, and the exchange rate of the US dollar against the ruble (since sales of oil and gas resources are carried out mostly in US dollars). The share of oil and gas revenues in the budget of the Russian Federation remained 30–50% until 2019 (Figure 4).

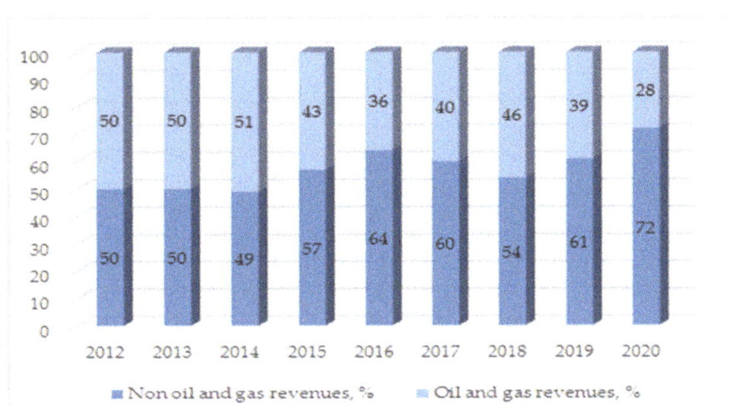

Figure 4. Share of oil and gas revenues in the budget of the Russian Federation, 2012–2020 [6].

The year 2020 brought significant challenges [30]: the rapid spread of the coronavirus (COVID-19) pandemic forced governments all around the world to establish lockdown measures, which significantly affected economic activity, employment, and people's livelihoods [31]. The economic crisis caused by the coronavirus pandemic has had a negative impact on many industries; however, some of them have proved to be quite resilient [32,33]. The world fuel and energy market experienced the greatest impact of COVID-19: the decline in economic activity has dramatically affected global energy demand, which, according to the Global Energy Perspective 2021 by McKinsey and Company, fell by 7% [30]. The fall in demand for fuel and energy resources in key sales markets led to a record drop in prices. At the end of the first quarter of 2020, the gas price had reached a 30-year low, and the oil price had reached a 22-year low. However, due to the partial lifting of restrictions in some countries, the demand for energy resources has partially returned [30]. According to the International Energy Agency's forecast published in the Global Energy Review 2021, global energy demand will grow by 4.6%, with the bulk of the demand coming from developing countries [34].

Furthermore, according to the OPEC forecast, world oil demand in 2021 will grow by 5.9 million barrels per day. It is predicted to grow in China, India, and some Asian countries. According to the forecast of the World Energy Agency, oil demand will increase by 5.5 million barrels per day, and it will recover mainly in the second half of the year [35].

With the global demand for energy resources, the share of gas will increase in the next few decades and peak in the late 2030s. The growth in oil demand will slow down, but oil will remain the most important energy resource for many years to come [30]. In this case, given the depletion of traditional fields with easily recoverable hydrocarbon reserves and the importance of mineral resources for the economy, resource provision is crucial for Russia. Despite the current macroeconomic conditions, the Arctic shelf is a promising area for providing Russia with raw materials, especially hydrocarbons [36,37].

3.2. Resource Potential Development and Labor Migration

As noted above, a significant share of Russian oil and gas resources is concentrated in the Arctic zone, specifically on the Arctic shelf. Projects for the development of offshore hydrocarbon fields in the Russian Arctic are technologically complex. The availability of the appropriate technologies is one of the key factors determining the commercial effectiveness of such projects. Moreover, complex technologies imply the need for relevant competencies of the workforce at all levels, from project managers to lower-level workers. Thus, it is impossible to implement complex mining technologies without qualified personnel.

It follows that the Arctic regions have great job opportunities for potential high-skilled labor immigrants. However, jobs in the field of raw material extraction in the Arctic are characterized by difficult working conditions, as mentioned above, including low temperatures, long and dark winters that provoke a depressive emotional state, a large number of physically stressful tasks, weak infrastructure, lack of social life, and so on. These factors make the work related to the development of the resource potential of the Arctic and its territories unattractive to the working population. At the moment, there is an acute personnel problem in the field of mineral resource extraction in the Arctic [38] (p. 1). As the analysis below shows, the Russian Arctic was characterized by an outflow of human resources at the time of this research.

There is unquestionably a close relationship between the migration of labor and the development of raw materials in the Arctic, which is clearly identified in a number of scientific works [38] (p. 3), [39,40]. Heleniak [39] (p. 2) demonstrated this for two Russian Arctic regions, Khanty-Mansiy and Yamalo-Nenets districts, which are key regions for the extraction of raw materials that are vital for the country: oil and gas. These regions were the only Russian Arctic territories that had constant migration inflows during the post-Soviet period due to high incomes, in contrast to the considerable outflows from other Arctic regions that do not possess rich natural resources. The same situation can be observed in the Nenets Autonomous district, which is rich in hydrocarbon resources. This is the only region in which migration inflows exceeded outflows throughout 2010–2019, as shown below (Figure 5).

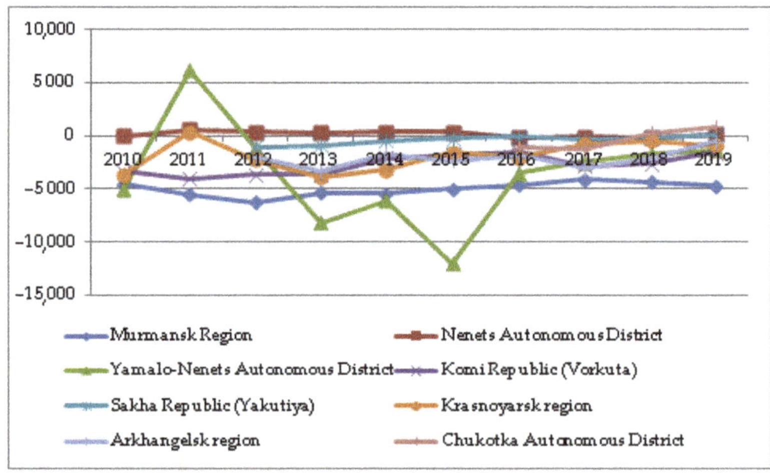

Figure 5. Dynamics of migration processes in the Arctic regions in 2010–2019, number of people.

In this regard, the task of attracting qualified personnel to work at Arctic mining enterprises, which are the main employers in the Arctic regions, is currently relevant. This task should be planned and solved primarily within industrial enterprises that perform the extraction of Arctic mineral resources. Such organizations should form material and non-material incentives for personnel that would increase the inflow of human resources from other regions to the Arctic. All of this should take place with active support from the state for the development of the Arctic regions and their infrastructure, thus increasing their attractiveness to migrant workers.

The next task in our study was to analyze the factors that influence the attractiveness of the Arctic region from the point of view of a potential labor immigrant.

3.3. Migration Processes in the Arctic Regions

The aim of this stage of the study was to identify the main trends associated with the inflow and outflow of the population in the Arctic regions. To this end, we chose an indicator of migration growth calculated as the difference between the number of arrivals and departures in the region. The dynamics of migration growth in the Arctic regions between 2010 and 2019 are shown in Figure 5. Data from state statistics bodies were collected for all the municipal areas of the Russian Arctic regions and used as the initial information for the analysis [41].

Figure 5 shows that the Yamalo-Nenets Autonomous district had the highest volatility of migration growth (calculated as the difference between migration increase and decline): its values ranged from 6249 people in 2011 to −11,972 people in 2015. These peak values are due to a significant inflow of labor resources to the city of Novy Urengoy in 2011 (5448 people), as well as the cities of Gubkinsky, Salekhard, and Nadym (1886, 1088, and 1071 people, respectively), and a significant outflow of the population across all municipal districts of the region in 2015. The maximum outflow of 5361 people was observed in the city of Novy Urengoy. Mass inflows and outflows of labor resources in such cities and municipalities are due to the launch of new oil and gas projects and the development of new fields, the closure of old ones, the movement of employees working in shifts, the outflow of young residents to promising regions of the country, and the departure of senior citizens to favorable climatic zones.

The lowest volatility of migration growth was in the Nenets Autonomous district, the Republic of Sakha, and the Murmansk region, which showed a consistent significant outflow of the population throughout the entire period, with an average of 5072 people per year.

Negative dynamics of migration growth were typical for the following Arctic regions (or the parts that are located in the Russian Arctic zone):

- Nenets Autonomous district;
- Arkhangelsk region (except for the last two years).

Positive dynamics of migration growth were observed in:

- Murmansk region;
- Komi Republic;
- Sakha Republic;
- Krasnoyarsk region;
- Chukotka Autonomous district.

"Above-zero" values of migration growth from 2010 to 2015 are observed only in the territory of the Nenets Autonomous district. This indicates that the inflows to the region were higher than the outflows over that period. However, since 2016, the situation has changed.

The results of the analysis show that migration processes in the Yamalo-Nenets Autonomous district have the highest volatility among the Arctic regions, which is primarily due to the ongoing mining activities in this region. At the time of the study, all Arctic territories were characterized by a stable outflow of the population. To identify the reasons for such trends, we analyzed key factors and socio-economic indicators that affect the influx of the population, particularly labor resources, to the Arctic regions.

3.4. Analysis of Methods and Indicators Used for Assessing Regional Attractiveness

In the literature, the attractiveness of a region to the working-age population is often described by the term "migration attractiveness". There are various methods and approaches to assessing the level of migration attractiveness of a region [42–49]. Niedomysl [42] described two alternative approaches to analyzing a location's attractiveness—assumption-based and statement-based—and their pros and cons. Portnov [43] identified employment and housing factors of interregional migration and proposed an approach to determine sustainable regional development policies aimed at a more balanced distribution of a country's population. Karachurina [44] focused her research on the attractiveness of centers and secondary cities in 74 Russian regions to internal migrants. Lundholm and others [45] examined interregional migration within the five Nordic countries—Sweden, Norway, Denmark, Finland, and Iceland—with a focus on the main motivating factors for moving.

Beglova and others [46] calculated the coefficient of migration attractiveness as a root of the ratio of the arrival coefficient to the departure coefficient, which reflects the excess of migration inflows over outflows. In addition, researchers studied the correlation between the coefficient of migration attractiveness and the economic, social, demographic, and environmental factors that determine it, resulting in a total of 16 quantitative indicators [46]. Similarly, in [47], the correlation between the migration balance and the migration attractiveness of cities was assessed using 18 socio-economic indicators.

Petrov et al. [48] investigated the relationship between migration growth rates and the quality of life of the population in the regions of Russia. The latter was used to assess the attractiveness of a region and was calculated on the basis of 12 indicators divided into 4 groups: physiological needs, safety needs, communication needs, and achievement needs. The researchers suggested assessing regional attractiveness in terms of the system of needs of migrants. Moreover, the study showed that migration growth in the Yamalo-Nenets Autonomous district, the Republic of Komi, Magadan, Murmansk, and some other regions was much lower than the national average. This is due to the fact that these regions are located in the Arctic zone of the Russian Federation and are characterized by special living conditions. Druzhinina et al. [49] focused their study on the influence of factors on the processes of external and internal migration in the northern territories of the Russian Federation. According to the authors, population density is the most attractive factor for internal migration, while external migration largely depends on the provision of communication services and the social security of the population. Additionally, providing the population with new housing is the most important area of socio-economic development for all migrants in the northern territories.

Much attention in the literature is paid to the issues of population migration and the attractiveness of territories. For decades, researchers have been studying the factors that contribute to the attractiveness of specific regions and territories to migrants. Thus, factors such as urban attractiveness, ecology, and proximity to amenities now play a more important role in the migration of young people than in the past [50]. The Organisation for Economic Co-operation and Development (OECD) names seven determinants that contribute to a country's attractiveness to high-skilled migration: quality of opportunities; income and tax; future prospects; family environment; skills environment; inclusiveness; and quality of life [51]. Each determinant includes several specific indicators, which can be both qualitative and quantitative. The term "talent attractiveness" is used to identify highly educated and talented people migrating in search of better living conditions. Its level is assessed in [52]. Ewers and Dicce [53] examined the relationship between highly skilled international migration and urban–regional development.

It is abundantly clear that the prospect of a better job, improved living conditions, and personal development are the principal motives that drive people to emigrate. "Power of attraction" is based on notable differences in social conditions, the situation in the labor market or, for example, the quality of life of society. Matuszczyk [54] analyzed 14 different indicators to measure migration attractiveness, such as the unemployment rate, GDP per capita, median of annual income (PPS), cost of living index, rent index, law enforcement index, severe material deprivation rate, and others. However, many researchers consider the unemployment situation to be one of the main factors influencing the resettlement of people [48].

The sustainable attractiveness of regions, including the Arctic zone, is influenced by economic and social circumstances within the territory. The latter can be estimated with the help of specific indicators. For this purpose, a number of methods exist and are widely discussed in the academic literature.

In 1993, Lipshitz [55] summarized the key factors and methods for assessing the development of a territory. In further research, Cziraky and others [56] proposed a multivariate statistical framework for assessing regional development, and Shahraki [57] investigated important factors in the process of regional development. Erlinda and others [58] provided an evaluation framework for the assessment of sustainable regional development using multiple criteria related to development scenarios established by stakeholders.

According to some research, assessing regional development is often performed using aggregate indexes. They involve various economic, social, and other indicators such as GRP, the number of beds in hospitals, investment attractiveness, income values, budget characteristics, and other factors [59,60].

Professor Svetunkov [61] highlights two main groups of methods used to evaluate the development of a region: integral indicators and econometric modeling.

Many researchers have assessed the attractiveness of a region, territory, or city using concepts such as its competitiveness [62–65]. The competitiveness of the region mainly involves the allocation of capital in the territory, the development of productive forces, the internal stability and strength of the role and influence of the region in external systems, and the ability to compete with homogeneous systems in their economic development and to offer a stable environment for business and residence. Some researchers have also included the living standards of the population.

Similar to the level of socio-economic development, researchers have proposed different approaches to assessing the level of competitiveness of a region, territory, or city based on various indicators or mathematical and statistical models [66–68].

International institutions and administrative bodies also participate in the calculation of territorial competitiveness indices. For example, the following techniques can be used:

The Human Development Index (HDI) is assessed by the United Nations Development Program. The HDI is a cumulative measure of key aspects of human development: a long and healthy life, knowledge, and suitable living standards [69].

The Regional Competitiveness Index (RCI) is assessed by the Directorate-General for Regional and Urban Policy of the European Commission. It aims to provide a consistent, comparable, and effective measurement of economic and social issues in the EU regions and is based on 11 factors, such as infrastructure, health, opportunities for education and business, technologies and innovation, and employment [66].

The level of competitiveness of the largest cities in the state of California is assessed by The Reason Public Policy Institute in the USA. It identifies the best cities to live in in the state. The rating is based on indicators that assess the location of the city, the temperature conditions, and services (medicine, transport, recreation, etc.) [70].

Almost all approaches studied by the authors assess the competitiveness of the region using a set of particular social and economic indicators combined into several key factors. The factors often include the standard of living, investment attractiveness, innovative activity, the level of development, the efficiency of resource use, and the financial, environmental, and organizational potential of the region.

The most commonly used indicators are the average per capita monetary income of the population; the profitability of gross output (works, services) of the region; the share of unprofitable organizations; the share of investments in fixed assets in the GRP; expenses of the consolidated budget per capita; the share of innovatively active organizations in their total number; exports; the share of small enterprises in the total number of registered enterprises; the share of graduated specialists, postgraduates, and doctoral students in the economically active population, etc.

The estimates obtained reflect various aspects of competitiveness at the regional level as well as the integral competitiveness of the region, allowing us to determine its strengths and weaknesses and serving as a basis for developing sustainable regional development strategies.

3.5. Correlation Analysis

To identify indicators that have a strong linear relationship with migration processes and the attractiveness of the Arctic regions, a number of economic and social indicators were selected with the help of experts.

Social indicators:

1. The population size, thousands of people;
2. The number of registered crimes per capita, pcs.;
3. Commissioning of residential buildings, thousands m^2 of total area of residential premises per person;
4. The volume of paid services to the population (in actual prices for the period), million rubles;
5. The birth rate, natural population growth (decline) per 1000 people;
6. The gross enrollment rate in preschool education as a percentage of the number of children aged 1–6 years.

Economic indicators:

7. Income per capita, rubles;
8. Average consumer expenditure per capita (per month), rubles;
9. Number of enterprises and organizations per 1000 people, pcs.;
10. Unemployment rate, percentage;
11. Share of the population with incomes below the subsistence minimum in the total population, percentage;
12. Cost of a fixed set of consumer goods and services, rubles.

We assessed the influence of each factor on migration processes by evaluating linear correlation coefficients between each of the above-mentioned indicators and the number of arrivals and departures in the region from 2010 to 2018. Table 2 presents the values of linear correlation coefficients calculated for the Murmansk region.

Table 2. Linear correlation coefficients for socio-economic indicators of the Murmansk region.

No.	Economic Indicators	Arrivals	Departures
1	Share of population with incomes below the subsistence minimum in the total population, percentage	−0.0541	0.0041
2	Unemployment rate, percentage	−0.3511	−0.3551
3	Income per capita, rubles	0.7635	0.6615
4	Average consumer expenditure per capita (per month), rubles	0.7960	0.7003
5	Number of enterprises and organizations per 1000 people, pcs.	−0.1367	−0.1183
6	Cost of a fixed set of consumer goods and services, rubles	0.8501	0.7368
No.	**Social Indicators**	**Arrivals**	**Departures**
7	Volume of paid services to the population (in actual prices for the period), million rubles	0.7594	0.6593
8	Gross enrollment rate in preschool education as a percentage of the number of children aged 1–6 years	0.9035	0.8057
9	Population size, thousands of people	−0.8079	−0.7027
10	Commissioning of residential buildings, thousands m^2 of the total area of residential premises per person	0.5216	0.4272
11	Number of registered crimes, pcs.	0.0955	0.0574
12	Birth rate, natural population growth (decline) per 1000 people	−0.6249	−0.5613

The correlation analysis led us to Equationte the following conclusions:

1. The share of the population with incomes below the subsistence minimum in the total population of the region has almost no linear relationship with migration indicators. This is due to the high volatility of the indicator. Similarly, the unemployment rate indicator has a low linear impact on migration, since it is characterized by the absence of any clearly defined linear trend that is inherent in indicators of migration inflows and outflows. Therefore, these indicators are not recommended for use in linear econometric migration models.
2. Income per capita has a stronger linear correlation with the number of arrivals than with the number of departures. This can be interpreted from the point of view of the

region's economic attractiveness: relocation to the Far North and Arctic regions is often due to material incentives, which, in this case, is the growth of income. A similar trend is observed for indicators that reflect consumer expenditure on goods and services.

3. The number of enterprises and organizations per 1000 people has a very low correlation with migration indicators and thus has little influence on the attractiveness of the region. This is also the case for the indicator "Number of registered crimes".

4. Migration inflows and outflows have a strong linear relationship with the indicators "Gross enrollment rate in preschool education" and "Population size" and a slightly weaker association with the indicator "Commissioning of residential buildings". At present, the correlation coefficient values of these indicators suggest that they are acceptable for use in simple linear econometric models.

5. "Birth Rate per 1000 people" has an insignificant linear relationship with migration. In other words, a natural increase or decrease in the population within the region is unlikely to be an incentive for potential migrants to relocate to the Arctic region and will not significantly contribute to a population influx from other regions.

6. In addition, analysis of the correlation of indicators shows that the population has an inverse linear relationship with income per capita with a correlation coefficient of -0.984.

7. In general, migration processes have a stronger linear relationship with social indicators than with economic ones, which may be a confirmation of the fact that migration of the population to/from the Arctic regions is not caused only by material incentives but, on the contrary, is mainly determined by social factors.

3.6. Econometric Modeling

In this research, we applied a contemporary economic and mathematical instrument that uses complex variables. It has been successfully used by some researchers to solve different economic problems [61]. The key principle of this tool is to combine two economic indicators into one model variable. This approach allows different aspects of a phenomenon to be addressed and its influence on parameters to be analyzed, which, in turn, could be a complex variable. Thus, we applied basic complex-valued model (1) with regard to migration processes.

$$y_{rt} + iy_{it} = (a_0 + ia_1) + (b_0 + ib_1)(x_{rt} + ix_{it}), \tag{1}$$

where y_r and y_i are components of an endogenous complex variable; x_r and x_i are components of an exogenous complex variable; and $a_0 + ia_1$ and $b_0 + ib_1$ are model coefficients.

Correlation between two complex indicators is evaluated as follows [61]:

$$r_{XY} = \frac{\sum (y_{rt} + iy_{it})(x_{rt} + ix_{it})}{\sqrt{\sum (x_{rt} + ix_{it})^2 \sum (y_{rt} + iy_{it})^2}} \tag{2}$$

If the real part of r_{XY} is close to 1, then the endogenous variable is linearly dependent on the exogenous one, while the imaginary part reflects the plot scatter of the regression model.

Econometric modeling of migration processes was carried out using data published by state statistics bodies using the Murmansk region as an example. We used indicators representing migration processes—the number of arrivals and departures from the region in the period—as the dependent complex variable in all econometric models. As factors that determine the variability of the considered indicators of migration, we chose seven socio-economic indicators that have high linear correlations with the dependent variables (Table 2). Then, we formed complex variables by grouping a pair of particular indicators that describe one process or phenomenon from two different aspects. The time series included seven observations, namely, the real values of the indicators from 2011 to 2017. Using the developed models, we predicted the values of migration inflows and outflows in the Murmansk region for the year 2018. The actual values in 2018 were used as benchmarks to assess the adequacy of the obtained forecasts.

Taking the above into account, we studied simple linear complex-valued models of type (1), as follows:

- "Income per capita" (Xr, rubles) and "Volume of paid services to the population" (Xi, million rubles) are variables in model 1. The complex coefficient of linear correlation between the exogenous and endogenous variables of the model is quite high: 0.731–0.048i.
- "Average consumer expenditure per capita" (Xr, rubles) and "Cost of a fixed set of consumer goods and services" (Xi, rubles) are variables in model 2. The complex coefficient of linear correlation between the exogenous and endogenous variables of the model is high: 0.814–0.005i.
- "Gross enrollment rate in preschool education, as a percentage of the number of children aged 1–6 years" (Xr) and "Commissioning of residential buildings" in thousands m^2 of the total area of residential premises per person (Xi) are the variables in model 3. The complex coefficient of linear correlation between the exogenous and endogenous variables of the model is high: 0.875–0.046i.
- "Gross enrollment rate in preschool education, as a percentage of the number of children aged 1–6 years" (Xr) and "Birth rate, natural population growth (decline) per 1000 people" (Xi) are variables in model 4. The complex coefficient of linear correlation between the exogenous and endogenous variables of the model is high: 0.895–0.066i.

Model 1 includes social and economic indicators as factors. Income per capita is the real part of the complex factor, and its imaginary part is the indicator of the volume of paid services provided to residents in the Arctic regions. This indicator reflects the degree to which high-level personal needs are satisfied in contrast to the basic ones.

The profit indicator is normally the main appeal to potential labor migrants to the region [71]. Additionally, high-level needs require appropriate social infrastructure. The amount of money spent indirectly indicates the availability of various social opportunities [61]. Therefore, we deemed the volume of paid services rendered to be a crucial factor in judging regional attractiveness and incorporated it as an imaginary component.

Model 2 is a simple linear regression model in which migration processes are the result of variation in two economic indicators that characterize the average expenses of residents of the Arctic regions. Consumer expenditure per capita and the cost of a fixed set of consumer goods and services both describe the process of spending personal funds but from different angles. In fact, they represent the average amount of money that the consumer possesses after all compulsory monthly expenses. This value is of interest to potential migrants to the Arctic regions, since it characterizes their material security if they move there.

Model 3 is a simple linear regression model in which the values of migration indicators depend on the social components of the attractiveness of the region; these components represent the provision of families with preschool educational institutions and housing. Model 4 is a modification of Model 3 and incorporates the social indicator of natural population growth instead of the indicator "Commissioning of residential buildings".

Figures 6–9 show graphs of the actual and calculated values of arrivals and departures in the Murmansk region. The calculated values were obtained using the above econometric complex-valued models.

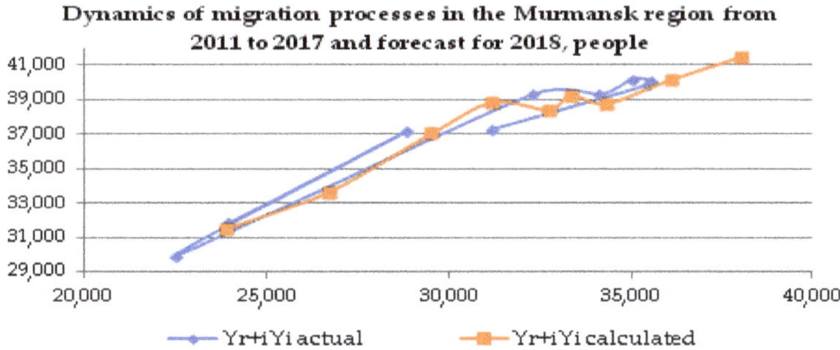

Figure 6. Actual and calculated values of migration indicators in the Murmansk region: Model 1.

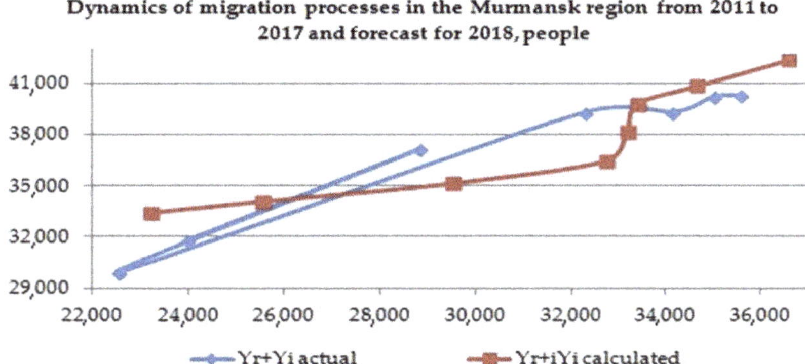

Figure 7. Actual and calculated values of migration indicators in the Murmansk region: Model 2.

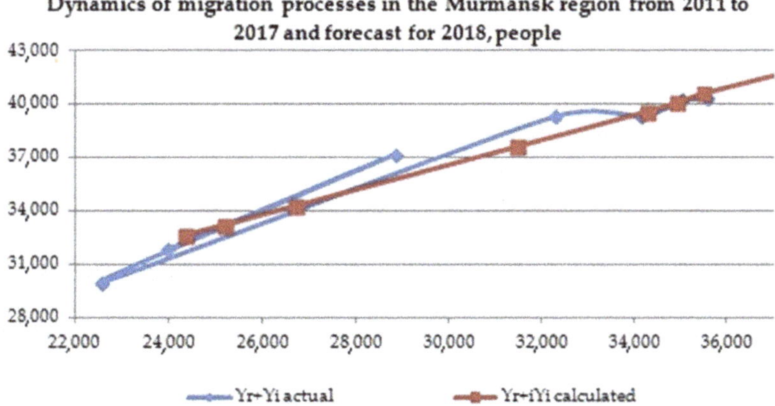

Figure 8. Actual and calculated values of migration indicators in the Murmansk region: Model 3.

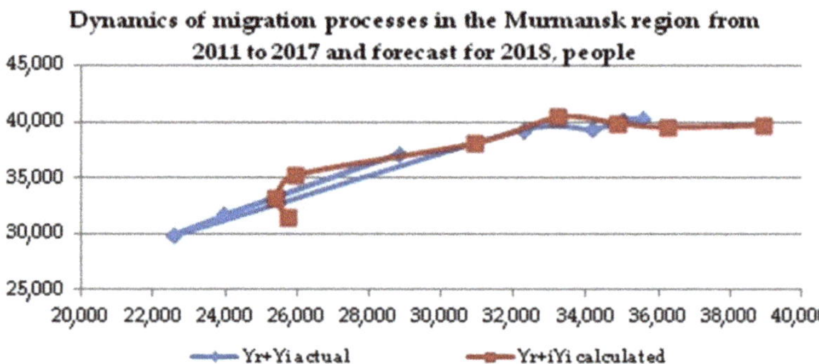

Figure 9. Actual and calculated values of migration indicators in the Murmansk region: Model 4.

The figures show that the predictions of all four econometric models closely match the initial data and are suitable for forecasting. In addition to the four models presented above, we performed a linear extrapolation of trends of migration inflows and outflows based on their dynamics from 2011 to 2017 in order to calculate forecast values for 2018. Table 3 shows the results of an analysis that compares different forecasts of arrivals and departures in the Murmansk region.

Table 3. Forecast and actual values of migration indicators in the Murmansk region for 2018.

Forecast Econometric Models		Forecast Values of Migration Indicators for 2018, People		Complex R-Squared
		Arrivals	Departures	
Extrapolation of trend	$y_r = 2098.9t + 21,937$ $y_i = 1461.3t + 31,005$	40,827.10	44,156.70	-
Model 1	$y_r + iy_i = (2.37 + 15,928.97i) +$ $+(0.92 + 0.63i)(x_r + ix_i)$	38,145.25	42,233.90	$0.535 + 0.002i$
Model 2	$y_r + iy_i = (6177.87 + 20,181.98i) +$ $+(1.05 + 0.1i)(x_r + ix_i)$	36,597.09	42,371.07	$0.663 + 2.63 \times 10^{-5}i$
Model 3	$y_r + iy_i = (-148,520.7 - 89,153.8i) +$ $+(2158.6 + 1519.2i)(x_r + ix_i)$	37,238.93	41,778.19	$0.765 + 0.002i$
Model 4	$y_r + iy_i = (-116,674.8 - 111,635.4i) +$ $+(1776.2 + 1788.9i)(x_r + ix_i)$	38,937.17	39,725.04	$0.801 + 0.004i$
Actual values of migration indicators for 2018, people		35,198	40,477	-

As indicated in the table, the forecasts of the arrivals and departures derived from the four studied complex-valued models are closer to the actual data than the results of a linear extrapolation of their trends for 2018.

Furthermore, the best forecasting results are obtained with model 2 for the indicator "arrivals" and model 4 for the indicator "departures". The fourth model has the highest complex R-squared. Thus, we can conclude that linear complex-valued models are suitable for predicting migration processes in the Arctic regions of Russia, despite their simplicity and inability to consider many factors.

3.7. Integral Estimation of the Attractiveness of Arctic Regions

As mentioned above, a large number of domestic studies have been devoted to assessing the attractiveness of territories. However, the literature contains little research assessing the attractiveness of the Arctic regions to labor migrants. However, it should be noted that living and working conditions in the regions of the Russian Arctic differ significantly. In this regard, one of the objectives of the study was to conduct a comparative analysis of four Arctic regions that are entirely included in the Arctic zone of the Russian Federation (Murmansk region and Nenets, Yamalo-Nenets, and Chukotka Autonomous districts) according to a number of key indicators. For this purpose, we calculated two integral indicators characterizing the economic and social attractiveness of the territories of the Russian Arctic using state statistics data.

To calculate the integral indicators of economic and social attractiveness of the Arctic regions, the authors used the values of six economic and six social indicators presented above for the period 2010–2018. To this end, four source data tables were generated for each region under consideration, one of which is shown below as an example (Table 4). The source data series were converted into values on a relative scale according to the "maximum–minimum" method. Its minimum and maximum values correspond to 0 and 1, respectively. The following Equations were used:

$$\frac{X_i - X_{min}}{X_{max} - X_{min}}, \tag{3}$$

$$1 - \frac{X_i - X_{min}}{X_{max} - X_{min}} \tag{4}$$

where X_i is the value of the indicator for region i, X_{max} is the maximum value of the indicator among the regions in the year under review, and X_{min} is the minimum value of the indicator among the regions in the year under review.

Table 4. Initial data for the Murmansk region for calculating integral indicators of attractiveness.

Indicators Values	2010	2011	2012	2013	2014	2015	2016	2017	2018
Social Indicators									
The population size, thousands of people	794.10	788.00	780.40	771.10	766.30	762.20	757.60	753.60	748.10
The number of registered crimes per capita, pcs.	0.02	0.02	0.02	0.02	0.02	0.02	0.02	0.02	0.02
Commissioning of residential buildings, thousands m² of total area of residential premises per person	0.01	0.04	0.03	0.03	0.03	0.03	0.04	0.08	0.06
The volume of paid services to the population (in actual prices for the period), million rubles	47.59	51.06	55.29	65.73	71.50	72.11	74.94	74.93	80.00
The birth rate, natural population growth (decline) per 1000 people	−0.20	−0.07	0.50	0.90	0.30	0.30	−0.30	−0.80	−1.50
The gross enrollment rate in preschool education as a percentage of the number of children aged 1–6 years	80.40	80.10	80.50	81.20	83.40	84.70	85.00	85.30	86.10
Economic Indicators									
Income per capita, rubles	24,047	25,303	28,932	32,912	34,149	36,848	36,116	37,108	41,564
Average consumer expenditure per capita (per month), rubles	15,640	17,262	19,526	23,404	26,547	27,113	27,469	28,744	30,699
Number of enterprises and organizations per 1000 people, pcs.	27.72	29.13	27.44	27.54	28.15	28.61	29.80	25.16	22.97
Unemployment rate, percent	8.60	8.60	7.70	7.20	6.70	7.80	7.70	7.00	6.80
Share of population with incomes below the subsistence minimum in the total population, percentage	19.10	13.20	13.60	11.10	10.80	10.90	12.70	12.80	9.90
Cost of a fixed set of consumer goods and services, rubles	6286.1	11,063	11,490	12,175	13,128	14,688	16,236	17,099	18,384

The authors used the "maximum –minimum" method because it avoids the excessive influence of each individual indicator on the integral one. Equation (3) is used when higher values of the indicator are preferred, and Equation (4) is applied otherwise.

The integral indicators of economic and social attractiveness were calculated separately for each year of the studied period according to Equations (5)–(7):

$$EA_j = \sum_{i=1}^{n} a_{ij} w_i, \tag{5}$$

$$SA_j = \sum_{i=1}^{m} c_{ij} z_i, \tag{6}$$

$$\sum_{i=1}^{n} w_i = 1; \ \sum_{i=1}^{m} z_i = 1, \tag{7}$$

where EA_j is the economic attractiveness of the region in year j; a_{ij}, c_{ij} are values of the i-th economic and social indicator in year j, respectively; Saj is the social attractiveness of the region in year j; w_i, z_i are the i-th economic and social indicators' weights, respectively; n and m are the number of economic and social indicators, respectively.

Table 5 presents the values obtained for each of the four regions, which were calculated based on 12 socio-economic indicators. We assume that the weight coefficients of all indicators are equal in order to avoid the excessive influence of any of the studied indicators on the integral assessment.

Table 5. Integral evaluation of economic and social attractiveness of the Arctic regions in 2010–2018.

Region	2010	2011	2012	2013	2014	2015	2016	2017	2018
	Economic Attractiveness								
Murmansk region	0.34	0.35	0.37	0.35	0.42	0.39	0.36	0.33	0.37
Nenets Autonomous district	0.58	0.51	0.60	0.53	0.57	0.50	0.44	0.44	0.50
Yamalo-Nenets Autonomous district	0.68	0.73	0.74	0.75	0.75	0.76	0.81	0.80	0.84
Chukotka Autonomous district	0.49	0.40	0.46	0.46	0.46	0.40	0.41	0.48	0.47
	Social Attractiveness								
Murmansk region	0.37	0.38	0.36	0.34	0.36	0.34	0.35	0.41	0.50
Nenets Autonomous district	0.22	0.22	0.40	0.47	0.25	0.30	0.29	0.22	0.16
Yamalo-Nenets Autonomous district	0.55	0.53	0.53	0.43	0.41	0.52	0.58	0.56	0.61
Chukotka Autonomous district	0.63	0.68	0.52	0.53	0.68	0.70	0.66	0.72	0.70

The data in this table show that the Yamalo-Nenets Autonomous district had the most economic stimulation in the 9 studied years. The economic activity is related to the high income per capita in the region, which, in turn, is conditioned by the presence of oil and gas extraction companies with good salaries and a small population. This region has had the best values of average consumer expenditure per capita, share of the population with incomes lower than the subsistence minimum in the total population, and unemployment rate. The Murmansk region has the lowest economic appeal. This is due to low incomes and consumer spending and high unemployment rates.

The Chukotka Autonomous district also has low economic attractiveness. It ranks last in terms of "Cost of a fixed set of consumer goods and services", which is significantly higher in the Chukotka district than in the other regions, and "Average consumer expenditure per capita". Although the "Unemployment rate" and "Share of population with incomes below the subsistence minimum in the total population" are quite low, these variables have had little effect on its attractiveness. "Income per capita" is also quite good.

However, the Chukotka district does not have a leading position in any of the indicators for the studied period, which determined its relatively low attractiveness.

In terms of social attractiveness, the Chukotka Autonomous district is highlighted by the set of studied indicators. When analyzing the individual values of indicators, it was found that this region was characterized by the lowest crime rate per person for almost all 9 years under review. Moreover, the region has a large number of square meters of commissioned residential buildings per person, which is several times higher compared to the Nenets Autonomous district, which has almost the same population size.

In addition, this region has consistently had the leading position among the four regions in terms of "Gross enrollment rate in preschool education" since 2010. In 2013, the "Volume of paid services per capita" in the Chukotka Autonomous district almost doubled compared to its value in 2012.

As a result, the region took the leading position for this indicator until 2018. In terms of "Birth rate", the Chukotka Autonomous district is significantly behind the Yamalo-Nenets Autonomous district, which has a population that is about 10 times higher.

Despite having the largest population of the four Arctic regions studied, which seems to be an attractive factor for young people, the Murmansk region ranks only third after the Chukotka and Yamalo-Nenets Autonomous districts according to the integral indicator of social attractiveness. This is due to its low (or even minimum) values of indicators such as "Commissioning of residential buildings per person", "Volume of paid services per capita", and "Birth rate", which is actually associated with a natural population decline. In addition, the Chukotka, Nenets, and particularly Yamalo-Nenets Autonomous districts show stable natural population growth.

The Nenets Autonomous district ranks last in terms of social attractiveness, primarily because it has the lowest values of "Population size" and "Volume of paid services per capita", as well as poor values of all the other indicators in comparison with other regions.

In the course of the study, the authors attempted to analyze the relationship between migration growth and the integral indicators of the attractiveness of the region. We hypothesized that changes in indicators of social and economic attractiveness have linear effects on the number of arrivals and departures in the region. However, the calculation of the linear complex correlation coefficient did not reveal a simple linear relationship between them, and this hypothesis was not confirmed. Further research can aim to identify non-linear models that reflect the relationship between migration flows in the regions of the Russian Arctic and their attractiveness to labor migrants.

4. Discussion

This study focused on the socio-economic assessment of the attractiveness of Arctic territories to potential migrants. We should note that the regions under consideration are significantly heterogeneous in terms of most indicators. The Murmansk region is characterized by the highest population, which is largely due to its milder climatic conditions, proximity to European countries, and developed infrastructure. The lowest population for the 9 studied years was in the Nenets Autonomous district, whose indicator values are 17–19 times lower than those in the Murmansk region. The rest of the indicators are also scattered for different Arctic regions. However, the territories under consideration have a common problem: the outflow of the working population to more favorable climatic zones.

The main limitations of the study are discussed below.

It is evident that the migration attractiveness of the Arctic region to the working population is determined by a range of indicators. However, not all indicators can be quantified. This is due to the fact that people rarely base their choices only on quantitative indicators in the decision-making process. In particular, it seems doubtful that a potential immigrant would analyze regional statistics on indicators such as GRP per capita, investment in fixed assets, or infant mortality rates before moving to a region. The decision to move is made on the basis of the individual's principles, life attitudes, beliefs, and other personal characteristics that cannot be quantified, as mentioned in [71].

Thus, the migration attractiveness of a region also depends on a range of non-measurable personal characteristics, such as tolerance to difficult weather conditions. Our research does not separate the influence of quantitative and qualitative parameters on migration processes, as the qualitative characteristics of the research object are difficult to formalize. This fact also prevents their use for assessing the attractiveness of the region together with socio-economic indicators. In addition, only measurable parameters can be analyzed with the proposed research methods. For these reasons, we limited our research to the investigation of quantitative characteristics of the object. Additional research covering qualitative characteristics in order to analyze their influence on migration processes in the Russian Arctic and assess the attractiveness of regions would provide further insight into the problem of migration outflows. Nevertheless, we consider it possible to identify key economic and social indicators that can be used to objectively assess the attractiveness of the Russian Arctic regions based on statistical data, which was performed in the framework of this study.

The second limitation is that this study focuses only on simple linear regression relationships between migration indicators and the indicators that cause their variation. In this regard, as dependent variables of the models, we studied only those indicators that have a close linear correlation with migration. The presented models do not account for other factors. In the future, we plan to apply complex-valued non-linear econometric models and include a wider range of indicators. In addition, it is possible to apply an individual approach to each region separately, adjusting the list of indicators for building regression models depending on their individual characteristics.

Finally, the research was limited by the relatively small amount of retrospective data for the following key reasons: (1) the investigated time series data must be stationary, and (2) statistical data are not publicly available for all analyzed regions. Since migration processes are very sensitive to changes such as legislative and legal regulations, social benefits, and economic incentives, we considered time series that were formed in relatively stable conditions, which limits the study timeframe. Moreover, we faced a data acquisition problem when trying to separate the statistical data for areas located beyond the Arctic Circle from the whole administrative region's dataset. For this reason, we had to analyze only those regions that are fully located beyond the Arctic Circle. In addition, some Arctic regions do not have any publicly available unified statistical information within the 2010–2019 timeframe, which also limited our research.

There is also one disputable point that should be mentioned. The choice of specific indicators for assessing the social and economic attractiveness of regions and building regression models may seem controversial. However, in the course of the study, the authors conducted a thorough analysis and selected attractiveness indicators based on the works of Russian and foreign researchers, specifics of the studied regions, and possibilities of accessing quantitative information for their assessment. According to our assessment, the list of selected indicators best characterizes the attractiveness of the Arctic region to migrants from other territories, especially Russian ones. Choosing equal weight coefficients for all indicators is debatable, although it prevents the excessive influence of any indicator on the result.

The objectivity of the research in selecting indicators for estimating the attractiveness of regions was achieved by using a number of independent experts and conducting correlation analyses to determine indicators that have a significant linear relationship with migration processes. The use of complex-valued econometric methods makes this study unique, since this is the first time that the tools of a complex-valued economy are used to simulate migration processes in the Arctic regions. The procedure for evaluating regional attractiveness is also new, as it involves a unique list of indicators identified during the research, which are specific to Arctic regions.

Thus, the results of this study are the six economic and six social indicators identified on the basis of a thorough literature review and expert surveys; the assessments of the linear relationship between each indicator and migration processes in the Arctic regions;

four complex linear regression models that can be used to predict the number of arrivals and departures in the region; and evaluation of the attractiveness of four Arctic regions in terms of their social and economic development.

The key findings of the research are as follows: (i) all Russian Arctic regions are characterized by a stable outflow of the population throughout the 10 studied years; (ii) migration processes are more linearly dependent on social indicators than on economic ones, which confirms the importance of the former; (iii) the four presented complex linear models produced better forecasting results than the simple extrapolation of the trends, so linear complex-valued models are suitable for predicting migration processes in the Arctic regions of Russia; (iv) the Chukotka Autonomous district has the highest social attractiveness, while economic incentives for migration are the highest for the Yamalo-Nenets Autonomous district.

The unexpected result of the study was that among all Arctic regions, the Chukotka Autonomous district took the lead in terms of social attractiveness. It should be emphasized that Chukotka is the most sparsely populated region of Russia and the most distant from the center. According to the Russian Statistical Agency, the area of the region is 4% of the total area of Russia, the population in 2021 is 49,527 people (0.03% of the population of Russia), and the population density is only 0.07 people/km^2. The region is very rich in mineral resources, so all socio-economic indicators are at a high level. The finding that the Yamalo-Nenets Autonomous district emerges at the top in terms of economic indicators is more predictable since oil and gas resources, the basis of the national budget, are concentrated in this region. The region is in 7th place for GRP among all regions of the country. However, due to the larger population (547.01 million people) and the population density (0.71 people/km^2) compared to Chukotka, the Yamalo-Nenets Autonomous district has a lower score in terms of social attractiveness.

The scientific contribution of this article to the study of migration processes in Arctic regions is the use of complex-valued econometric models to predict future migration flow. These studies were conducted for the first time, and the calculations proved to be effective and have potential practical applications.

We suggest that this method be tested on a larger dataset on migration.

5. Conclusions

The social and economic development of the Arctic region depends directly on the number of its residents, migration changes, and its migration attractiveness. The analysis of the region's migration attractiveness helps to identify the reasons for population outflow and, therefore, makes it possible to influence and eliminate them.

The study of the migration attractiveness of the Arctic regions was carried out in several stages. The first stage was devoted to a detailed analysis of the foreign and domestic literature on the problems and prospects of the development of the Northern territories, the migration attractiveness of the region, and various ways of assessing it. Analysis of migration processes in the Russian Arctic showed that there was an outflow of population from all regions over the past decade. The authors assume that there is a relationship between the migration processes currently taking place in the Arctic regions and their attractiveness to the working population.

In the second stage of the study, key social and economic indicators that contribute to the region's attractiveness to potential migrants were identified. For this purpose, the authors involved experts on the problems of economic development of the Arctic territories. The result of the expert survey was a list of 12 quantitative socio-economic indicators characterizing the attractiveness of the Arctic region.

To assess the dependence of migration processes on each of the selected socio-economic indicators, the authors conducted a correlation analysis using official data from state statistics bodies. It was revealed that not all indicators have a linear relationship with migration growth; simple relationships are attributed to a greater association between the number of arrivals and departures in the region and the level of its social development. This confirms that the motives for moving to the Far North are not just economic incentives.

Migration processes were modeled using the tools of a complex-valued economy. In particular, simple linear regression complex-valued models were used that reflect the relationship between a pair of dependent and a pair of independent model variables. The number of arrivals and the number of departures in the Arctic region were selected as a pair of dependent variables. For the independent variable of the linear complex-valued model, four pairs of indicators were proposed that reflect the social and economic attractiveness of the region. Thus, the authors formed four linear complex-valued models in order to test them for their ability to form good predictions of migration indicators. Calculations performed on the Murmansk region as an example showed that these models are suitable for forecasting, since the values of the number of arrivals and departures in the region calculated for 2018 are close to the actual data.

In the final stage of the study, a comparative analysis of the attractiveness of the Arctic regions (Murmansk region and Nenets, Yamalo-Nenets and Chukotka Autonomous districts) was performed. The authors evaluated the integral indicators of the social and economic attractiveness of the region on the basis of 12 indicators by calculating the weighted averages. It was found that the most attractive region over a 10-year period in terms of the economic situation is the Yamalo-Nenets Autonomous district, whereas the Chukotka Autonomous district is the most attractive in terms of social conditions. However, it should be noted that the Yamalo-Nenets Autonomous district is the country's oil and gas center, which implies significant tax revenues at all budget levels, as well as a high level of income of the population. The Chukotka Autonomous district is a region with a minimum population density, which affects the statistical indicators in terms of their increase.

Further research in this area will focus on modeling migration processes in all Arctic regions using linear and non-linear econometric complex-valued models.

Author Contributions: Conceptualization A.C. and P.K.; Methodology A.C., P.K. and N.R.; Validation P.K.; Investigation A.C. and A.N.; Formal Analysis A.C., N.R. and A.N.; Writing—Original Draft Preparation A.C., N.R. and A.N. All authors have read and agreed to the published version of the manuscript.

Funding: This research received no external funding.

Institutional Review Board Statement: Not applicable.

Informed Consent Statement: Not applicable.

Data Availability Statement: Publicly available datasets were analyzed in this study. This data can be found here: https://eng.gks.ru.

Conflicts of Interest: The authors declare no conflict of interest.

Appendix A

Dear colleague, please fill in the information about yourself and select five economic and five social indicators from the list, which, in your opinion, are advisable to use to assess the migration attractiveness of the Russian Arctic region. Put a cross or any other sign in a free field.

Full name____

Position____

Place of work______________________________________

The List of Indicators

No.	Social Indicators:
1.	The population size, thousands of people
2.	The number of registered crimes per capita, pcs.
3.	The number of doctors per 10,000 population, people
4.	The number of hospital beds for 24 h hospitals per 10,000 population, people
5.	The capacity of outpatient clinics per 10,000 population, visits per shift
6.	The birth rate, natural population growth (decline) per 1000 people
7.	Commissioning of residential buildings, thousands m^2 of total area of residential premises per person
8.	The volume of paid services to the population (in actual prices for the period), million rubles
9.	Putting into operation of sports and recreation complexes, units
10.	Commissioning of hotels, places
11.	The gross enrollment rate in preschool education, as a percentage of the number of children aged 1–6 years

No.	Economic Indicators:
1.	Number of enterprises and organizations per 1000 people, pcs.
2.	Average annual number of employees of organizations, thousand people
3.	Income per capita, rubles
4.	Average consumer expenditure per capita (per month), rubles
5.	Retail trade turnover (in actual prices), mln. rubles
6.	Unemployment rate, percent
7.	Recognized as unemployed, thousands of people
8.	The volume of work performed in the type of economic activity "Construction", mln. rubles
9.	Share of population with incomes below the subsistence minimum in the total population, percent
10.	Public catering turnover, mln. rubles
11.	Cost of a fixed set of consumer goods and services, rubles

References

1. Gulas, S.; Downton, M.; D'Souza, K.; Hayden, K.; Walker, T.R. Declining Arctic ocean oil and gas developments: Opportunities to improve governance and environmental pollution control. *Mar. Policy* **2017**, *75*, 53–61. [CrossRef]
2. Andreassen, N. Arctic energy development in Russia—How "sustainability" can fit? *Energy Res. Soc. Sci.* **2016**, *16*, 78–88. [CrossRef]
3. Gladkiy, Y.N.; Eidemiller, K.Y.; Samylovskaya, E.A.; Sosnina, M.N. Conceptual theories and ideologies of sustainable development of the arctic in the era of changing technological paradigms. *IOP Conf. Ser. Earth Environ. Sci.* **2019**, *302*, 012069. [CrossRef]
4. Ellis, J. Law of the sea convention. *Essent. Concepts Glob. Environ. Gov.* **2014**, 113–114.
5. Moe, A. The dynamics of Arctic development. In *Asia and the Arctic*; Sakhuja, V., Narula, K., Eds.; Springer: Singapore, 2016; pp. 3–13. [CrossRef]
6. Federal Budget and Budgets of the Budget System of the Russian Federation for 2020 (Preliminary Results). Ministry of Finance of the Russian Federation Official Internet Site. Available online: https://minfin.gov.ru (accessed on 22 April 2021).
7. Hintsala, H.; Niemela, S.; Tervonen, P. Arctic potential—Could more structured view improve the understanding of Arctic business opportunities? *Polar Sci.* **2016**, *10*, 450–457. [CrossRef]
8. Litvinenko, V. The role of hydrocarbons in the global energy agenda: The focus on liquefied natural gas. *Resources* **2020**, *9*, 59. [CrossRef]
9. Tcvetkov, P.; Cherepovitsyn, A.; Makhovikov, A. Economic assessment of heat and power generation from small-scale liquefied natural gas in Russia. *Energy Rep.* **2020**, *6*, 391–402. [CrossRef]
10. Vasilenko, N.; Khaykin, M.; Kirsanova, N.; Lapinskas, A.; Makhova, L. Issues for development of economic system for subsurface resource management in russia through lens of economic process servitization. *Int. J. Energy Econ. Policy* **2020**, *10*, 44–48. [CrossRef]
11. Kobylkin, D.N. The resources of the Arctic shelf are our strategic reserve. *Energy Policy* **2019**, 6–9. Available online: https://energypolicy.ru/wp-content/uploads/2020/02/%D0%BD%D0%BE%D1%8F%D0%B1%D1%80%D1%8C-2019.pdf (accessed on 22 April 2021).

12. Cherepovitsyn, A.E.; Lipina, S.A.; Evseeva, O.O. Innovative approach to the development of mineral raw materials of the Arctic Zone of the Russian Federation. *J. Min. Inst.* **2018**, *232*, 438–444. [CrossRef]
13. Smits, C.C.A.; Justinussen, J.C.S.; Bertelsend, R.G. Human capital development and a Social License to Operate: Examples from Arctic energy development in the Faroe Islands, Iceland and Greenland. *Energy Res. Soc. Sci.* **2016**, *16*, 122–131. [CrossRef]
14. Vasiltsov, V.S.; Vasiltsova, V.M. Strategic planning of Arctic shelf development using fractal theory tools. *J. Min. Inst.* **2018**, *234*, 663–672. [CrossRef]
15. Tsvetkova, A.; Katysheva, E. Present problems of mineral and raw materials resources replenishment in Russia. In Proceedings of the 19th International Multidisciplinary Scientific GeoConference SGEM, Albena, Bulgaria, 30 June–6 July 2019; STEF92 Technology: Sofia, Bulgaria, 2019; Volume 19, pp. 573–578. [CrossRef]
16. Novoselov, A.; Potravny, I.; Novoselova, I.; Gassiyb, V. Selection of priority investment projects for the development of the Russian Arctic. *Polar Sci.* **2017**, *14*, 68–77. [CrossRef]
17. Tolvanen, A.; Eilu, P.; Juutinen, A.; Kangas, R.; Kivinen, M.; Markovaara-Koivisto, M.; Naskali, A.; Salokannel, V.; Tuulentie, S.; Similä, J. Mining in the Arctic environment—A review from ecological, socioeconomic and legal perspectives. *J. Environ. Manag.* **2019**, *223*, 832–844. [CrossRef]
18. Cochrane, J.J.; Freeman, S.J. Working in Arctic and Sub-arctic conditions: Mental health issues. *Can. J. Psychiatry* **1989**, *34*, 884–890. [CrossRef] [PubMed]
19. Tripartite sectoral meeting on occupational safety and health and skills in the oil and gas industry operating in polar and Subarctic climate zones of the northern hemisphere. In Proceedings of the Note on the Proceedings of International Labour Office, Geneva, Switzerland, 26–29 January 2016; Available online: https://www.ilo.org/wcmsp5/groups/public/---ed_dialogue/---sector/documents/meetingdocument/wcms_495260.pdf (accessed on 30 January 2020).
20. ISO/DIS 35101(en). Petroleum and Natural Gas Industries—Arctic Operations—Working Environment. Available online: https://www.iso.org/obp/ui/#iso:std:iso:35101:dis:ed-1:v1:en (accessed on 31 January 2020).
21. Andersen, T.; Poppel, B. Living conditions in the Arctic. *Soc. Indic. Res.* **2002**, *58*, 191–216. [CrossRef]
22. Poppel, B. *SLiCA: Arctic Living Conditions: Living Conditions and Quality of Life among Inuit, Saami and Indigenous Peoples of Chukotka and the Kola Peninsula*; Rosendahls-Schultz Grafisk & Nordic Council of Ministers: Copenhagen, Denmark, 2015; p. 426. Available online: https://books.google.ru/books?id=ffLIBgAAQBAJ&lpg=PP1&hl=ru&pg=PA4#v=onepage&q&f=false (accessed on 28 September 2019).
23. Petersen, H.; Poppel, B. *Dependency, Autonomy, Sustainability in the Arctic*; Ashgate: Farnham, UK, 1999; p. 375.
24. Briere, M.D.; Gajewski, K. Human population dynamics in relation to Holocene climate variability in the North American Arctic and Subarctic. *Quat. Sci. Rev.* **2020**, *240*, 106370. [CrossRef]
25. Desjardins, S.P.A.; Friesen, T.M.; Jordan, P.D. Looking back while moving forward: How past responses to climate change can inform future adaptation and mitigation strategies in the Arctic. *Quat. Int.* **2020**, *549*, 239–248. [CrossRef]
26. Alekseeva, M.B.; Bogachev, V.F.; Gorenburgov, M.A. Systemic diagnostics of the Arctic industry development strategy. *J. Min. Inst.* **2019**, *238*, 450–458. [CrossRef]
27. Navigator of Popular Professions in the Arctic Zone of the Russian Federation 2021–2035. Available online: https://arctic-navi.hcfe.ru/#/ (accessed on 23 April 2021).
28. Nikishin, A. Arctic: The Territory of Leadership. Parliamentary Newspaper. Publication of the Federal Assembly of the Russian Federation 2017. Available online: https://www.pnp.ru/economics/arktika-territoriya-liderstva.html (accessed on 22 April 2021).
29. Investment Portal of the Arctic Zone of Russia. Official Internet Site. Available online: https://arctic-russia.ru (accessed on 22 April 2021).
30. Global Energy Perspective 2021. McKinsey and Company. Available online: https://www.mckinsey.com/~{}/media/McKinsey/Industries/Oil%20and%20Gas/Our%20Insights/Global%20Energy%20Perspective%202021/Global-Energy-Perspective-2021-final.pdf (accessed on 15 April 2021).
31. Rouleau, J.; Gosselin, L. Impacts of the COVID-19 lockdown on energy consumption in a Canadian social housing building. *Appl. Energy* **2021**, *287*, 116565. [CrossRef]
32. Ilinova, A.; Dmitrieva, D.; Kraslawski, A. Influence of COVID-19 pandemic on fertilizer companies: The role of competitive advantages. *Resour. Policy* **2021**, *71*, 102019. [CrossRef]
33. Fadeev, A.; Larichkin, F.; Afanasyev, M. Arctic Offshore Fields Development: New Challenges & Opportunities at the Current Post-Pandemic Situation (Conference Paper). *IOP Conf. Ser. Earth Environ. Sci.* **2020**, *554*, 012008. [CrossRef]
34. Global Energy Review 2021, IEA, Paris. Available online: https://www.iea.org/reports/global-energy-review-2021 (accessed on 20 April 2021).
35. 2021: A Year of Timid Hopes, Fragile Agreements and New Sources. Available online: https://globalenergyprize.org/ru/2021/01/28/2021-god-robkih-nadezhd-hrupkih-soglashenij-i-novyh-istochnikov (accessed on 21 April 2021).
36. Fadeev, A.; Lipina, S.; Zaikov, Z. Innovative approaches to environmental management in the development of hydrocarbons in the Arctic shelf. *Polar J.* **2021**. Available online: https://www.researchgate.net/publication/349965780_Innovative_approaches_to_environmental_management_in_the_development_of_hydrocarbons_in_the_Arctic_shelf (accessed on 19 April 2021). [CrossRef]
37. Sergeev, V.; Ilin, I.; Fadeev, A. Transport and Logistics Infrastructure of the Arctic Zone of Russia. *Transp. Res. Procedia* **2021**, *54*, 936–944. [CrossRef]

38. Wang, L.; Huang, J.; Cai, H.; Liu, H.; Lu, J.; Yang, L. A study of the socioeconomic factors influencing migration in Russia. *Sustainability* **2019**, *11*, 1650. Available online: https://www.mdpi.com/2071-1050/11/6/1650/htm (accessed on 19 April 2021). [CrossRef]

39. Heleniak, T. Migration in the Arctic. Arctic Yearbook 2014. Available online: https://arcticyearbook.com/images/yearbook/2014/Scholarly_Papers/4.Heleniak.pdf (accessed on 16 April 2021).

40. Skrupskaya, Y. Migration in the Arctic Region. Higher School of Economics Publ. House: Moscow, Russian Federation, 2020. Working Paper WP7/2020/02. (Series WP7 "Mathematical Methods for Decision Making in Economics, Business and Politics"). Available online: https://www.hse.ru/data/2020/06/17/1607980066/WP7_2020_02______.pdf (accessed on 16 April 2021).

41. Federal State Statistics Service. Available online: https://eng.gks.ru (accessed on 20 July 2019).

42. Niedomysl, T. Migration and place attractiveness. *Geogr. Reg.* **2006**, *68*, 46. Available online: http://uu.diva-portal.org/smash/get/diva2:168343/FULLTEXT01 (accessed on 12 May 2020).

43. Portnov, B.A. Modeling the migration attractiveness of a region. In *Desert Regions*; Portnov, B.A., Hare, A.P., Eds.; Springer: Berlin, Germany, 1999; pp. 111–131. Available online: https://link.springer.com/chapter/10.1007/978-3-642-60171-2_6#citeas (accessed on 2 June 2020).

44. Karachurina, L.B. Attractiveness of centers and secondary cities of regions for internal migrants in Russia. *Reg. Res. Russ* **2020**, *10*, 352–359. [CrossRef]

45. Lundholm, E.; Garvill, J.; Malmberg, G.; Westin, K. Forced or free movers? The motives, voluntariness and selectivity of interregional migration in the Nordic countries. *Popul. Space Place* **2004**, *10*, 59–72. [CrossRef]

46. Beglova, E.I.; Nasyrova, S.I.; Yangirov, A.V. Assessment of migration attractiveness of Russian Federation federal districts. *Int. J. Civ. Eng. Technol. (IJCIET)* **2018**, *9*, 323–338. Available online: http://www.iaeme.com/MasterAdmin/Journal_uploads/IJCIET/VOLUME_9_ISSUE_12/IJCIET_09_12_036.pdf (accessed on 16 December 2019).

47. Anisimova, E.A.; Glebova, I.S.; Khamidulina, A.M.; Karimova, R.R. Correlation of migration level and city attractiveness. *Int. Bus. Manag.* **2016**, *10*, 5577–5580. [CrossRef]

48. Petrov, M.B.; Kurushina, E.V.; Druzhinina, I.V. Attractiveness of the Russian regional space as a living environment: Aspect of the migrants' behavioural rationality. *Econ. Reg.* **2019**, *15*, 377–390. [CrossRef]

49. Druzhinina, I.V.; Kurushina, E.V.; Kurushina, V.A. Attractiveness of the Arctic zone and the northern territories of Russia for migrants. *Int. J. Ecol. Econ. Stat.* **2017**, *38*, 152–163. Available online: http://www.ceser.in/ceserp/index.php/ijees/article/view/5308 (accessed on 17 August 2019).

50. Schäfer, P.; Just, T. Does urban tourism attractiveness affect young adult migration in Germany? *J. Prop. Investig. Financ.* **2018**, *36*, 68–90. [CrossRef]

51. Migration Policy Affects the Attractiveness of OECD Countries to International Talent. 29 May 2019. Available online: http://www.oecd.org/newsroom/migration-policy-affects-attractiveness-of-oecd-countries-to-international-talent.htm (accessed on 18 June 2019).

52. Tuccio, M. Measuring and assessing talent attractiveness in OECD countries. In *OECD Social, Employment and Migration Working Papers No. 229*; OECD Publishing: Paris, France, 2019; Available online: https://www.oecd.org/migration/mig/Measuring-and-Assessing-Talent-Attractiveness-in-OECD-Countries.pdf (accessed on 13 March 2020).

53. Ewers, M.C.; Dicce, R. High-skilled migration and the attractiveness of cities. In *High-Skilled Migration: Drivers and Policies*; Czaika, M., Ed.; University Press Scholarship Online: Oxford, UK, 2018. [CrossRef]

54. Matuszczyk, K. The migration attractiveness index 2017. *Migr. Attract. Index Bull.* 2018. Available online: http://ceedinstitute.org/attachments/420/39855ff647bf393ca4fba0ad823fce5a.pdf (accessed on 11 August 2019).

55. Lipshitz, G. The main approaches to measuring regional development and welfare. *Soc. Indic. Res.* **1993**, *29*, 163–181. Available online: http://www.jstor.org/stable/27522688 (accessed on 16 March 2020). [CrossRef]

56. Cziraky, D.; Sambt, J.; Rovan, J.; Puljiz, J. Regional development assessment: A structural equation approach. *Eur. J. Oper. Res.* **2006**, *174*, 427–442. [CrossRef]

57. Shahraki, A.A. Regional development assessment: Reflections of the problem-oriented urban planning. *Sustain. Cities Soc.* **2017**, *35*, 224–231. [CrossRef]

58. Erlinda, N.; Fauzi, A.; Sutomo, S.; Putri, E.I.K. Assessment of sustainable regional development policies: A case study of Jambi province, Indonesia. *Econ. World* **2016**, *4*, 224–237. [CrossRef]

59. Mederly, P.; Novacek, P.; Topercer, J. Sustainable development assessment: Quality and sustainability of life indicators at global, national and regional level. *Foresight* **2003**, *5*, 42–49. [CrossRef]

60. Shi, Y.; Ge, X.; Yuan, X.; Wang, Q.; Kellett, J.; Li, L.; Ba, K. An integrated indicator system and evaluation model for regional sustainable development. *Sustainability* **2019**, *11*, 2183. [CrossRef]

61. Svetunkov, S.G. *Complex-Valued Modeling in Economics and Finance*; Springer Science + Business Media: New York, NY, USA, 2012; 318p. [CrossRef]

62. Meyer-Stamer, J. *Systematic Competitiveness and Local Economic Development—Discussion Paper*; Mesopartner: Duisberg, Germany, 2008.

63. Rucinska, S.; Rucinsky, R. Factors of regional competitiveness. In Proceedings of the 2nd Central European Conference in Regional Science—CERS, Nový Smokovec, Slovakia, 10–13 October 2007; pp. 902–911. Available online: https://www.researchgate.net/publication/228452752_Factors_of_regional_competitiveness (accessed on 15 February 2020).

64. Zhang, W.; Deng, F.; Liang, X. Comprehensive evaluation of urban competitiveness in Chengdu based on factor analysis. In Proceedings of the Ninth International Conference on Management Science and Engineering Management; Xu, J., Nickel, S., Machado, V., Hajiyev, A., Eds.; Springer: Berlin/Heidelberg, Germany, 2015; p. 362. [CrossRef]
65. Malecki, E. Jockeying for position: What it means and why it matters to regional development policy when places compete. *Reg. Stud.* **2004**, *38*, 1101–1120. [CrossRef]
66. Annoni, P.; Dijkstra, L.; Gargano, N. The EU Regional Competitiveness Index 2016. European Commission, Directorate-General for Regional and Urban Policy, WP 02/2017. Available online: https://ec.europa.eu/regional_policy/sources/docgener/work/201701_regional_competitiveness2016.pdf (accessed on 18 October 2019).
67. Xu, J. Research the evaluation index system of regional competitiveness in Henan province. *Chin. Stat.* **2012**, *3*, 58–60.
68. Zhou, Y.; Lu, Z. Research on Evaluation of Regional Economic Competitiveness Based on Fuzzy Analytic Hierarchy Process. *Reg. Educ. Res. Rev.* **2019**, *1*, 24. [CrossRef]
69. United Nations Development Programme. Human Development Reports. Available online: http://hdr.undp.org/en/content/human-development-index-hdi (accessed on 27 January 2020).
70. Segal, G.F.; Moore, A.T.; Nolan, J. *California Competitive Cities: A Report Card on Efficiency in Service Delivery in California's Largest Cities*; Reason Foundation: Los Angeles, CA, USA, 2002.
71. Iakovleva, I.A.; Sharok, V.V. Social factors that make work in Arctic region attractive. In *Advances in Social Science, Education and Humanities Research, Proceedings of the International Conference on Communicative Strategies of Information Society—CSIS, Saint-Petersburg, Russia, 26–27 October 2018*; Atlantis Press: Amsterdam, The Netherlands, 2018; Volume 289, pp. 286–291. [CrossRef]

Article

An Investigation into Current Sand Control Methodologies Taking into Account Geomechanical, Field and Laboratory Data Analysis

Dmitry Tananykhin [1,*], Maxim Korolev [2], Ilya Stecyuk [2] and Maxim Grigorev [1]

[1]   Petroleum Faculty, Saint Petersburg Mining University, 199106 Saint Petersburg, Russia; makcum1298@mail.ru

[2]   Oil and Gas Production Department, LLC RN-Purneftegaz, 629830 Gubkinskiy, Russia; korolevhik@yandex.ru (M.K.); iastetsyuk@png.rosneft.ru (I.S.)

*   Correspondence: dmitryspmi@mail.ru

Abstract: Sand production is one of the major issues in the development of reservoirs in poorly cemented rocks. Geomechanical modeling gives us an opportunity to calculate the reservoir stress state, a major parameter that determines the stable pressure required in the bottomhole formation zone to prevent sand production, decrease the likelihood of a well collapse and address other important challenges. Field data regarding the influence of water cut, bottomhole pressure and fluid flow rate on the amount of sand produced was compiled and analyzed. Geomechanical stress-state models and Llade's criterion were constructed and applied to confirm the high likelihood of sanding in future wells using the Mohr–Coulomb and Mogi–Coulomb prototypes. In many applications, the destruction of the bottomhole zone cannot be solved using well mode operations. In such cases, it is necessary to perform sand retention or prepack tests in order to choose the most appropriate technology. The authors of this paper conducted a series of laboratory prepack tests and it was found that sanding is quite a dynamic process and that the most significant sand production occurs in the early stages of well operation. With time, the amount of produced sand decreases greatly—up to 20 times following the production of 6 pore volumes. Finally, the authors formulated a methodological approach to sand-free oil production.

Keywords: sanding; sand control; poorly consolidated reservoir; prepack test; slotted liner; geomechanical modeling

Citation: Tananykhin, D.; Korolev, M.; Stecyuk, I.; Grigorev, M. An Investigation into Current Sand Control Methodologies Taking into Account Geomechanical, Field and Laboratory Data Analysis. *Resources* **2021**, *10*, 125. https://doi.org/10.3390/resources10120125

Academic Editor: Antonio A. R. Ioris

Received: 17 November 2021
Accepted: 10 December 2021
Published: 13 December 2021

Publisher's Note: MDPI stays neutral with regard to jurisdictional claims in published maps and institutional affiliations.

1. Introduction

The process of sand production is often associated with the development of poorly cemented reservoirs. The first reservoir equilibrium stress state is already reached during the drilling process, and becomes more severe with further well operation. As a result, the destruction of rocks in the bottomhole zone occurs when stresses exceed their tensile strengths [1–3]. This leads to an increased concentration of suspended rock particles in produced liquid, causing submersible and surface equipment malfunctions. Resulting in a decreased well operation factor due to an increase in the frequency and duration of repairs and, as a result, in operating costs [4–7].

There are three main initiation mechanisms of rock destruction. Two of them consist of violating rock integrity by exceeding their compressive or tensile strengths with shear and tensile stresses, respectively. The dynamics of sand production as a result of tensile stresses is, as a rule, short-term, rapidly decaying and local in character and does not lead to significant difficulties during well operation. The third mechanism is associated with volumetric destruction of pore space and is currently poorly studied due to the complexity of the physical processes and the difficulties associated with clear formalization of the task due to multiple influencing factors [1].

As a result of this process, a plastic zone that grows with time is formed around the perforations, associated with the appearance of permanent deformations, the mechanical and reservoir properties of which differ from the remote part of the formation.

Calculation methods based on geomechanical data should be used to prevent the occurrence of critical stresses, at which the destruction process of the bottomhole zone intensifies. One of these methods is selecting the optimal drawdown for sand free operation of the well. The prediction of critical drawdown is formalized by a problem solved using geomechanical modeling [8–15]. Modeling the stability of the bottomhole formation zone (BHZ) makes it possible to predict potential complications during well operation associated with the mechanical properties of the rocks. Such models are used to determine the optimal well completion and magnitude of the sand-free drawdown along with the location and orientation of perforations [16–18].

In order to build the model, the following data are necessary: well logging data, the results of core studies (compressive strength, static and dynamic Young's moduli, Poisson's ratio), as well as operating and drilling data.

The purpose of this research is to increase the turnaround time of the well due to the development of an algorithm for selecting the optimal operating parameters of the well, in conditions of sand removal from the formation by taking into account geomechanical, field and laboratory data analysis.

Technologies for the Operation of Wells Complicated by Sand Production

There are two general technological approaches used in combating the sand production process:

(a)   Preventing the ingress of mechanical inclusions;
(b)   Allowing and working with the consequences rock particle ingress into the wellbore [19,20].

Both approaches are actively being developed, which is confirmed by numerous publications on the use of various technologies in many Russian (and other) fields: Russkoye, Messoyakhskaya group of fields, Van-Yeganskoye, Medvezhye, Komsomolskoye, Vankorskoye, etc. The photo of the region (depositional environment) where the oilfield discussed in this article is located can be seen in Figure 1.

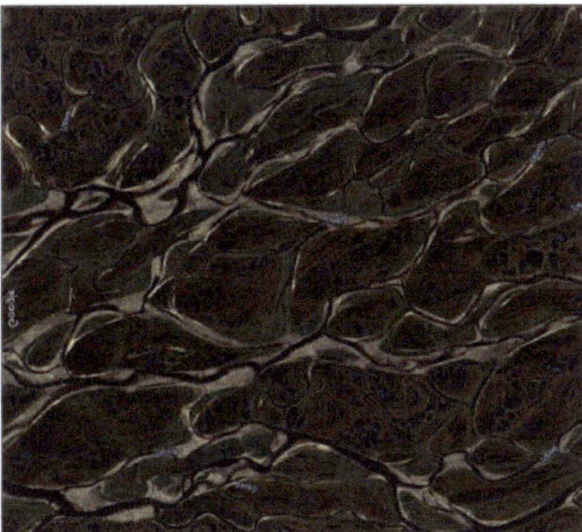

Figure 1. Satellite photo of the area.

Reservoir deposits are identified within the upper part of the Pokurskaya site and are characterized by tidal Upper Cretaceous sediments of the Cenomanian stage, represented by weakly compacted rocks: sands, sandstones, silts, siltstones and argillites (mudstones). The deposits are characterized by explicit facies heterogeneity. The above-mentioned technologies show variable efficiency, and their application is accompanied by significant disadvantages, for example:

- The use of screens causes stress destabilization in the bottomhole zone;
- There is an increase in the extra skin factor ranging from 2 to 10;
- There is a need for their periodic replacement/cleaning (due to erosion wear);
- The use of chemical compositions for fixing the bottomhole zone can reduce the permeability (in some cases up to 70%) due to clogging of highly permeable channels (since the injected composition enters them first), and they also operate for a limited period of time;
- Gravel packing is not always possible (for example, in horizontal wells), and where used, imposes restrictions on the completion of the well;
- Specific gravel pack assemblies require either carefully graded gravel or specially prepared gravel (which is more expensive in terms of its applicability in horizontal wells).

There is also an operational method for limiting sand production—regulating the technological parameters of the well operation, which consists of reducing the depression to the minimum permissible values in order to prevent the ingress of rock particles into the well, but its disadvantage is quite clear—an artificially low flow rate [21–24].

In the case of high-viscosity oil, these factors are exacerbated many times over due to the low productivity index (PI) of the well, which has made the sand management approach the subject of some interest [25–27].

This approach consists of two aspects: careful and constant monitoring of the operating parameters of individual wells and optimization of risks (in the form of predictive calculations and modeling) that inevitably appear when rock is removed from the formation without control.

Given these aspects, in developing this approach, it was understood that thorough consideration of each well was necessary in order to obtain a general situational understanding [28–32].

Additionally, some predictive analytics methods were studied, consisting of calculations regarding:

- Predicting the initialization time of the sand development process;
- The volume of sand production;
- The ability of the rock particles to migrate in the bottomhole zone.

The above-mentioned technique requires the analysis of a vast amount of data, including both the formation properties and well parameters.

Therefore, we formulated an approach based on the studying the influencing parameters on the sand production process: pH of formation water, oil and liquid production rate, water cut, number of shutdowns and starts of wells, aperture of installed downhole screen, method of well completion, reservoir and bottomhole pressure, drawdown, well arrangement, particle size distribution and many others.

2. Materials and Methods

2.1. Geomechanical Modeling

Many researchers [33–43] find a notably high influence of fine fractions (<50 μm) on the well operation (mainly plugging screens), as a result, the material associated with the formation of sand arches was worked out, but significant results were not achieved in by analyzing the literature (except for the connection of the aforementioned arches with the process of the natural decrease in the number of suspended particles in the first few days of well operation).

Some investigations look at the use of chemical compositions for the selective retention of fine fractions; however, no field test results have been conducted [14,15,44–48].

Interdependences were investigated within the framework of a field with high viscosity oil, currently under development between the following parameters: the number of suspended particles, the influence of the sand production process on the operation of individual wells and the study of the effectiveness of screens of certain standard sizes and structures [49,50]. It was found that with different production rates and water-cuts, the interdependence of the number of suspended particles on certain parameters may or may not be observed, as, for example, with the flow rate: its increase or decrease does not affect the number of suspended particles in the fluid flow, which is unexpected (since the fluid flow with a higher speed should entrain more rock particles from the formation). It was found that for wells with a liquid flow rate of <100 m^3/day, there is a significant dependence for the number of suspended particles on water cut (for wells with a flow rate >100 m^3/day, there is no dependence). The effect of water cut on the process of formation destruction has been noted numerous times, which is some confirmation of the fact that phase flow is one of the key parameters that should be taken into account when working with a poorly cemented reservoir.

The authors analyzed the operating experience of one of the facilities and constructed a distribution of the well stock, categorizing wells as either "complicated" or "uncomplicated" according to the following criteria: flow rates (Figure 2), water cut (Figure 3) and target bottomhole pressure (Figure 4). It can be seen from the graphs that operational complications (failures due to erosive wear or clogging with mechanical inclusions of downhole pumping equipment are mostly observed in the well stock with a flow rate of less than 100 m^3/day and a water cut of less than 50%).

Coupling of wells is not always applicable due to the differentials in the tubing diameter, equipment and other factors, which influence the sanding process. The number of wells for consideration for the first category (0–50 m^3/day) is three times fewer than the second. Making conclusions based on the beforementioned data seems questionable since a two-time increase in liquid flow rate leads to a rise in sanding. Nevertheless, a further increase in flow rate does not lead to complications in the well.

It is worth noting that many authors have found during their investigations that the amount of sand carried out increases along with an increase in water cut. Nevertheless, the graph above does not show this effect, since when the water cut is above 50%, there are no complicated wells at all.

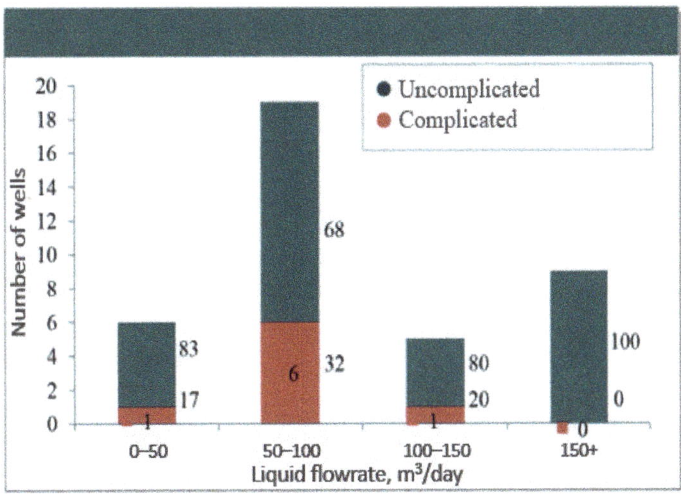

Figure 2. Distribution of sand-prone well stock by fluid flow rate.

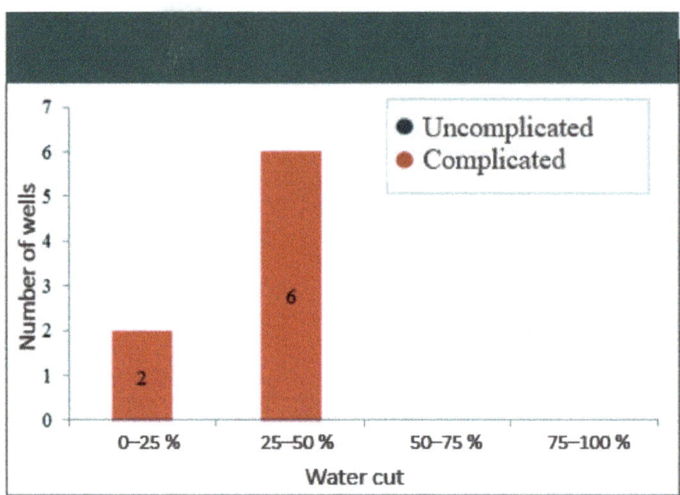

Figure 3. Distribution of sand-prone well stock by water cut.

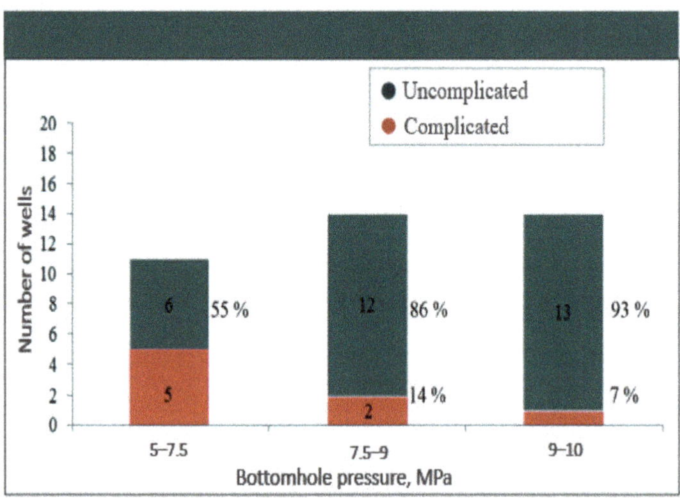

Figure 4. Distribution of sand-prone well stock by bottomhole pressure.

With a decrease in bottomhole pressure (on average, a higher depression of the reservoir), a dependence of a higher level of sanding and complication can be observed.

This phenomenon is associated with the low bearing capacity of the flow and with high viscosity of the oil, which is explained by the lower sedimentation rate of the sand particles. Well operation is carried out according to the target bottomhole pressure control program, which is determined by the requirements of rational oilfield.

The trend towards lower drawdowns in the bottomhole formation zone is confirmed by the analysis of Figure 4, where it can be seen that the least number of failures and complications in the well stock with a target bottomhole pressure of more than 9 MPa, with the initial formation pressure—10.6 MPa.

Thus, the main issue of scientific and practical interest is the prediction of the onset of reservoir destruction and further determination of the critical bottomhole pressure (which leads to the production of sand together with the formation fluid) and, ultimately, to finding the optimal dynamics of bottomhole pressure lowering. There are two basic ways to solve this issue—practical modeling (laboratory tests) via sand retention tests (SRT) and Prepack tests and mathematical modeling. Physical modeling gives good results and provides a lot of information; however, preparing these experiments is time consuming, especially with bulk modeling included.

A literature review [16–19] makes it possible to recommend geomechanical modeling as a tool for assessing the stability of the bottomhole formation zone during operation in the conditions of weakly consolidated sandstones. The stability model is adapted to the data of caliper, imager, mini-frac and modular dynamic tests (MDT) studies. The model-building sequence for one-dimensional geomechanical modeling for PK formations is shown in Figure 5.

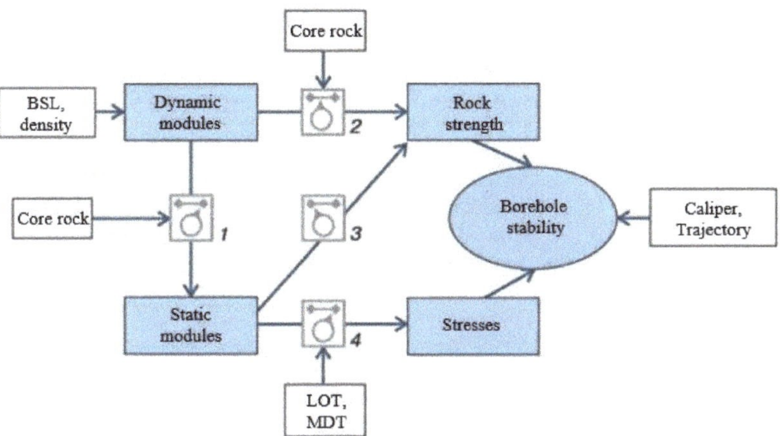

Figure 5. General scheme of geomechanical modeling. BSL—Broadband sonic logging, LOT—frac-test, MDT—stress-test with bottomhole tester.

The 1D model of stability of the bottomhole formation zone according to the criteria of Mogi–Coulomb and Mohr–Coulomb is based on the current parameters of the formation and according to the data of the well operation:

- Reservoir pressure;
- Vertical stress;
- Minimum and maximum horizontal stresses;
- Adhesion strength of the rock;
- Angle of friction;
- Borehole azimuth;
- Well profile;
- Biot's poroelastic constant;
- Poisson's ratio.

The bottomhole pressure and the angular position around the circumference of the wellbore are the specified parameters in this model. Detailed algorithm for this modeling procedure is presented in [9].

The results of the calculations of the 1D model are used to determine the admissible value of depression, at which the fracture of the bottomhole formation zone will not occur. The next stage is a calculation of the optimal step for lowering the bottomhole pressure to value, when the well is brought to the target operating mode (at analogous fields— 0.3–0.5 MPa/day). A significant advantage of the proposed approach is the ability to assess the critical depression value even in the absence of core studies of the mechanical properties of the rock.

As a result, the following geomechanical modeling algorithm was developed:

1. Construction of the one-dimensional model of mechanical properties using well logging and standard correlations;
2. Calculation of stresses and adaptation of the minimum stress to the data on mini-frac;
3. Adaptation of the maximum horizontal stress and strength along the profile of the caliper;
4. Calculation of the critical depression profile based on the correlations set in the software and adaptation of strength based on the development and operation history of previously perforated intervals;
5. Forecast of critical depressions for intervals.

2.2. Prepack Test Design

The methodology for the current prepack tests series was developed to simulate reservoir conditions in the lab with different screens, flowrate conditions and drawdowns. The same differentials as in Figures 1–3 were chosen to be variables in the tests. Thus, tests were run with different water/gas cuts (30, 50, 90%), different drawdowns (gradP1 and gradP2, which were four times higher than gradP1).

Slotted liner was chosen to be tested in this series due to its simplicity, availability on the market and prevalence among Russian oil and gas companies. Screens with aperture sizes of 100, 150, 200, 500, 700 and 1000 μm (mcm) were tested. Initial test runs showed that the 1000 mcm screen was inappropriate for the testing conditions. A schematic representation of the testing facility is shown in Figure 6. The coreholder is equipped with a cuff, and the crimp pressure was set to 2.04 MPa.

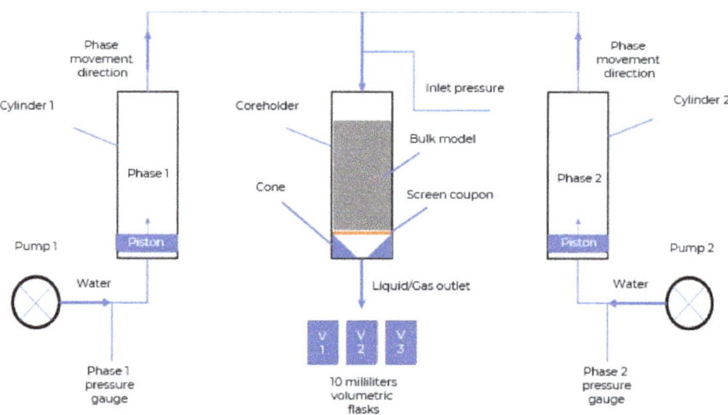

Figure 6. Prepack testing facility.

Bulk models were made with intention to reach porosity, permeability and particle size distribution (PSD) mirroring that of reservoirs. The original PSD curve was taken from one of the oilfields and corresponded to the PK1 formation, shown in Figure 7. Sand from oilfield was used as a material to make bulk models. It was pre-extracted with solvent flushing (until solvent was transparent), dried at 60 degrees Celsius and then sorted using sieves (section sizes of 100, 125, 160, 215, 250, 315 and 500 μm).

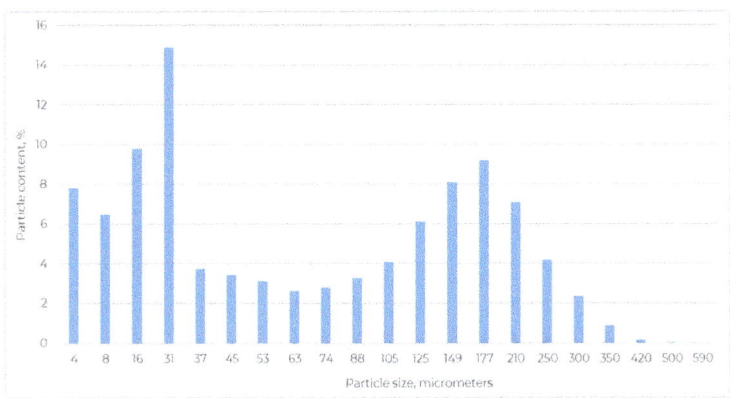

Figure 7. Reservoir's PSD curve.

Bulk models (Figure 8) were made with moist tamping technique, while brine was used as liquid to dampen the models. The diameter of the model is 3 cm, height varied from 5 to 7 cm. Vacuum treatment in a jar with mineral oil was then applied to reach the initial saturation parameters for brine and oil. Mineral oil of the appropriate viscosity (75 mPa*s) was later used as a model of oil in the experiments.

Figure 8. Bulk model with screen filter disk installed.

Samples of liquid were taken three times in a row using volumetric flasks (10 mL) to get dynamic data on the number of suspended particles. Suspended solids concentration (SPC) was then calculated with the mass method. Later, the sand was extracted and analyzed for PSD with a laser in-line particle analyzer.

3. Results

3.1. Geomechanical Modeling

Large reservoirs under the development of LLC "RN-Purneftegaz" were selected as test subjects for geomechanical modeling; namely, the PK1 oil and gas-condensate field reservoir, as well as reservoirs PK18 and PK19-20 of the oil- and gas-condensate field. Here, PK is the name of the formation, and the numbers refer to the number of a single layer in the entire formation.

The developed reservoirs are uncemented sandstone, so the equipment operates under conditions of increased abrasive wear. The removal of mechanical inclusions (mainly sand) is very significant from 3 mg/L to 2050 mg/L (average 109 mg/L) for the PK1 reservoir and 5.5–1080 mg/L (average 81 mg/L) for the PK19-20. Table 1 shows the geomechanical properties of the reservoirs under consideration within the Pokurskaya site, used for the calculation.

Table 1. Reservoir parameters for oil fields.

Parameter	Oilfield 1	Oilfield 2	Oilfield 3
Reservoir pressure, MPa	10.3	12.0	10.8
Rock strength, MPa	5.6	5.6	6.5
Vertical stress, MPa	23.0	22.9	21.1
Maximum horizontal stress, MPa	17.7	21.5	17.5
Minimum horizontal stress, MPa	16.0	19.6	16.4
Rock cohesion strength, MPa	0.22	0.29	0.24
Friction angle, deg	24	27	32
Well's azimuth, deg	210	210	329
Well deviation from vertical axis (zenith angle), deg	89.9	89.5	89.7
Biot's constant	0.8	1	1
Poisson's ratio	0.31	0.2	0.32

The simulation results are presented in Figures 5–7, where it can be seen that, according to the Mogi–Coulomb and Mohr–Coulomb criteria, the fracture of the bottomhole formation zone will occur even with a minimum pressure drop of 0.1 MPa for all the reservoirs studied (blue line—rock strength, red—current stress). This is also confirmed by the Leid criterion, if $\Delta\sigma 1$ and $\Delta\sigma 3 > 0$, rock destruction should be expected (Table 2). These results indicate that it is imperative to substantiate the technology to prevent the destruction of rocks in the bottomhole formation zone or to deal with sand production in the well.

Table 2. Additional parameters.

Parameter	Oilfield 1	Oilfield 2	Oilfield 3
$\Delta\sigma 1$	91.1	89.6	81.7
$\Delta\sigma 3$	84.4	200.6	−1.85

The previous statements are also confirmed by Lade's criterion (Table 2).

According to Lade's research and modeling, if $\Delta\sigma 1$ and $\Delta\sigma 3 > 0$, rock destruction will occur. In Table 2, $\sigma 1$ is the significant principal effective stress and $\sigma 3$ is the minor principal effective stress.

These results indicate that it is imperative to substantiate the technology to prevent the destruction of rocks in the bottomhole formation zone or deal with sand production in the well.

The simulation results are presented in Figures 9–11, where we can observe that, according to the Mogi–Coulomb and Mohr–Coulomb criteria, the fracture of the bottomhole formation zone will occur even with a minimum pressure drop of 0.1 MPa for all the presented reservoirs (blue line—rock's strength, red—current stress).

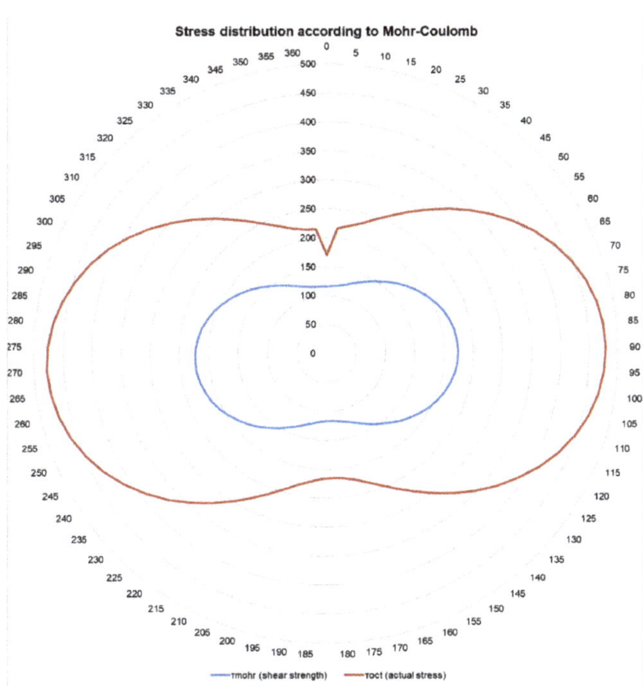

Figure 9. Calculated stresses for Field 1 (blue line—shear strength, red line—actual stress).

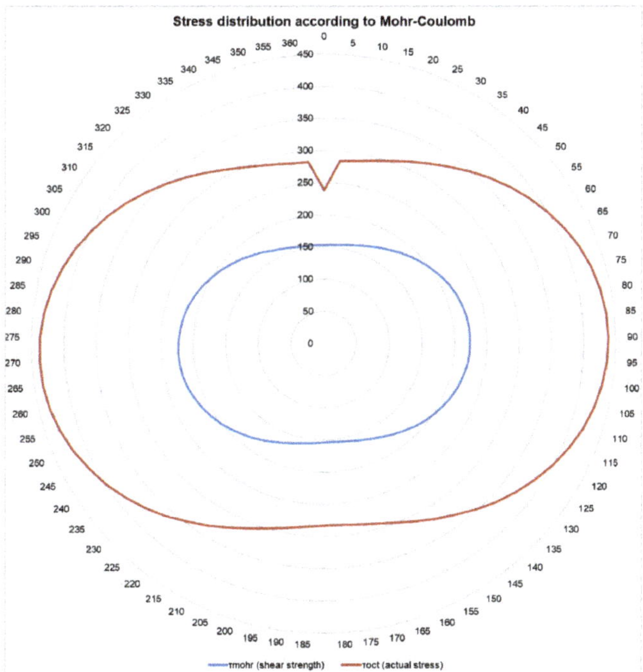

Figure 10. Calculated stresses for Field 2 (blue line—shear strength, red line—actual stress).

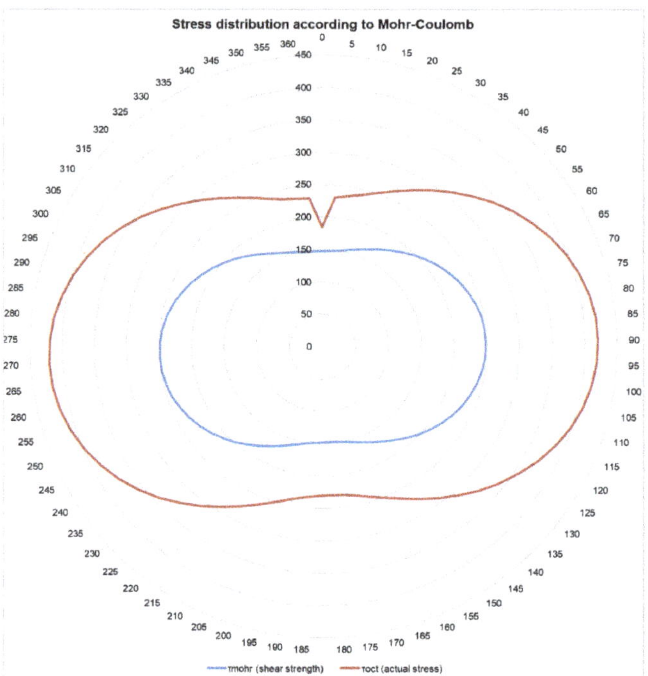

Figure 11. Calculated stresses for Field 3 (blue line—shear strength, red line—actual stress).

Sand control with technological restrictions cannot be applied with such reservoir conditions. In this case, it is crucial to ensure that a proper screen or other sand control method will be suitable and will work with maximum efficiency. An inappropriate screen can lead to severe sand influx into the well or decrease permeability in the bottomhole.

3.2. Prepack Tests

Since samples were taken sequentially three times at the beginning of the experiment, we had the opportunity to track the change in the number of carried particles for the first 30 mL of the pumped fluid. In almost every experiment, SSC was smaller in the latter stages than during stage 1. This indicates and confirms theories that the most severe sand influx happens during well stimulation and later the amount of sand decreases dramatically. The remaining fluid was collected in a separate container and was not analyzed further, but visual observation showed that the amount of suspended particles in it was minimal, being almost transparent.

Overall, the amount of pumped liquid in Oil/Water experiments was always 250 mL and changed from 25 to 175 mL in experiments with gas. An example of the data obtained is shown in Figure 12 below.

For example, for experiment "O/W 30/70 gradP1" SPC at stage 2, there was only 40% of SPC in first stage, which later decreased to 13% of SPC at stage 1.

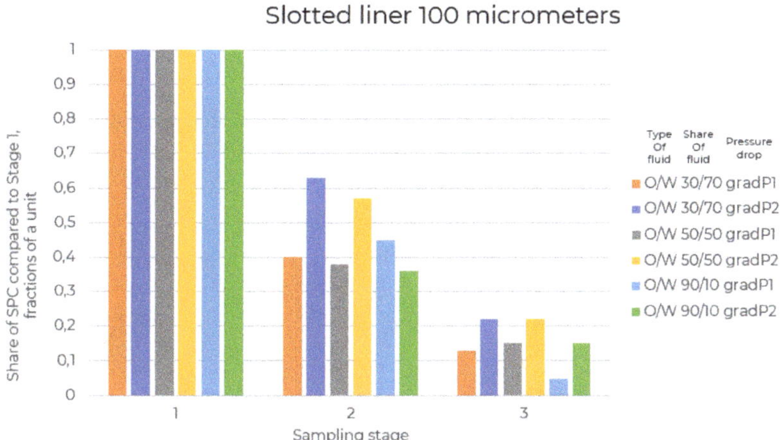

Figure 12. Results of the experiments with oil and water in different water cuts (shares are respective).

4. Discussion

Theoretical and laboratory studies had been carried out to identify the reasons for the removal of mechanical inclusions from the bottomhole formation zone, as well as methods to prevent the destruction of the reservoir formation. The findings of this research make it possible to develop a methodology for an algorithm of geomechanical modeling and subsequently to recommend the optimal parameters for bringing a well into operation.

The results of 1D-geomechanical modeling confirm the hypothesis about the destruction of the bottomhole formation zone at the objects of the Pokurskaya site both during drilling and maintenance and workover of wells and during well operation. The analysis of field well's failures showed that 84% of mechanical impurities or erosion failures occurred on the first or second voyage of equipment during the process of bringing the well into operation with lower bottomhole pressure at 0.5–1 MPa/day. Furthermore, data also suggest that 61.5% of failures occurred after well shutdowns (by production limitation, workovers, etc.). Therefore, it is necessary to take into account the geomechanical properties of the rock when planning the development of such reservoirs.

In cases where it is not possible to solve the problem of sand influx by technical means, sand control technologies must be employed. The best way to thoroughly investigate the efficiency of the proposed technology is testing in lab conditions by simulation reservoir parameters.

5. Conclusions

Using the methodology proposed in the article, the authors noticed a significant decrease in the amount of sand produced after filtration of 3–6 pore volumes. As a result of the implementation of a complex methodology, which included literature analysis, field data analysis, geomechanical modeling and lab testing, the authors developed the following recommendations:

1. Carrying out 1D and 3D geomechanical modeling in order to clarify the drilling parameters, i.e., permissible bottomhole pressure over reservoir pressure on the formation and the rock penetration rate during the drilling;
2. The well completion method should be selected from the operating experience of similar objects, using the endings in assemblies with downhole screens;
3. Bringing the well into operation should be carried out with a minimum gradient of lowering bottomhole pressure—0.2–0.5 MPa/day—which is confirmed by the experience of similar objects' operation.

4. During the operation of the well, considering the possibility of using technologies designed to prevent the removal of mechanical inclusions from the formation.

6. Patents

One of the basic elements of this work is a computer program written by the authors "A program for calculating stability criteria and rupture pressures during the operation of wells complicated by sand occurrence" (RU 2020611693).

Author Contributions: D.T.—General expertise and lab tests designing; M.K.—Field data gathering, analysis and modeling; I.S.—Field data analysis; M.G.—Modeling, lab tests designing and performing of the experiments. All authors have read and agreed to the published version of the manuscript.

Funding: The research was performed at the expense of the subsidy for the state assignment in the field of scientific activity for 2021 № FSRW-2020-0014.

Institutional Review Board Statement: Not applicable.

Informed Consent Statement: Not applicable.

Data Availability Statement: All data included in the article are available for public usage.

Acknowledgments: We would like to acknowledge the Saint Petersburg Mining University for the opportunity to work with their equipment and perform experiments and allowing us to use the Research Equipment Sharing Center of the Mining University for probe analysis for the PSD curves.

Conflicts of Interest: The authors declare no conflict of interest.

References

1. Tananykhin, D.S. Justification of the Technology of Fixing Weakly Cemented Sandstones in the Bottomhole Zone of Oil and Gas Wells by a Chemical Method. Ph.D. Thesis, Saint Petersburg Mining University, Saint Petersburg, Russia, 2013.
2. Bondarenko, V.A. Increasing the efficiency of securing the bottomhole formation zone in order to reduce sand production (on the example of deposits in the Krasnodar region). Ph.D. Thesis, Kuban State Technological University, Krasnodar, Russia, 2014.
3. Matanovic, D.; Cikes, M.; Moslavac, B. *Sand Control in Well Construction and Operation*; Springer: Berlin/Heidelberg, Germany, 2012; 204p. [CrossRef]
4. Mahmud, H.B.; Leong, V.H.; Lestariono, Y. Sand production: A smart control framework for risk mitigation. *Petroleum* **2020**, *6*, 1–13. [CrossRef]
5. Al-Awad, M.N.; El-Sayed, A.A.H.; Desouky, S.E.D.M. Factors affecting sand production from unconsolidated sandstone Saudi oil and gas reservoir. *J. King Saud Univ. Eng. Sci.* **1998**, *11*, 151–174. [CrossRef]
6. Litvinenko, V.S.; Dvoynikov, M.V.; Trushko, V.L. Elaboration of a conceptual solution for the development of the Arctic shelf from seasonally flooded coastal areas. *Int. J. Min. Sci. Technol.* **2021**, in press. [CrossRef]
7. Grachev, S.I. *Development of Oil Fields with Horizontal Wells: A schoolbook*; Grachev, S.I., Samoilov, A.S., Eds.; Tyumen Technical University: Tyumen, Russia, 2015; 144p, ISBN 978-5-9961-1067-4.
8. Joshi, S.D. Augmentation of well productivity with slant and horizontal well. *J. Petrol. Techn.* **1988**, *40*, 729–739. [CrossRef]
9. Zare-Reisabadi, M.; Kaffash, A. Sanding Potential Evaluation Based on a New True-Triaxial Failure Criterion. In Proceedings of the First International Conference of Oil, Gas, Petrochemical and Power Plant, Tehran, Iran, 16 July 2012.
10. Al-Ajmi, A.M.; Zimmerman, R.W. Stability analysis of vertical boreholes using the Mogi–Coulomb failure criterion. *Inter-Natl. J. Rock Mech. Min. Sci.* **2006**, *43*, 1200–1211. [CrossRef]
11. Rahman, K.; Khaksar, A.; Kayes, T.J. Minimizing Sanding Risk by Optimizing Well and Perfora-tion Trajectory Using an Integrated Geomechanical and Passive Sand-Control Approach. In Proceedings of the SPE Annual Technical Conference and Exhibition, Denver, CO, USA, 21–24 September 2008. [CrossRef]
12. Rogachev, M.K.; Mukhametshin, V.V.; Kuleshova, L.S. Improving the efficiency of using resource base of liquid hydrocarbons in jurassic deposits of western Siberia. *J. Min. Inst.* **2019**, *240*, 711–715. [CrossRef]
13. Fjær, E.; Ruistuen, H. Impact of the intermediate principal stress on the strength of heterogeneous rock. *J. Geophys. Res.* **2002**, *107*, ECV-3. [CrossRef]
14. Al-Shaaibi, S.K.; Al-Ajmi, A.M.; Al-Wahaibi, Y. Three-dimensional modeling for predicting sand production. *J. Pet. Sci. Eng.* **2013**, *109*, 348–363. [CrossRef]
15. Hill, A.D.; Zhu, D. The relative importance of wellbore pressure drop and formation damage in horizontal wells. In Proceedings of the SPE Europec/EAGE Annual conference and exhibition, Vienna, Austria, 9–12 June 2008.
16. Dikken, B.J. Pressure drop in Horizontal wells and it's effect on production performance. *J. Pet. Technol.* **1990**, *42*, 1426–1433. [CrossRef]
17. Chen, N.H. An explicit equation for friction factor in pipe. *J. Am. Chem. Soc.* **1979**, *18*, 296–297. [CrossRef]

18. Ouyang, L.-B.; Sepehr, A.; Khalid, A. General wellbore flow model for horizontal, vertical, and slanted well completions. *Spe J.* **1998**, *3*, 124–133.

19. Yue, P.; Du, Z.M.; Chen, X.F.; Tang, C. The Pressure Drop Model of Liquid Flow with Wall Mass Transfer in Horizontal Wellbore with Perforated Completion. *Math. Probl. Eng.* **2014**, *2014*, 9. [CrossRef]

20. Chernyshov, S.; Репина, В.А.; Krysin, N.; Macdonald, D. Improving the efficiency of terrigenous oil-saturated reservoir development by the system of oriented selective slotted channels. *J. Min. Inst.* **2020**, *246*, 660–666. [CrossRef]

21. Novy, R.A. Pressure drops in horizontal wells: When can they be ignored? In Proceedings of the SPE Annual Technical Conference and Exhibition in Washington, Houston, TX, USA, 3–6 October 1993.

22. Edelman, I.; Ivantsov, N.; Shandrygin, A.; Makarov, E.; Zakirov, I. Approaches to development of high-viscosity oil fields in Arctic conditions using the example of Russkoe field. In Proceedings of the SPE Arctic and Extreme Environments Conference and Exhibition, Moscow, Russia, 18–20 October 2011.

23. Betekhtin, A.N.; Kostin, D.K.; Tikhomirov, E.V.; Nikolaev, M.N.; Lyapin, V.V.; Misbakhov, R.Z. Laboratory Studies of Sand Control Systems Used for Heavy Oil Production from Unconsolidated Rocks. In Proceedings of the SPE Russian Petroleum Technology Conference, Moscow, Russia, 16–18 October 2017. [CrossRef]

24. Aksenova, N.A.; Ovchinnikov, V.P.; Anashkina, A.E. (Eds.) *Technology and Technical Means of Well Completion with Unstable Reservoirs: Monograph*; Trinity International University: Tyumen, Russian, 2018; 134p.

25. Kuncoro, B.; Ulumuddin, B.; Palar, S. Sand control for unconsolidated reservoirs. In Proceedings of the Simposium Nasional IATMI 2001, Yogyakarta, India, 3–5 October 2001.

26. Van den Hoek, P.J.; Kooijman, A.P.; De Bree, P.; Kenter, C.J.; Zheng, Z.; Khodaverdian, M. Horizontal-Wellbore Stability and Sand Production in Weakly Consolidated Sandstones. *SPE Drill. Completion* **2000**, *15*, 274–283. [CrossRef]

27. Wang, H.; Gala, D.P.; Sharma, M.M. Effect of fluid type and multiphase flow on sand production in oil and gas wells. In Proceedings of the SPE Annual technical conference and exhibition, San-Antonio, TX, USA, 9–11 October 2017.

28. Andrews, J.S.; Joranson, H.; Raaen, A.M. Oriented Perforating as a Sand Prevention Measure—Case Stud-ies from a Decade of Field Experience Validating the Method Offshore Norway. In Proceedings of the Offshore Technology Con-ference, Houston, TX, USA, 5–8 May 2008.

29. Dvoynikov, M.; Buslaev, G.; Kunshin, A.; Sidorov, D.; Kraslawski, A.; Budovskaya, M. New concepts of hydrogen production and storage in Arctic region. *Resources* **2021**, *10*, 3. [CrossRef]

30. Poroshin, M.A.; Tananykhin, D.S.; Grigoriev, M.B. Analysis of laboratory methods for studying the sanding process in the development of oil fields. *Eurasian Sci. J.* **2020**, *12*.

31. David, C. Water weakening triggers mechanical instability in laboratory fluid substitution experiments on a weakly-consolidated sandstone. In Proceedings of the 50th US Rock Mechanics/Geomechanics Symposium, Houston, TX, USA, 26–29 June 2016.

32. Xiao, J.; Wang, J.; Sun, X. Fines Migration: Problems and Treatments. *Oil Gas Res.* **2017**, *3*, 123. [CrossRef]

33. Wang, X.; Osunjaye, G. Advancement in openhole sand control applications using shape memory polymer. In Proceedings of the SPE Annual Technical Conference and Exhibition, Dubai, United Arab Emirates, 26–28 September 2016.

34. Shagiakhmetov, A.M.; Raupov, I.R.; Terleev, A.V. Investigation of selective properties of the gel-forming composition for the limitation of water inflow to carbonate reservoirs conditions. *Int. J. Civ. Eng.* **2019**, *10*, 485–492.

35. Ramezanian, M.; Emadi, H.; Wang, H. A modified design for gravel packing with expandable rubber beads. In Proceedings of the SPE Western Regional Meeting, San Jose, CA, USA, 23–26 April 2019.

36. Ramezanian, M.; Emadi, H. Gravel-Packing experiments with oil-swelling rubber particles. *SPE Drill. Completion* **2020**, *35*, 252–261. [CrossRef]

37. Yuan, Y.; Goodson, J.; Johnson, M.; Gerrard, D. In-situ mechanical and functional behavior of shape memory polymer for sand management applications. In Proceedings of the Brazil Offshore Conference and Exhibition, Macae, Brazil, 14–17 June 2011.

38. Ahmadi, M.H.; Alizadeh, S.M.; Tananykhin, D.; Hadi, S.K.; Iliushin, P.; Lekomtsev, A. Laboratory evaluation of hybrid chemical enhanced oil recovery methods coupled with carbon dioxide. *Energy Rep.* **2021**, *7*, 960–967. [CrossRef]

39. Carrejo, N.; Horner, D.; Johnson, M. Shape memory polymer as a sand management alternative to gravel packing. In Proceedings of the Canadian Unconventional Resources Conference, Calgary, AB, Canada, 15–17 November 2011.

40. Wu, B.; Choi, S.K.; Feng, Y.; Denke, R.; Barton, T.; Wong, C.Y.; Boulanger, J.; Yang, W.; Lim, S. Evaluating Sand Screen Performance Using Improved Sand Retention Test and Numerical Modelling. In Proceedings of the Offshore Technology Conference Asia, Kuala Lumpur, Malaysia, 22–25 March 2016.

41. Kotb, O.; Haftani, M.; Nouri, A. An Investigation into Current Sand Control Testing Practices for Steam Assisted Gravity Drainage Production Wells. *Eng* **2021**, *2*, 435–453. [CrossRef]

42. Ballard, T.; Beare, S.; Wigg, N. Sand Retention Testing: Reservoir Sand or Simulated Sand-Does it Matter? In Proceedings of the SPE International Conference & Exhibition on Formation Damage, Lafayette, LA, USA, 24–26 February 2016.

43. Haftani, M.; Kotb, O.; Nguyen, P.H.; Wang, C.; Salimi, M.; Nouri, A. A Novel sand control testing facility to evaluate the impact of radial flow regime on screen performance and its verification. *J. Pet. Sci. Eng.* **2020**, *195*, 107903. [CrossRef]

44. Soroush, M.; Roostaei, M.; Mohammadtabar, M.; Hosseini, S.A.; Mahmoudi, M.; Sadrzadeh, M.; Ghalambor, A.; Fattahpour, V. Design for Reliability: Purpose Driven Sand Control Methods for Cased and Perforated Wells. In Proceedings of the SPE Annual Technical Conference & Exhibition, Denver, CO, USA, 5–7 October 2020.

45. Mayorga Cespedes, E.A.; Roostaei, M.; Uzcátegui, A.A.; Soroush, M.; Izadi, H.; Hosseini, S.A.; Schroeder, B.; Mahmoudi, M.; Gomez, D.M.; Mora, E.; et al. Sand Control Optimization for Rubiales Field: Trade-Off Between Sand Control, Flow Performance and Mechanical Integrity. In Proceedings of the SPE Latin American and Caribbean Petroleum Engineering Conference, Bogota, Coulombia, 17–19 March 2020.
46. Mahmoudi, M.; Roostaei, M.; Fattahpour, V.; Uzcatequi, A.; Cyre, J.; Sutton, C.; Fermaniuk, B. Standalone Sand Control Failure: The Role of Wellbore and Near Wellbore Hydro-Thermo-Chemical Phenomenon on the Plugging and the Flow Performance Impairments of the Standalone Sand Screen. In Proceedings of the SPE Thermal Well Integrity and Design Symposium, Baniff, Canada, 27–29 November 2018.
47. Rogachev, M.K.; Nguyen Van, T.; Aleksandrov, A.N. Technology for Preventing the Wax Deposit Formation in Gas-Lift Wells at Offshore Oil and Gas Fields in Vietnam. *Energies* **2021**, *14*, 5016. [CrossRef]
48. Aleksandrov, A.N. Simulating the Formation of Wax Deposits in Wells Using Electric Submersible Pumps. In *Advances in Raw Material Industries for Sustainable Development Goals*; Aleksandrov, A.N., Kishchenko, M.A., Nguyen, V.T., Eds.; CRC Press: London, UK, 2021; pp. 283–295. [CrossRef]
49. Rajaoalison, H.; Zlotkowski, A.; Rambolamanana, G. Mechanical Properties of Sandstone using non-Destructive Method. *J. Min. Inst.* **2020**, *241*, 113. [CrossRef]
50. Jafarpour, H.; Moghadasi, J.; Khormali, A.; Petrakov, D.G.; Ashena, R. Increasing the stimulation efficiency of heterogeneous carbonate reservoirs by developing a multi-bached acid system. *J. Pet. Sci. Eng.* **2019**, *172*, 50–59. [CrossRef]

Article

Arctic Shelf Oil and Gas Prospects from Lower-Middle Paleozoic Sediments of the Timan–Pechora Oil and Gas Province Based on the Results of a Regional Study

Oleg Prischepa, Yury Nefedov * and Victoria Nikiforova

Department of Geology of Oil and Gas, Saint-Petersburg Mining University, 22 line, 2, 199106 St. Petersburg, Russia; prischepa_om@pers.spmi.ru (O.P.); nikiforova_vs@pers.spmi.ru (V.N.)
* Correspondence: nefedov_yuv@pers.spmi.ru

Abstract: The Timan–Pechora oil and gas province (TPP), despite the good geological and geophysical knowledge of its central and southern regions, remains poorly studied in the extreme northwestern part within the north of the Izhma–Pechora depression and the Malozemelsk–Kolguev monocline, and in the extreme northeast within the Predpaikhoisky depression. Assessing the oil and gas potential of the Lower Paleozoic part of the section is urgently required in the northwestern part of the TPP, the productivity of which has been proven at the border and in the more eastern regions of the province (Pechora–Kolva, Khoreyverskaya, Varandei–Adzva regions), that have been evaluated ambiguously. A comprehensive interpretation of the seismic exploration of regional works was carried out, with the wells significantly clarifying the structural basis and the boundaries of the distribution of the main seismic facies' complexes. The capabilities of potentially oil- and gas-producing strata in the Silurian–Lower Devonian were studied. An analysis of migration routes in transit strata used for basin modeling in order to reconstruct the conditions of oil and gas formation that are common in the land and water areas of the Arctic zone of the TPP was carried out. Modeling allowed us to reach an understanding of the formation of large zones with possible accumulations of hydrocarbons, including the time at which the formation occurred and under what conditions, to establish space–time links with possible centers of generation to identify migration directions and, based on a comparison with periods of intensive generation of hydrocarbons both directly located within the work area and beyond (noting the possible migration), to identify zones of the paleoaccumulation of hydrocarbons. The body of existing literature on the subject made it possible to outline promising oil and gas accumulation zones, with the allocation of target objects for further exploration in the Lower Paleozoic part of the section.

Keywords: Arctic shelf; Timan–Pechora oil and gas province; seismic exploration; hydrocarbon potential; Russian Arctic

Citation: Prischepa, O.; Nefedov, Y.; Nikiforova, V. Arctic Shelf Oil and Gas Prospects from Lower-Middle Paleozoic Sediments of the Timan–Pechora Oil and Gas Province Based on the Results of a Regional Study. *Resources* **2022**, *10*, 3. https://doi.org/10.3390/resources11010003

Academic Editor: Ben McLellan

Received: 16 November 2021
Accepted: 14 December 2021
Published: 28 December 2021

1. Introduction

More than 5500 deep oil and gas wells with a drilling volume of 10 million meters have been drilled within the Timan–Pechora sedimentary basin (TP SB). Only 54 wells have been drilled in the water area of the sedimentary basin, most of which have only exposed Mesozoic deposits. Drilling within the limits of the land of the TP SB is also uneven. In the northern half of the Izhma–Pechora synclise, which is the area of this study, it does not exceed 0.01 m/km^2 (Figure 1), and it does not exceed 0.3 m/km^2 in the Malozemelsko–Kolguevskaya monocline, which indicates the lack of knowledge of this region and the prospects for the discovery of new oil and gas objects (Figure 1).

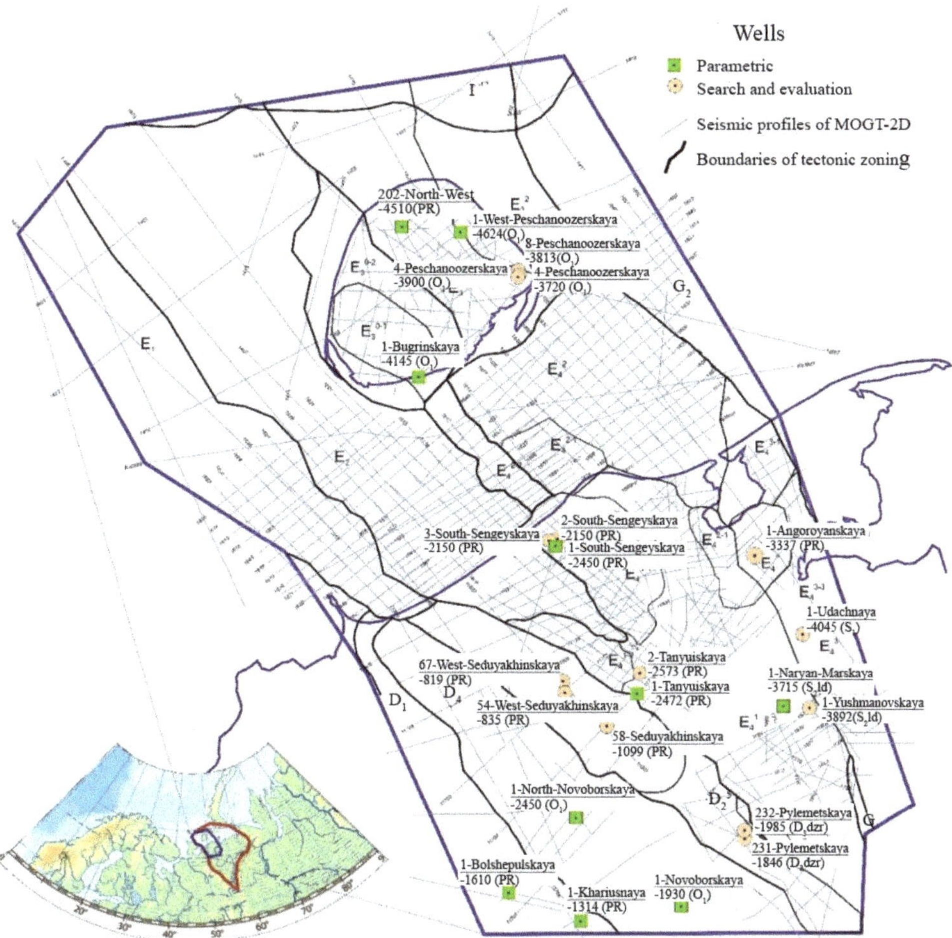

Figure 1. Scheme of seismic study [1].

In the Pechora–Kolva aulacogen, located in the east, it increases to 200 m/km² in some areas. The Khoreyver depression of the Khoreyversko–Pechoromorskaya syneclise is characterized by an average drilling depth of 50–100 m/km², but its coastal part has been studied to a much lesser extent—the drilling depth is only 10 m/km². In the water area on the eastern half of Kolguev Island (East Kolguev structural zone), the drilling density is close to 50 m/km². The most significant drilling operations on land were carried out by state-owned enterprises from the 1970s to the early 1990s of the last century [2,3].

A promising direction for further study of the region is the study of the storage reservoir and flow behavior in the Ordovician shale formation lying directly under the Silurian oil and gas source clay strata, which is a high-quality impermeable layer. Studies of the capacity of similar shale strata, from the point of view of the possibility of CO_2 storage capacity and flow behavior in the shale formation, have been carried out by the Journal of Petroleum Science and Technology 2022 [4].

2. The Economic Significance of the Resource Development of the Hydrocarbon Potential of the Arctic Zone of the Russian Federation

The hydrocarbon (HC) potential of the Arctic zone of the Russian Federation (AZRF) is considered as a strategic reserve for the reproduction of the mineral resource base of oil and gas in the long term. A detailed geological study and the industrial development of the Arctic regions, especially the water areas, are associated with a number of problems related to natural and climatic features, the complex geological structure of deposits, technological features, and the imperfection of the regulatory framework—the current unfavorable economic situation in the global and domestic markets for fuel and energy resources [5].

According to the results of the latest quantitative assessment, the volume of forecast resources of the Russian Arctic is estimated at more than 270 billion TOE (tonne of oil equivalent), including about 48.5 billion tons of oil and condensate and over 220 trillion m^3 of natural gas. At the same time, land resources are estimated at 156 billion TOE (58% of the total volume), of which a significant part is accounted for by gas, while the share of the water areas accounts for 42% of all resources, mainly gas and gas condensate. The largest hydrocarbon potential of the land of the Arctic zone, if we consider the subjects of the Russian Federation, is estimated to be in the Yamalo–Nenets Autonomous District (120.5 billion TOE, including 98.8 trillion m^3 of gas), followed by the Nenets Autonomous District (the northern part of the Timan–Pechora oil and gas province). In addition, with regards to the water area, the southern parts of the Kara and Barents Seas have the greatest potential, which are where all the deposits identified on the Arctic shelf and, accordingly, the oil, gas, and condensate reserves are located.

The raw hydrocarbon potential of the Arctic Seas of Russia is represented today by 19 deposits, most of which are located within the lips and bays of the Kara Sea; some are shelf extensions of onshore objects [5,6], and a smaller one is in the Pechora Sea on the continuation of the TPP. The volume of proven oil reserves is relatively small at 454 million tons, while the volume of projected oil resources is estimated at 12.8 billion tons, of which only 5% is concentrated in currently licensed areas. The volume of proven gas reserves is much larger at 9.2 trillion m^3, 70% of which comprises fields with valid licenses; the volume of the forecasted natural gas resource base is 86.5 trillion m^3 [7–9].

The development of the HC raw material base of the AZRF, first of all, is associated with the implementation of large-scale projects for the exploration, production, and transportation of oil and gas on land, with the prospect of further using the created coastal industrial infrastructure in order to involve the marine hydrocarbon potential in industrial turnover and in the development of the Northern Sea Route. The coastal parts of the northern seas are now considered as outposts of testing and implementation of the latest technologies in the most difficult conditions [8]. The Pechora Sea (the southern part of the Barents Sea) is considered as one of the most important bridgeheads for advancing into the Arctic water area on the basis of the already identified and predicted hydrocarbon potential [10–12].

Since 2013, a project for the industrial development of the Prirazlomnoye oil field has been implemented in the Pechora Sea, the operator of which is PJSC "Gazprom Neft Shelf". The project can be called successful in terms of being able to test a huge number of development technologies and logistics solutions in difficult ice conditions with the development of the interaction of various services and support vessels.

The Shtokman gas condensate field has been prepared for development in the Barents Sea, but the start of development was postponed due to technological difficulties, the need for huge investments, and economic constraints caused by a significant reduction of the gas market in Europe for the Russian Federation and the withdrawal from projects of strategic partners. Among the achievements of domestic marine technology in the Arctic waters of Russia, one worthy of note is the Varandey terminal, which was built by the forces of

PJSC "Lukoil" in the Pechora Sea and has an oil shipment volume of up to 12.5 million tons/year [13].

At the same time, the development of the hydrocarbon resource base on the Arctic shelf has significant limitations, and all the significant achievements of recent years in the preparation of new reserves have been reduced to the additional exploration of offshore extensions of land deposits. Domestic programs for the development of the raw material base of the Arctic seas are very impressive: according to the Ministry of Natural Resources, a significant amount of seismic work has been planned until 2023, and the number of exploratory wells drilled in the Arctic waters should reach 127 units [14]. The real pace of exploration in the water part of the Russian Arctic is limited by the huge cost of offshore drilling, the small capacity of its own drilling fleet, and sanctions and restrictions [15].

3. Materials and Methods

The study methodology consisted of a number of sequential steps [16–19], including:

- construction of a geological model of the studied area, including tracing the distribution of Lower Paleozoic deposits;
- identification of possible oil and gas systems;
- recovery of the dive history;
- restoration of the primary hydrocarbon potential (before the beginning of catagenesis);
- assessment of the scale of hydrocarbon generation;
- assessment of possible migration directions from the oil and gas system;
- identification of natural reservoirs (collectors and cap rocks) in the Lower Paleozoic part of the section;
- analysis of possible hydrocarbon accumulation zones and types of traps in them [18,20–24].

The Kingdom and TEMIS 3D software systems (VNIGRI licenses) were used to process seismic survey data and build a basin model. To calculate the hydrocarbon potential of the oil and gas system, balance equations were used, obtained on the basis of numerous experimental data from VNIGRI [25,26] and linked to the degree of transformation of organic matter (catagenesis) [27–32].

The greatest contribution to the clarification of the hydrocarbon potential of the studied area was made by such parts of this study as:

Refinement of the structural model (a series of structural maps) based on the interpretation and linking of the profile seismic survey on the land part of the site, in the transitional shallow water zone (depths less than 25 m), on Kolguev Island, and in the relatively deep water zone (more than 25 m).

Clarification of the boundaries of the development and thickness of the Lower Paleozoic part of the section (separately the Lower-Middle Ordovician terrigenous strata, Lower Silurian carbonate strata, Lower Devonian terrigenous-carbonate strata, Middle Devonian-Lower Frasnian terrigenous strata, and Middle-Upper Frasnian (so-called "Domanik") carbonate-siliceous-clay strata).

Identification of characteristic features inherent in oil and gas source deposits (organic carbon content of more than 0.5% per rock), with an emphasis on the Lower Paleozoic part of the section.

Identification of lithological and facies features and boundaries of the distribution of the oil and gas system, traditional for the Timan–Pechora OGP—the Domanik oil and gas source formation.

Conduction of a laboratory study by pyrolysis of core samples from the Ordovician, Lower Silurian, and Middle Frasnian parts of the section.

Determination of the stratigraphic interval of the distribution of Lower Silurian sediments enriched with organic matter and tracing their distribution within the studied area.

Determination of the lithological composition (microscopic and petrographic study of samples) and lithotypes, typical for sediments enriched with organic matter.

Study of the petrophysical features of Ordovician and Middle Frasnian terrigenous deposits in order to identify reservoirs.

Paleotectonic reconstructions along the lines of individual seismic profiles in order to determine the time of immersion of the source sediments to the depths of the "oil window".

Mapping the current concentration of organic carbon and the catagenetic maturity of organic matter.

Recalculation of the amount of organic carbon in Silurian and Upper Devonian rocks at the beginning of catagenesis.

Clarification of possible ways of hydrocarbon migration through the transit terrigenous strata of the Lower and Middle Devonian.

Construction of a basin model in the TEMIS and assessment of the hydrocarbon potential of the studied area.

To assess the residual oil and gas potential of the Silurian strata, an integrated approach was applied using quantitative models of oil and gas generation developed by Neruchev S.G. [25,26] for the main genetic types of organic matter, which can be used as a basis for assessing the oil and gas formation of any potential oil and gas source strata.

Sapropel OM is characterized by the highest oil-producing potential (I-II types of kerogen according to international terminology).

Based on these models, in which the weight percentage of generated liquid and gaseous hydrocarbons in the mass of the initial substance is calculated.

In this study, the technology of TEMIS for basin modeling is used. The essence of modeling is to restore the geological processes that determine oil and gas formation [27–32].

The resource assessment by the volume-genetic method is based on the capabilities of a rock enriched with organic matter to generate hydrocarbons. This method evaluates the generation potential of the oil and gas source strata and the degree of its realization. The method of calculating the generated hydrocarbons by the oil-producing stratum located in the center of an oil and gas formation includes the calculation of the current content (mass) of organic carbon (C_c) in oil and gas-producing rocks, an estimate of the mass of generated hydrocarbons per unit mass of C_c, and an estimate of the total amount of generated hydrocarbons by the oil and gas-producing rock [25,26].

The theories and technologies of basin modeling have been successfully tested for a number of other regions [31,33–39].

4. Source Material

The research area is characterized by a very reduced range of industrial oil and gas content, established only on Kolguev Island, despite the fact that in the northern part of the land area of the TPP or the eastern part of the research area, the range of industrial oil and gas content is very significant—with deposits from the Lower Silurian to the Middle Triassic [40–42].

In the sedimentary cover within the studied area, the Cambrian-Lower-Middle Ordovician terrigenous, Middle Ordovician-Lower Devonian terrigenous-carbonate, Middle Devonian-Frasnian terrigenous, Domanik-Tournaisian carbonate, Upper Visean-Lower Permian carbonate, and the Upper Permian and Triassic terrigenous sediments that promise oil and gas complexes are distinguished from the bottom-up. The principal difference from the eastern regions is the reduced volume of the sedimentary cover from the Middle Devonian to the Jurassic deposits as well as the presence of graben-like deflections filled with Cambrian-Lower Ordovician terrigenous deposits (Figure 2) [30,43,44].

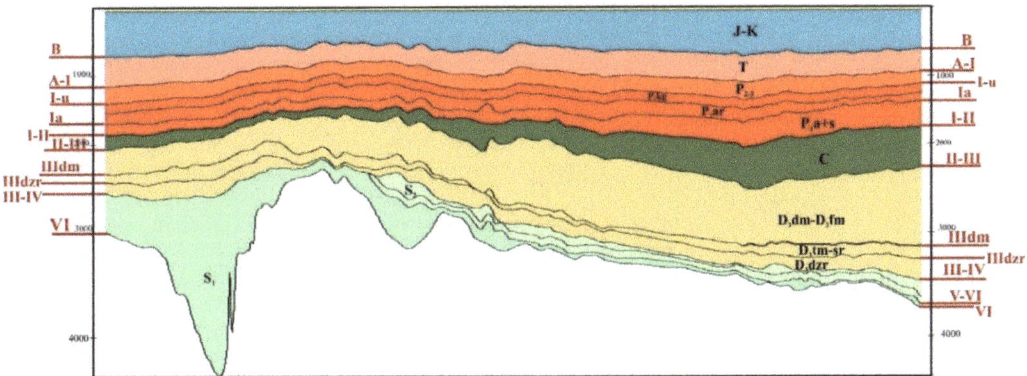

Figure 2. Geological and geophysical profile illustrating the structure of the sedimentary cover of the work area [40].

5. Prospects of Oil and Gas Potential

In accordance with the oil and gas geological zoning of the Timan–Pechora oil and gas province, the research area includes the Seduyakhinsko–Kipiyevsky oil and gas region (OGR), belonging to the Izhma–Pechora oil and gas region (OGR) (only a small fragment is presented in the territory of the study), and the West–Kolguevsky, Naryan–Marsky, and East–Kolguevsky OGR, which are the part of the Malozemelsko–Kolguevskaya OGR [45–47].

5.1. Izhma–Pechora Oil and Gas Region

Industrial accumulations have not been established in the north of the Izhma–Pechora OGR; to the south, in the territory of the Komi Republic, they were detected in a narrow stratigraphic range in the Upper Devonian carbonate deposits. In the extreme southwestern part of the Izhma–Pechora OGR, the Seduyakhinsko–Kipiyevsky OGR stands out.

The Seduyakhinsko–Kipiyevsky OGR is tectonically confined to the Seduyakhin disjunctive shaft complicated by the Seduyakhin–Yangyt bridge. The industrial oil and gas potential of the section within the OGR has not been established.

Prospects may be associated with the pre-Domanik terrigenous part of the section in the zone of the Charkayu–Pylemetsk deep fault, which controls the wedging of the Upper Devonian Dzhersky deposits.

5.2. Malozemelsko–Kolguevskaya Oil and Gas Region

The Malozemelsko–Kolguevskaya oil and gas region is one of the most extensive areas of the northern part of the Timan–Pechora province; the area of its marine part alone (together with Kolguev Island) is 25.6 thousand km^2. In tectonic terms, it corresponds to the monocline of the same name, inclined to the north and northeast; its regional slope is 15–17 m/km. Sedimentary cover structures are expressed in the form of steps and shafts. Most of them correspond to ledges or projections of the foundation. The amplitudes of the shafts, which have a north-western strike similar to other tectonic elements that make up the monocline, vary from 50 to 250 m; their length reaches 100–140 km.

The sedimentary thickness of the monocline increases in the northeast and northwest, from 500–1000 to 4500–5500 m. It is distinguished by differences in the lateral distribution of the lower part of the sedimentary cover and the upper part before and after the Visean deposits. There is a wedging of deposits of most horizons to the southern border of the region, the stratigraphic breaks in the pre-Permian interval.

There are three districts within the oil and gas region: West–Kolguevsky, East–Kolguevsky, and Naryan–Marsky.

The East–Kolguevsky OGR in the study area is tectonically confined to the West–Kolguevsky trough, overlapped by the West–Kolguevskaya monocline. No accumulations of oil and gas were detected within the district. A distinctive feature is the sharply reduced thickness of most parts of the Paleozoic section and the predominance of coastal and coastal-marine terrigenous sediments for complexes with a predominantly carbonate composition to the east.

Prospects for oil and gas potential can be associated, first of all, with the terrigenous Upper Permian-Triassic reservoirs formed by sandstones of deltaic and alluvial-lacustrine origin, as well as non-structural traps in the terrigenous formations of the Silurian-Lower Permian interval of the section [46,47].

The Naryan–Marsky OGR tectonically corresponds to the Malozemelskay monocline (in the study area, it is isolated within the Sengei structural nose and the Naryan–Marsky stage, complicated by the South Sengei and East Seduyakhin ledges) and the Udachnaya stage.

The prospects of oil and gas potential are associated with both the Paleozoic and Mesozoic parts of the sedimentary cover section. Zones of oil and gas accumulation formed within promising oil and gas complexes are characterized by inheritance, which is associated with the peculiarities of the tectonic development of the territory [2,46].

The East–Kolguevsky OGR tectonically corresponds to the East–Kolguevsky structural region.

The oil and gas potential of the Lower Permian and Lower Middle Triassic deposits has been established within the OGR. The Peschanoozerskoye and Tarkskoye deposits on Kolguev Island have been identified. The highest productivity was noted in the Triassic sandstones of the Peschanoozerskoye deposit at depths of 1300–1600 m. The prospects of the search are mainly associated with low-amplitude and medium-sized structures in the development zones of natural reservoirs with a mainly terrigenous composition.

6. Development of the Structural Frame

In the waters of the Pechora Sea and in the area of its junction with the land of the Timan–Pechora province, a modern complex of seismic exploration regional works was carried out, which made it possible to significantly refine the geological and geophysical model of the structure of the northwestern part of the Timan–Pechora oil and gas province and its marine continuation in the waters of the Pechora Sea.

Interpretation of seismic data and borehole materials was made for the purpose of structural constructions on reflecting horizons (RH): VI (foundation surface), V-V1 (acoustic roof surface of Ordovician terrigenous deposits), IV (roof of Silurian deposits), III-IV (roof surface of Silurian-Lower Devonian deposits of different ages), III2 (surface of Middle Devonian-Lower Frasnian deposits), IIIdm (solely Domanik deposits), IIIfm1 (acoustic surface of Famennian deposits), IIv (roof surface of Visean deposits of the lower carboniferous), Ia (horizon, separating the carbonate and terrigenous part of the Lower Permian), A-I (the roof of the Lower Triassic), and B (the horizon in the Jurassic sediments). The seismogeological model created in the PC Kingdom was the basis for clarifying the tectonic and oil and gas geological zoning of the northern part of the Timan–Pechora province, including the waters of the Pechora Sea (Figure 3).

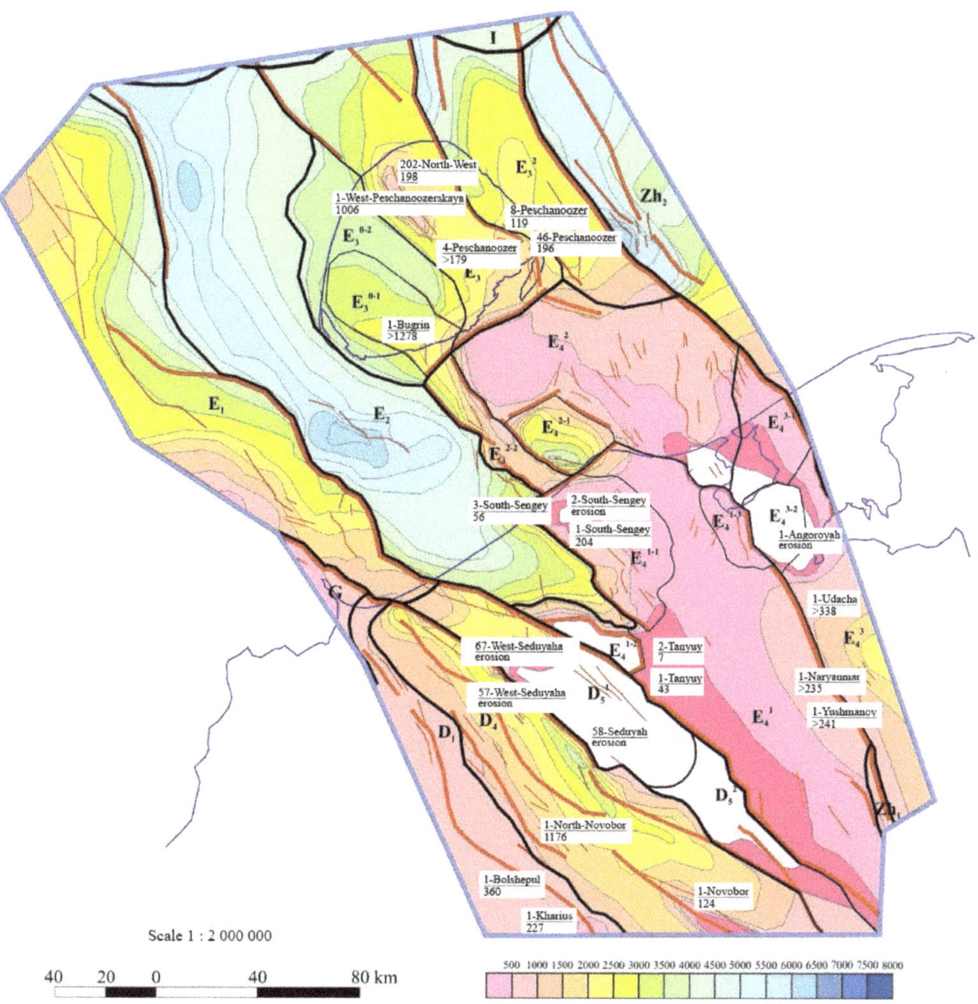

Figure 3. Tectonic scheme of the Timan–Pechora province.

The boundaries of the main tectonic elements were significantly clarified (Izhma–Pechora syneclise (north), the Malozemelsko–Kolguevskaya monocline, the Pechora–Kolva aulacogen (north), the Khoreyversko–Pechoromorskaya syneclise, the North Pechoromorskaya monocline, the Pripaikhoysko–Priyuzhnonovozemelsky trough, and the Gulyaev–Varandeyskaya structural-tectonic zone) primarily in the water part of the research area, and their subordination was established by the nature of the relationship. The discrepancy between the structural plans on the surface of the sedimentary cover and the surface of Permian-carbonates was then established, which determined significant differences in the control of oil and gas content of different structural stages (Figures 4 and 5).

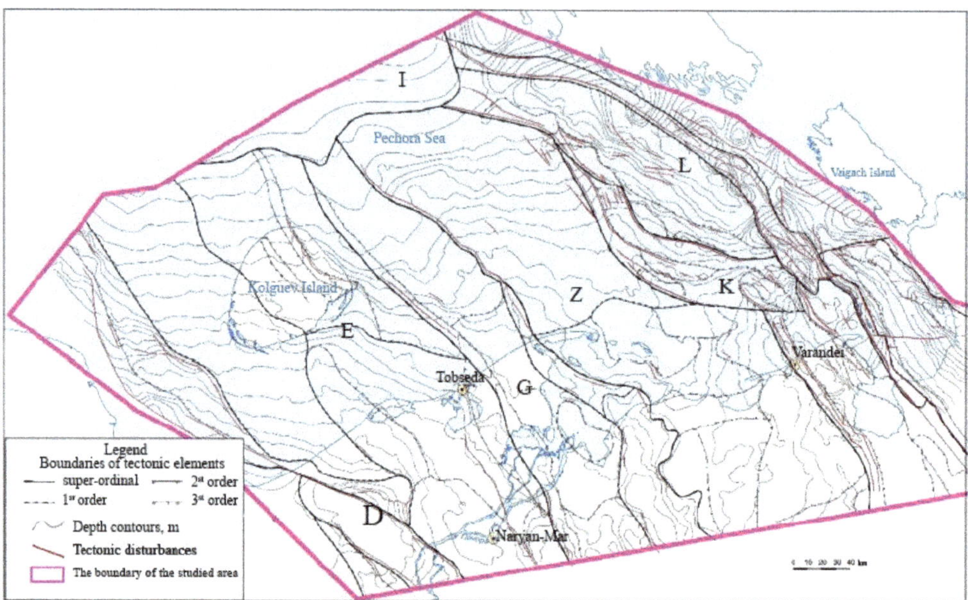

Figure 4. Structural and tectonic scheme of the northern part of the Timan–Pechora province (on the roof of the Permian-carbonates of RH-Ia).

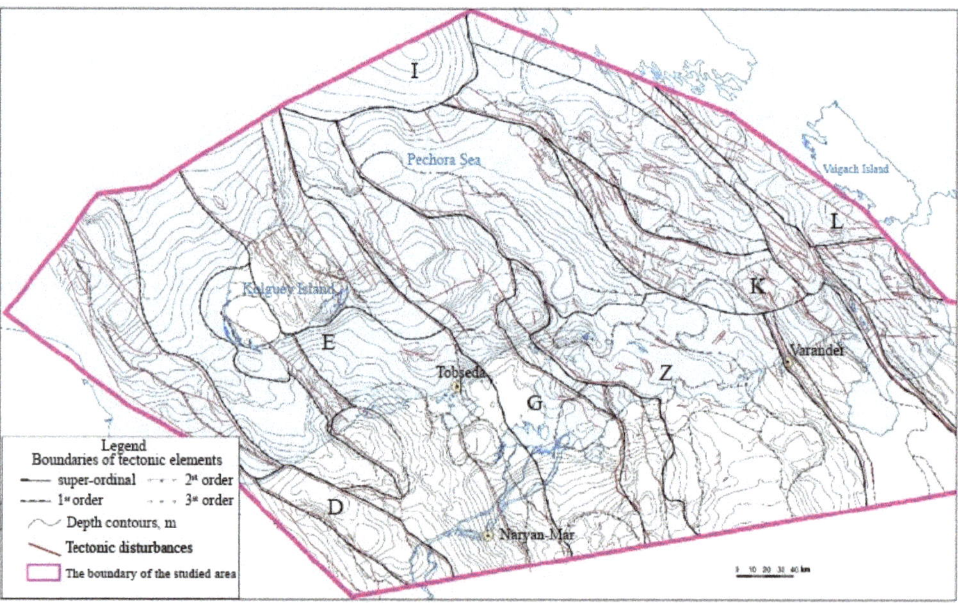

Figure 5. Structural and tectonic scheme of the northern part of the Timan–Pechora province (on the surface of the foundation of RH-Va).

The correlation of well sections along the lines of regional profiles allowed us to identify the features of the development of oil and gas complexes.

The restoration of the history of immersion and the isolation of oil and gas systems (the evolution of HC generation centers), noting the catagenetic zonality for the Barents Sea and North Kolguev generation centers, allowed us to draw conclusions about the zonality in the distribution of the phase composition of fluids in the water part of the basin (mainly oil fields were identified in the eastern part; gas and gas condensate fields were identified in the western part). It is assumed that there could be several centers of oil and gas formation, including the possible influence of the Domanik-Famennian oil and gas source rocks, located to the north of Kolguev Island.

Regional structural constructions made it possible to clarify the tectonic and oil and gas geological zoning (Figure 6) [46,47].

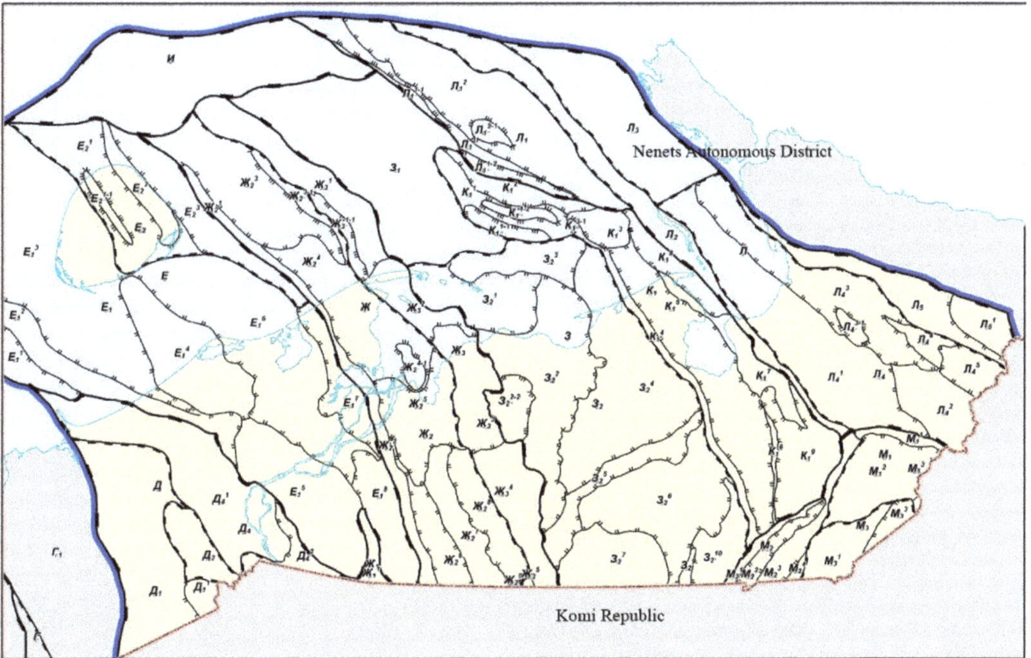

Figure 6. Scheme of the tectonic zoning of the northern part of the Timan–Pechora province and the adjacent water area.

7. Promising Oil and Gas Complexes and Oil and Gas Generating Strata of the Lower Paleozoic Part of the Section

The composition, structure, and ratio of rocks comprising natural reservoirs (NR) are determined by their genesis: the origin of certain types of precipitation, the frequency of sedimentation, the nature of secondary transformations. The genetic approach is an effective method of forecasting reservoirs as a whole and its individual elements. The most significant formation with generation potential is the Domanik-Famennian-Tournaisian complex. It also includes the Domanik oil and gas generating strata, which is fundamental for the province; its potential for oil is estimated at 70% of its contribution to the formation of TPP deposits, and more than 50% for that of gas.

In the sedimentary cover within the studied area, the Cambrian Lower-Middle Ordovician terrigenous, Middle Ordovician-Lower Devonian terrigenous-carbonate, Middle Devonian-Frasnian terrigenous, Domanik-Tournaisian carbonate, Upper Visean-Lower Permian carbonate, and the Upper Permian and Triassic terrigenous sediments that promise oil and gas complexes are distinguished from the bottom to the top [5,39].

7.1. Cambrian-Lower-Middle Ordovician Terrigenous Complex

The basal Lower-Middle Ordovician terrigenous complex is established on the Bugrinsky dome in the East Kolguevsky block and is assumed to be in the West Kolguevsky trough on the basis of the general patterns of development of quartz sandstone formations in posthumous post-Baikal subsidences known in the south in the Izhma-Pechora syneclise. The thickness of this complex in the depocenter of the deflection, which reaches up to 2.0 km on the tops of the Sengei Mountain and the South Sengei Dome, is reduced to tens of meters.

Terrigenous Ordovician deposits are characterized by data from drilling wells in the areas: No. 1–Bugrinskaya, No. 1—Bolshepulskaya, No. 1, 2—Khariusny, No. 1—Novoborskaya. They are mainly represented by terrigenous sediments of considerable capacity. The development within the studied area is limited mainly by the West Kolguevsky paleodeflection as well as by the paleotrough located along the Timansky ridge.

Within the water area, the oil and gas complex (OGC) is represented in the volume of the Upper Cambrian Lower-Middle Ordovician deposits (Sedjel formation of the Tremadocian stage, the Nibel formation of the Arenigian stage, and the conditionally isolated Lanvirnian stage). The deposits of the complex with angular disagreement lie on the foundation of the Proterozoic age and are represented by sandy-clay deposits of the littoral and supralithoral. It is assumed that the development of the complex in the modern water area is wider than on land. The presence of the Middle and Upper Ordovician is established in the West Kolguev trough, where the presence of Lower and Middle Devonian is also assumed in its uncompensated central-part formations of coastal-marine and deltaic sandstones. The thicknesses of the complex vary from 1.5–2.0 km in the West–Kolguevsky trough, to 0.5–1.5 km on its slopes. It is completely absent on the Bugrinsky buried dome (north of the Bugrinsky stage) and on the local elevations of the Sengei shaft and the Udachnaya stage. The lithological composition of the complex naturally changes from terrigenous-carbonate to the west and south of Kolguev Island, to mainly carbonate in the north of the Udachnaya stage. A characteristic feature of the sedimentation conditions is the compensated filling of the basin, which was provided by the demolition of detrital material from the Timan and the elevated Malozemelsky block. The most powerful section of the undifferentiated Sediol and Nibel formations of the Lower Ordovician, and possibly the Middle Ordovician terrigenous deposits with a total capacity of at least 1278 m, was opened by well No. 1—Bugrinskaya on Kolguev Island. In the upper part of the section, the role of significantly clayey, carbonate-clay bundles with a thickness of 5–40 m separating sand layers with a thickness of 5–120 m is noticeably increasing.

7.2. Middle Ordovician-Lower Devonian Terrigenous-Carbonate OGC

The province is allocated with deposits in the volume of the Upper Middle-Upper Ordovician, the Silurian and Lower Devonian, and in the volume of the Lower and Upper Silurian and Lower Devonian of the studied area. The deposits of the Lower Silurian (Lower Devonian), represented by carbonate and terrigenous rocks, were discovered by drilling only in the Naryan–Marsky, Udachnaya, and South–Sengei areas. The boundary of the complex is erosive and corresponds to the surface of the Lower Devonian, Silurian, and Ordovician deposits of different ages. The complex is very limited in distribution within the land of the studied area, and more significantly, in the water area. At the same time, its volumes are incomplete due to the partial erosion of the Lower Devonian, preserved only in the Western Kolguevsky trough and in the northeastern part of the Udachnaya stage [48].

Silurian deposits within the research area were discovered only on the Malozemelsko–Kolguevskaya monocline, where it has a predominantly carbonate composition and a total thickness from 0 to 500 m. In the volume of the Silurian system, the lower and upper sections are distinguished.

The deposits of the Lower Devonian are represented by fragmentary deposits of the Lohkovsky stage, presumably in the volume of the Ovinparmian horizon only in the extreme eastern sections of the research site. Its presence in the section is proved by drilling data on the Peschanoozerskoye area in wells No. 4 (thickness, 46 m) and No. 46 (thickness, 256 m). The rocks are represented by carbonate-clay and sulfate-carbonate-clay deposits in lagoons in the West Kolguevskaya depression and the Peschanomorskoye structural zone.

7.3. Middle Devonian-Frasnian Terrigenous OGC

The complex in the work area has a reduced stratigraphic volume. According to seismic data, there are practically no deposits of the Middle Devonian within the studied object. The Middle Devonian sandstones are assumed to be present only in the West Kolguevsky trough. The Lower Frasnian subcomplex (Dzhersky horizon) is almost ubiquitous and has been opened by wells of the Tanuyskaya, Pylemetskaya, and South Sengei areas. The thickness of the horizon reaches 150–200 m. Timan–Sargaev carbonate-terrigenous deposits cover the entire territory of the considered area, but due to significant sandiness they sometimes lose their qualities of a regional cover, preserving them in the east and locally in the West Kolguevsky trough (Figure 7).

Above the section lies the Domanik-Tournaisian terrigenous-carbonate OGC, which is not the subject of study.

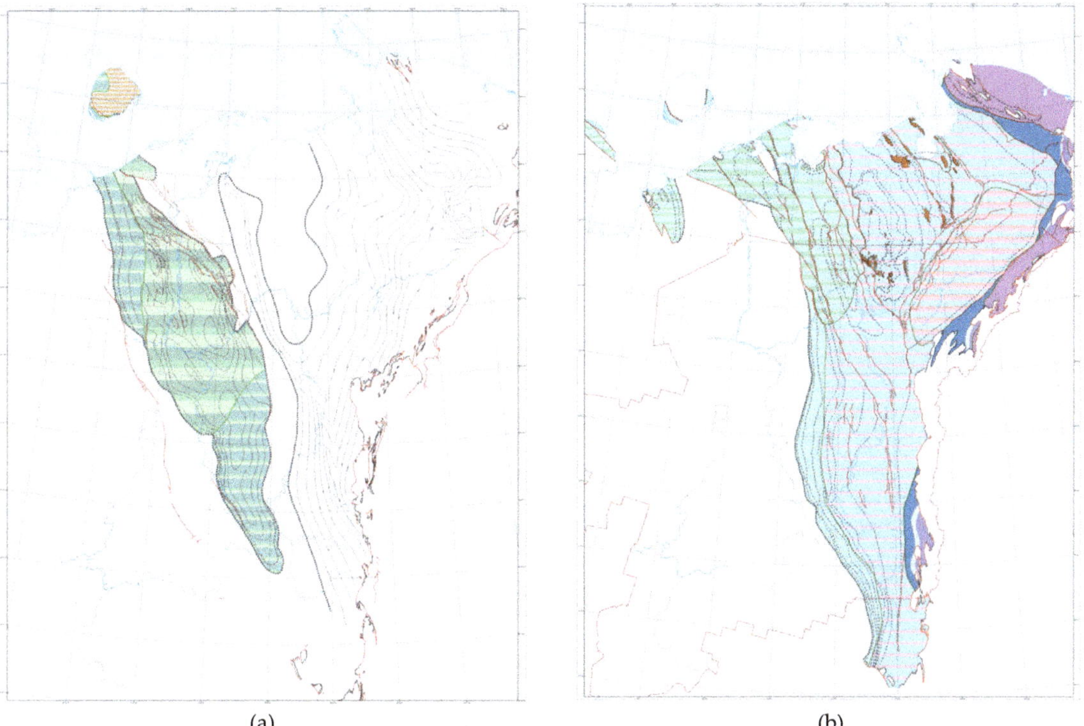

(a)   (b)

Figure 7. Lithological-facies scheme: (**a**) of the Cambrian Lower-Middle Ordovician terrigenous complex of the TPP; (**b**) of the development of the Middle Ordovician-Lower Devonian carbonate-terrigenous complex of the TPP.

Sedimentation conditions and patterns of distribution of natural reservoirs and oil-generating strata.

The composition, structure, and ratio of the rocks comprising the NR are due to its genesis: the origin of certain types of precipitation, the frequency of sedimentation, and the nature of secondary transformations. The genetic approach is an effective method of forecasting reservoirs as a whole and its individual elements (Figures 8 and 9).

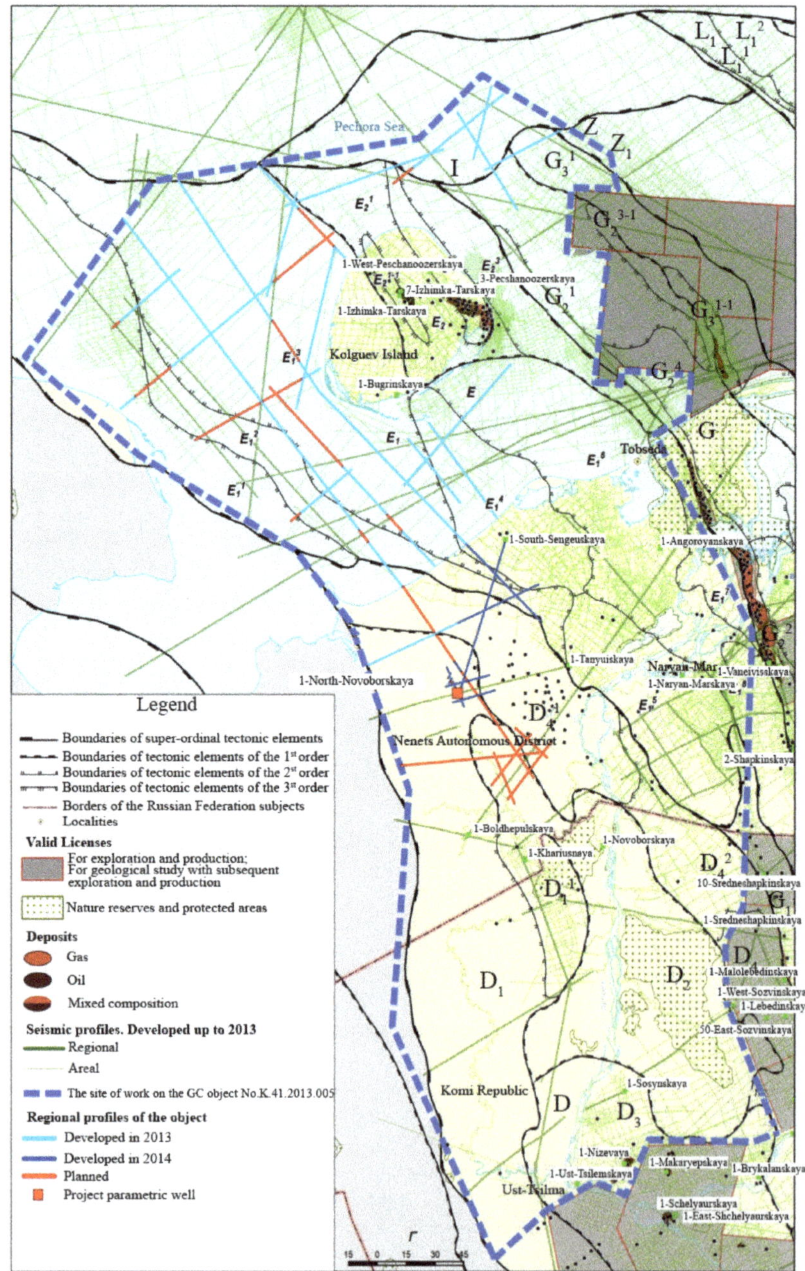

Figure 8. The scheme of seismic exploration of the research area.

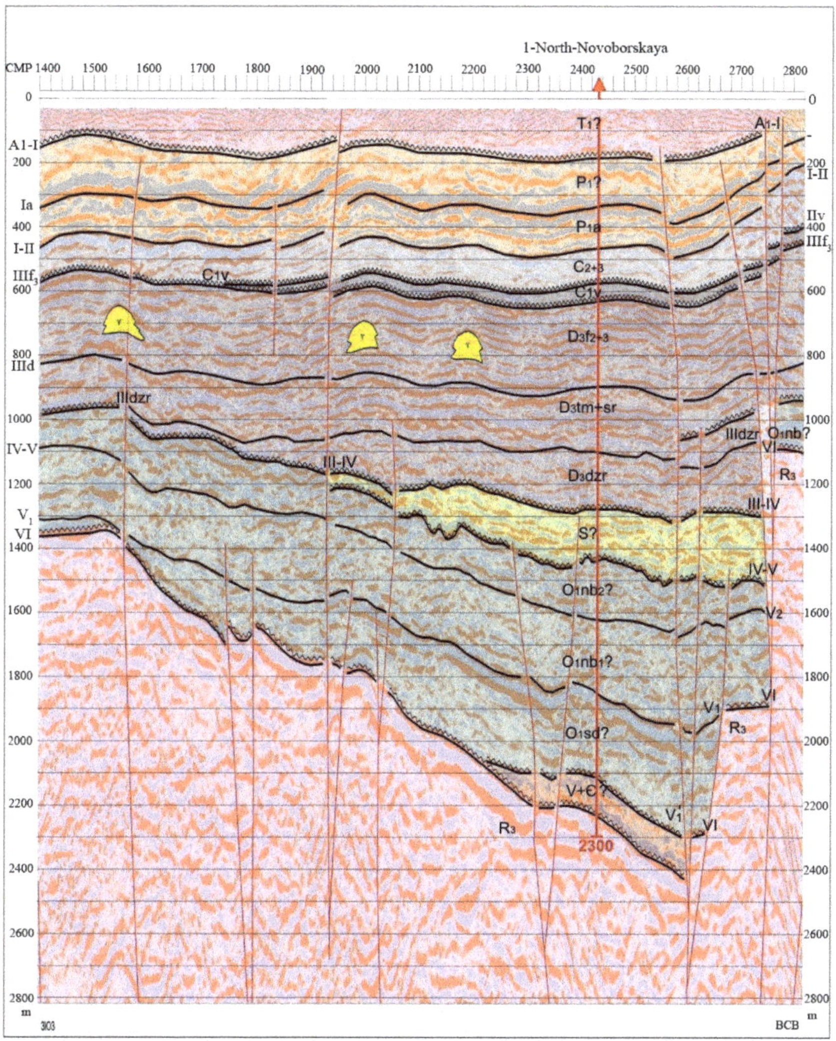

Figure 9. Seismogeological section within the basin with the largest thicknesses of the Lower-Middle Ordovician deposits.

On the basis of a comprehensive interpretation of geological and geophysical data within each oil and gas complex, the patterns of distribution of reservoir rocks and fluid traps forming natural reservoirs of various scales have been revealed.

7.4. Natural Reservoirs of the Lower-Middle Ordovician Terrigenous Oil and Gas Prospective Complex

The reservoir strata accumulated in the mode of compensated sedimentation. The alternation of layers with different filtration-thickness characteristics is rhythmic in nature due to the seasonal cyclicity of their sedimentation and/or the uneven speed, and possibly the multidirectional nature of the shoreline at the initial stages of filling the basin. Bundles with optimal reservoir potential are confined to the tops of rhythms corresponding to the formation of "washed" precipitation during relatively calm periods of sedimentation.

In the sandstones of the Lower Ordovician interval, similar in genesis to the south of the Izhma-Pechora syneclise (well No. 1—Sosyanskaya, No. 40—Khabarikhinskaya, No. 1—Dvoynikovaya), sandstone reservoirs are characterized by an open porosity coefficient of 20% (according to the sections). According to the results of laboratory studies of the core, the open porosity goes up to 14%, and the permeability ranges from 3.3 to 6.8 m up to 519 mD.

At the last stages of compensated filling, the Ordovician bay morphologically represented a peneplenized littoral-supralithoral zone. The leveled bottom relief and impeded water exchange over large areas caused the formation of substantial clay deposits (the Nibel formation) with sufficient thickness, which have satisfactory screening properties and can act as clay screens with a local and possibly a zonal distribution. It should be noted that in the clay column, there may be low-power sand–silt interlayers associated with the seasonal activation of sedimentation, and/or interlayers of carbonate–clay composition caused by the periodic or seasonal increase in salinity of the basin waters. In both cases, the reliability of the cover is reduced. These interlayers can act as a reservoir with a low filtration capacitance potential. The NR of the Lower-Middle Ordovician terrigenous complex can be characterized by a three-layer structure.

Noting the power of the scattering thickness (up to 300–400 m), with the exception of areas where there are no Nibel deposits, hydrocarbon deposits can be confined to high-amplitude structures. Since the Sediol formation strata most likely do not have their own oil-generating potential, the deposits are most likely near fault zones, where younger Silurian (Upper Silurian) deposits containing a sufficient total amount of organic matter are hypsometrically lower than the Lower Ordovician reservoirs.

In the volume of the regionally isolated Upper Ordovician-Lower Devonian terrigenous-carbonate OGC within the object of study, the Silurian carbonate and Lower Devonian carbonate-terrigenous subcomplexes were identified. There are no carbonate deposits of the Upper Ordovician in the studied area.

7.5. Natural Reservoirs of the Silurian Carbonate Subcomplex

The oldest deposits of the complex, uncovered by wells in the area under consideration and in adjacent areas, are of Lower Silurian age.

Facies of the littoral zone with mixed sedimentation, represented by variegated gray colored carbonate-terrigenous sediments, are found on the South Sengei ledge and on the Tanuyskaya area.

In the variegated gray colored carbonate-terrigenous formations of the Lower Silurian, the reservoirs are of quartz sandstone layers.

The presence of reservoirs in the Silurian deposits of the object was also confirmed further north in well No. 3—South-Sengei, where 11 reservoir layers with an effective capacity from 2.4 m to 5 m, and porosity from 5–8% to 12–16% were identified.

A comprehensive analysis of drilling materials and the results of field-geophysical studies allowed us to identify two main groups of development zones of the Silurian natural reservoirs. The first is the development zone of terrigenous-carbonate low-medium-thickness reservoirs under the regional Timan–Sargaev cover, spatially connected with high projections of the foundation (South Sengei and Tanuyskaya areas). On the Tanuyskaya area, Silurian rocks most likely form a hydrodynamically connected (single) reservoir with Dzhersky sandstones. On the arches of the paleomorphological ledges, the hydrodynamic activity of the waters was quite active, which led to the accumulation of desiccated sediments and significantly reduced the fluid-resistant properties of the cover.

The second group (II) is the development zone of carbonate low-medium-thickness reservoirs under the regional Timan–Sargaev and intraformational Silurian (Upper Silurian) cover. It is spatially connected with the development zone of shelf facies in the eastern part of the object and in its water area.

7.6. Natural Reservoirs of the Lower Devonian Carbonate-Terrigenous Subcomplex

The Lower Devonian interval of the section is absent from the sections of the wells of the studied area. The forecast of the complex reservoirs was carried out on the basis of data from seismic materials analyzed from the point of view of regional patterns of development of the Timan–Pechora paleobasin.

The nature of sedimentation in the Lower Devonian was due to extensive regional regression.

The zones of possible reservoir development are associated with areas of increased thickness (depocenters) of the Lower Devonian sediments, which (according to the basic principles of the paleogeomorphological method) can be considered as the main directions for forecasting sand formations (lenses, layers). The reservoir characteristics of sandstones can be significantly reduced by the abundant clay and/or carbonate-clay cement expected in the sediments of reservoirs with difficult water exchange.

In the north-western regions of the object, the possible presence of local, low-thickness, and mainly terrigenous NR of a complex type, which are lenticular and possibly mantle-shaped, is predicted.

The screening strata of the local type in the Lower Devonian complex are most likely of poor quality, associated with their significant carbonate content and areas, possibly comprising sulphate.

7.7. Natural Reservoirs of the Middle Devonian-Lower Frasnian Terrigenous Oil and Gas Complex

The Middle Devonian-Frasnian sub-Domanik oil and gas complex in the work area has a reduced stratigraphic volume. According to seismic data, there are no deposits of the Middle Devonian within the studied object. The Lower Frasnian subcomplex is almost ubiquitous and has been opened by wells of the Tanuyskaya, Pylemetskaya, and South-Sengei areas.

By its origin, the Lower Frasnian reservoir is polyfacial: coastal-marine and shallow-shelf. The reservoir potential is mainly associated with the Dzhersky and Lower Timan intervals, which, in the absence of the Middle Devonian, acquire the value of a search object [49,50].

The Dzhersky terrigenous deposits formed under coastal conditions with sufficiently active hydrodynamics in the shallow shelf that existed in most of the studied territory. The Dzhersk formations are characterized by the high sandiness of the section as well as the presence of laterally sustained and most likely mantle-shaped sand bodies with good reservoir potential. Reservoirs are mainly pored. The open porosity can vary from 5–6% to 15–18%. The permeability is intergranular.

The Dzhersky and Lower Timan sand layers, overlain by clays of the Upper Timan–Sargaev age, form a single hydrodynamically connected reservoir. The reservoir potential of this stratum is confirmed by the results of testing in wells.

8. Basin Modeling

The natural geological bodies of the formation level—where the processes of oil and gas formation that took place (and/or are taking place) are oil and gas source formations or horizons (OGSH)—are responsible for the oil and gas potential of various regions. The total oil and gas-producing potential of a particular oil and gas basin (OGB) should be estimated by the amount of OM contained in OGSH. The OGSH are characterized by certain concentrations of OM, types of OM, and a certain volume, i.e., thickness and area of development, as well as the maturity of OM (catagenesis). In order for OGSH (or their combination) to become a focus of oil and gas formation (OGF), the minimum values of HC migration densities from them should be at least 50 thousand tons/km^2 (50 million nm^3/km^2) under ideal accumulation conditions (the immediate vicinity of the reservoir and its optimal thickness). In general, the border focal density of migration can be assumed to be 100 thousand tons/km^2 (100 million m^3/km^2). In order to identify the localization of OGF, it is necessary to determine the area of development and the thickness of the OGSH as well as the concentration of OM (C_{org}) in them; it is also necessary to identify the degree of catagenesis of OM, and with the use of the created models of the generation-migration of HC, to calculate the scale of migration for the type of OM corresponding to this OGSH.

On the territory of the TP SB, the Domanik-Tournaisian formation and the Silurian formations (29% and 25% for oil, respectively) play the main role in terms of the scale of oil and gas formation; Lower Permian formations (12% for oil and 22% for gas) can be attributed to less significant ones.

The Domanik-Tournaisian formation was continued from the continent to the sea on the basis of facies constructions and seismic data, according to the results of which reef bodies were identified, outlining the zone of the pre-reef facies of the sublatitudinal strike. Thin layers of domanikoid rocks are found in the section of the clay-carbonate deposits of well No. 1—Pakhancheskaya, which is typical for some areas of the reef facies near the reef body [50]. These interlayers are enriched with OM and are not of interest from the standpoint of large-scale oil and gas formation, but they indicate the relative proximity of the development of domanikoids of the formation scale. The thickness of the Domanik OGSH increases in general from west to east—from 20 to 300 m—as the age range of the formation expands. The average concentrations of Corg vary in the range of 0.6–1.5%. In the work area, the Domanik-Tournaisian OGSH is characterized by the following degrees of catagenesis of OM: MK21 and MK22 [49].

The mudstones of the Lower Ordovician and Upper Cambrian are characterized by shallow-sea facies formation conditions and sapropel organic matter. At the same time, a large proportion of algal OM is noted in Ordovician samples. The degree of transformation of rocks according to bituminology remains relatively low, but the correctness of these definitions is questionable, especially given the large spread of analysis results.

The modern geothermal regime of the TP SB is characterized by the relatively low intensity of the thermal field; the average geothermal gradients in the section do not usually exceed 2.5 °C/100 m [25]. In the TPB, the so-called abbreviated ("subdonets") zonality of the catagenesis of OM [49] was established due to the intense geothermal regime in the past (paleogradients of the order of 5 °C/100 m), which is characteristic of Paleozoic basins. Table 1 shows the paleodepth dimension of such a scale compared with measurements of the reflectivity of vitrinite R_a (measurements carried out in oil) and R_o (measurements carried out in air) (Table 1) [49,50].

Table 1. The dimension of the Paleozoic catagenetic scale of the Timan–Pechora basin and its comparison with the optical indicators of vitrinite [35,36].

Diving Depth Intervals, km	Catagenesis Substages	Gradations of Catagenesis	The Reflectivity of Vitrinite	
			R_a, %	R_o, %
Up to 1.0	Protocatagenesis	PK_{1-2}		
1.0–1.5		PK_3	Up to 7.0	0.45 (up to 0.50)
1.5–2.4		MK_1	7.0–7.6	0.50–0.65
2.4–2.8		MK^1_2	7.6–7.9	0.65–0.73
2.8–3.2	Mesocatagenesis	MK^2_2	7.9–8.2	0.73–0.83
3.2–3.8		MK_3	8.2–9.0	0.83–1.12
3.8–4.1		MK_4	9.0–9.8	1.12–1.51
4.1–4.4		MK_5	9.8–10.7	1.51–1.98
4.4–5.0	Apocatagenesis	AK_1	10.7–11.5	1.98–2.45
5.0–6.5		AK_{2-3}	11.5–14.0	2.45–5.50

The scheme of the distribution of the centers of oil and gas generation of hydrocarbons in the water area of the Timan–Pechora basin is shown in Figure 10.

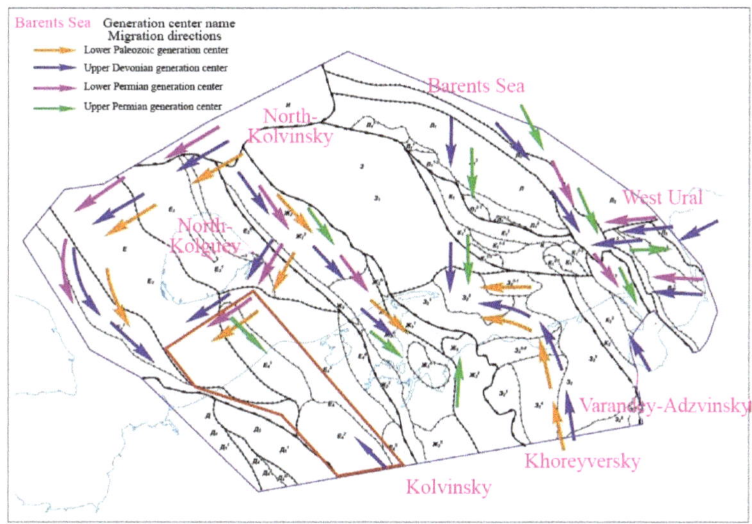

Figure 10. The main centers of generation of the north of the TPP and the directions of migration.

The thicknesses of the OGSH of the Silurian complex in the work area vary in the range of 50–400 m, while the direction of thickness changes and their gradients are aligned with the seismic map of the total thicknesses of O_2-D_1. Average concentrations of C_{org} in the OGSH (S_{1-2}) vary in the range of 0.25–0.60% (isocarbs 0.3; 0.5). The organic matter in the Silurian OGSH is characterized by the degree of catagenesis of MK_3 (Figure 11).

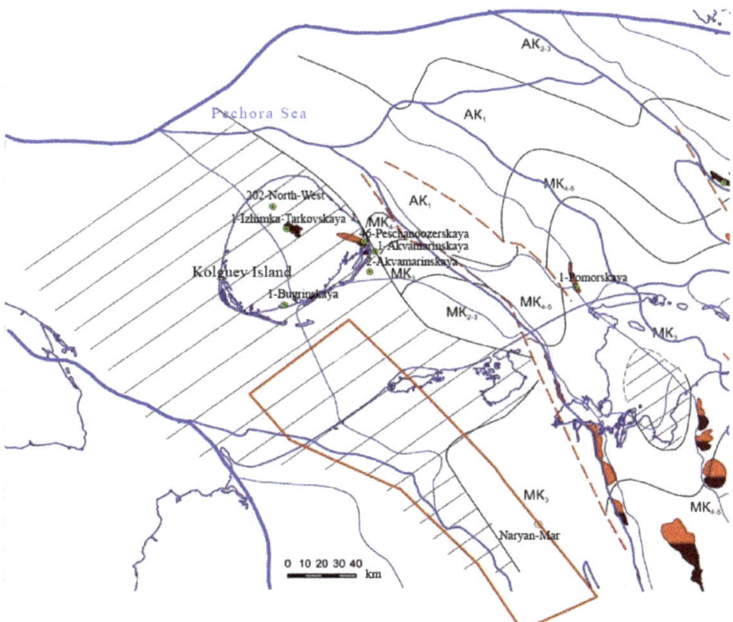

Figure 11. Scheme of catagenesis of OM on the roof of Silurian deposits.

In most of the territory of the TBP, the catagenetic scale has a shortened, so-called subdonets character corresponding to the maximum paleogradient of 5 °C/100 m. The scale is justified by numerous data on the definitions of the reflectivity of vitrinite.

The data of analytical geochemical studies made it possible to create a basin model for the studied region. The modeling of the processes of generation, migration, and accumulation of hydrocarbons was carried out using the TemisFlow software package. The geological history of the basin is modeled with backstripping technology, which involves restoring the history of the basin's dives by sequentially removing structural horizons. The modeling of the geochemical history is carried out based on the kinetic model of hydrocarbon transformation based on the work of the French Institute of Petroleum (IFP). The applied modeling method makes it possible to link disparate geochemical and structural materials to assess the qualitative characteristics of oil and gas source rocks and the degree of realization of their potential, as well as to obtain approximate but reasonable quantitative data on the scale of generation and migration of hydrocarbons.

As a result of the work carried out to assess the oil and gas source potential in the Lower-Middle Paleozoic section of the MKM, the main prospects for the formation of possible hydrocarbon accumulations are associated with Silurian oil-producing strata. It was not possible to separate the Lower and Upper Silurian deposits at the research site, and the characteristics of their oil and gas production potential are given together.

To the west of the fault separating the structures of the Pechora–Kolva Aulacogen from the Malozemelsko–Kolguevskaya monocline, a band of the absence of oil and gas source strata is mapped. The scale of the thicknesses of the OGSH of the Silurian complex in the work area is 50–400 m, while the direction of change in thicknesses and their gradients are aligned with the seismic map of the total thicknesses of O_2-D_1. Average concentrations of C_{org} in the OGSH (S_{1-2}) vary in the range of 0.25–0.60%.

To create the model, the constructed structural maps for the main reflecting horizons and thickness maps between them were used, the degree of erosion as a result of large interruptions in sedimentation was estimated, and prospective source strata, forecast reservoirs, and fluid barriers of the lower structural floor of the MKM were estimated (Figure 12).

At the initial stage of the work, the 1D modeling method was also used to evaluate the catagenetic transformation of the OM according to deep-drilling data. The modeling of the temperature history of catagenetic changes in rocks was based on T.K. Bazhenova's opinion that in Paleozoic basins, there was a reduced catagenetic zonality due to an increased paleogradient of temperatures of up to 4.5–5.0 °C/100 m, as well as data on the modern thermal gradient for mobile foundation blocks (which was the MKM) at 2.4–2.7 °C/100 m [6,46].

To assess the reliability of fluid traps and the generation potential of oil source rocks, the results of the logging data analysis for deep wells extrapolated to the territory of the entire site were used. The surfaces of fluid traps and oil source rocks were calculated by the method of convergence in the gap between the main structural horizons, by a proportional division of the thickness between the main deep surfaces. The correction of the depth reference was made based on the borehole data, where it is possible. To assess erosion, it was assumed that the maximum apparent thickness of the eroded complex was almost equal to the thickness of the complex before erosion. Furthermore, the amount of erosion was calculated as the difference between the maximum thickness and the current thickness. At the same time, the paleoterrain at the time of the formation of the analyzed complex was noted, and the thickness of the filling of paleotroughs was excluded from the calculation, for which the correlation of additional horizons was performed [40,48,50].

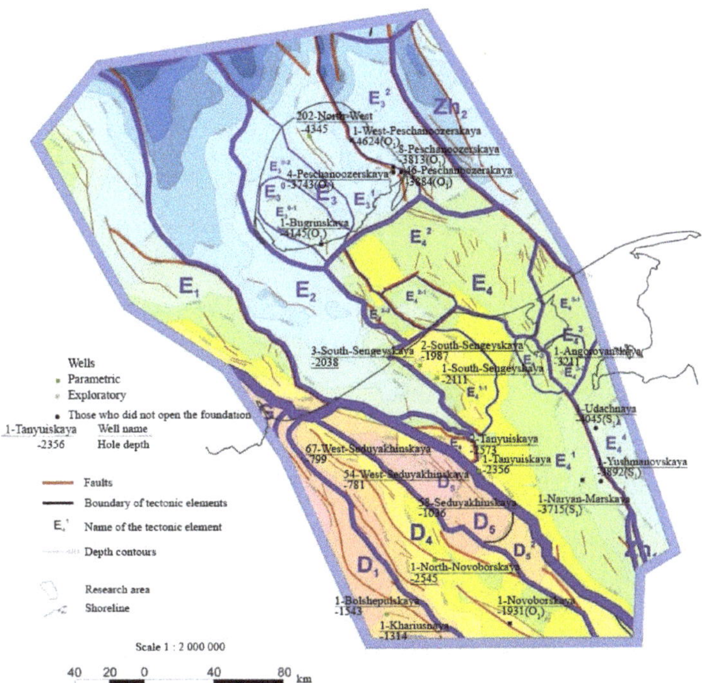

Figure 12. Structural map of the foundation (RH VI) with tectonic zoning [41].

Tectonic elements: Suborder: G—Timan ridge, D—Izhma–Pechora syneclise, E—Malozemelsko–Kolguevsky megablock (monocline), Zh—Pechora–Kolva megablock (aulacogen), I—North Pechora Sea monocline; 1st order: D_1—Neritsky stage, D_4—Novoborsko–Sozvinskaya structural zone, D_5—Seduyakhinsko–Malolebedinsky disjunctive megalithic bank, E_1—Korginskaya stage, E_2—West–Kolguevskaya depression, E_3—Kolguevsky block, E_4—Malozemelsky block, Zh_1—Pechora–Kozhvinsky graben, Zh_2—Denisovsky block; 2nd order: D_{51}—Seduyakhinsky disjunctive shaft, D_{52}—Seduyakhinsko–Yantygsky bridge, E_{30}—West Kolguev uplift, E_{31}—Kolguev structural zone of horsts and grabens, E_{32}—Peschanomorskoye structural zone, E_{41}—Naryan–Marsky stage, E_{42}—Udachnaya stage, E_{43}—Kharitseysko–Shapkinskaya stage; 3rd order: E_{30-1}—Bugrinsky dome, E_{30-2}—West–Bugrinsky stage, E_{41-1}—South Sengei dome, E_{41-2}—East Seduyakhinsky ledge, E_{41-3}—Nerut graben, E_{42-1}—Sengei graben, E_{42-2}—Sengei horst, E_{43-1}—Khareysky stage, E_{43-2}—South Anorgayakhsky dome.

The basin model was carried out on the basis of the data on the geochemical characteristics (including lithology, type of organic matter, and the history of immersion).

It was revealed that the organic matter in the Silurian OGSH in the work area is characterized by a degree of catagenesis from MK_2 to MK_3, while only in the extreme northern periphery is it up to MK_{4-5}. The main center of generation is shown in Figure 13.

The identification and substantiation of oil source strata in the modeling process was a difficult task due to the small size of the work area and its marginal position in the oil and gas basin. Belonging to the reservoir, fluid trap or oil source strata was determined based on the lithological characteristics of the strata and the content of C_{org}. Bundles in the Upper Silurian and Upper Devonian Domanik deposits have been tested as oil and gas source strata [3,47,49].

The Upper Silurian deposits in the work area are a layer of clay-carbonate interlayer, which was studied in natural outcrops along the Dolgaya River. An analysis of the GC curve of well No. 1—Naryan–Marsky allowed us to take the thickness of the interlayers with the oil source potential as 1/3 of the total thickness of the Upper Silurian deposits. The Lower Devonian sediments developed locally; they do not have widespread distribution. Their development is associated with the most loaded sections of the pre-Timan structural stage.

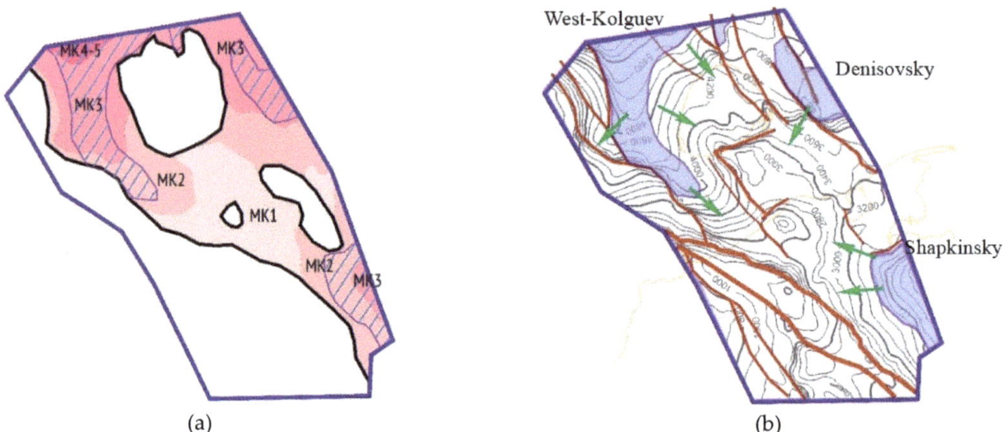

Figure 13. (**a**) Scheme of catagenetic zonality of Upper Silurian deposits [47]; (**b**) Scheme of the development of the center of generation and the migration direction of the HC of the Malozemelsko–Kolguevskaya monocline and the zones of its junction with adjacent territories [42].

9. Conclusions

Based on the interpretation of the results of seismic surveys carried out in the northwestern part of the TPP, including its water areas, a generalization of the results of deep drilling that included parametric well No. 1—Severo-Novoborskaya, structural maps, and thickness maps of all seismic facies complexes developed in the territory and the water area—which were used in basin modeling with the Temis software package for the purpose of reconstructing the conditions of immersion and the transformation of potential oil and gas formations—has been significantly refined, the most promising of which is the Silurian OGSH.

According to the created basin model, the total mass of hydrocarbon matter generated only by the oil source strata of the work area is 3.17 billion TOE. In addition, the modeling allowed us to get an idea of the time and conditions needed for the formation of large zones for the possible accumulation of hydrocarbons. Zones of HC paleoaccumulation are predicted based on their space–time relationships with possible centers of generation, on comparisons with periods of the intensive generation of HC both from space directly located within the work area and beyond (noting the possible migration), and on migration directions. It is concluded that it is necessary to search for transit routes of HC migration from the gas formation center located in the water area and the HC generation center located within the Pechora–Kozhvinsky megalithic bank.

According to the constructed model, the total generated mass of the HC substance within the work area is 3.17 billion tons of HC. This value displays the mass of hydrocarbons generated only by the site's own oil source strata, and it does not take into account hydrocarbons that could migrate from the outside, from the Pechora–Kolva aulacogen, or from the area north of Kolguev Island from the Domanik OGSH.

The conducted research allowed us to identify promising zones of oil and gas accumulation (PZOGA), spatially confined to the western slope of the West Kolguevskaya depression, the Sengei ledge, the zones of articulation of the Naryan–Marsky stage with the Nerut graben, and the Seduyakhinsky shaft. These PZOGA are considered as the primary objects of exploration for hydrocarbon deposits in the Middle Ordovician-Lower Devonian and Middle Devonian-Lower Frasnian oil and gas complexes. The most important geological feature of the development of these complexes is the formation of extended traps of the stratigraphic type (sometimes with tectonic shielding), which allows us to hope for the formation of large zones of hydrocarbon accumulation located in the frontal part of the migration flow from the submerged areas of the region.

Author Contributions: O.P.: clarification of oil and gas content prospects of lower-middle Paleozoic sediments of the Timan–Pechora oil and gas province. Analysis of the development of oil and gas source strata, updating the assessment of the oil and gas potential (oil and gas resources) of the Arctic zone of TPP, writing the section "Prospects of oil and gas potential", "Basin modeling" identification of tectonic boundaries of cratons and platform plates according to the tectonic scheme of the consolidated foundation, and "Conclusions". Y.N.: compilation of the seismic profiles of the Arctic zone of TPP, processing of geophysical data, writing the sections "The economic significance of the resource development of the hydrocarbon potential of the of the Russian Federation" and "Source material". V.N.: processing of geochemical data, writing the sections "Development of the structural frame" and "Promising oil and gas complexes and oil and gas generating strata of the Lower Paleozoic part of the section". All authors have read and agreed to the published version of the manuscript.

Funding: The research was carried out with the support of a subsidy for the implementation of the state task in the field of scientific activity for 2021 No. FSRW-2020-0014.

Data Availability Statement: Not applicable.

Conflicts of Interest: The authors declare no conflict of interest.

References

1. Prischepa, O.; Borovikov, I.; Grokhotov, E. Oil and gas potential of the little-studied part of the north-west of the Timan-Pechora oil and gas province according to the results of basin modeling. *J. Min. Inst.* **2021**, *247*, 66–81. [CrossRef]
2. Prischepa, O.; Nefedov, Y.; Grokhotov, E. Prospects for oil and gas potential of the Timan-Pechora province's water area extension based on the results of a comprehensive generalization of regional geophusical studies. *Sci. J. Russ. Gas Soc.* **2020**, *2*, 6–15.
3. Prishchepa, O.M.; Ilyinsky, A.A.; Zharkov, A.M. *Assessment of Resource Potential and Directions of Study of Unconventional Sources of Hydrocarbon Raw Materials of the Russian Federation—Methodological Problems of Geological Exploration and Scientific Research Works in the Oil and Gas Industry: All-Russian Meeting*; VNIGRI: St. Petersburg, Russia, 2013.
4. Prischepa, O.M. Averyanova Conceptual base and terminology of hydrocarbons of shale strata and low-permeability reservoirs. *Geol. Geophys. Dev. Oil Gas Fields* **2014**, *6*, 4–15.
5. Jia, B.; Chen, Z.; Xian, C. Investigations of CO_2 storage capacity and flow behavior in shale formation. *J. Pet. Sci. Eng.* **2022**, *208*, 109659. [CrossRef]
6. Alekseev, A.D.; Zhukov, V.V.; Strizhnev, K.V.; Cherevko, S.A. Study of hard-to-recover and unconventional objects according to the principle of reservoir factory in the formation. *J. Min. Inst.* **2017**, *228*, 695–704.
7. Farukshin, A.A.; Ayrapetyan, M.G.; Zharkov, A.M. The Geological Uncertainties Evaluation Features of Late Stage Development Deposits in Lower Cretaceous Oil and Gas Bearing Formations of Western Siberia. In Proceedings of the Geomodel, Gelendzhik, Russia, 7–11 September 2020; pp. 1–5.
8. Kaminsky, V.; Zuykova, O.; Medvedeva, T.; Suprunenko, O. Hydracarbon Potential of the Russian Continental Shelf: Present-Day Condition and the Developmen Prospects Mineral Resources of Russia. *Econ. Manag.* **2018**, *1*, 4–9.
9. Mardashov, D.; Islamov, S.; Nefedov, Y. Specifics of well killing technology during well service operation in complicated conditions. *Period. Tche Quim.* **2021**, *17*, 782–792. [CrossRef]
10. Carayannis, E.; Ilinova, A. Russian Arctic Offshore Oil and Gas Projects: Methodological Framework for Evaluating Their Prospects. *J. Knowl. Econ.* **2019**, *11*, 1403–1429. [CrossRef]
11. Cherepovitsyn, A.; Lipina, S.; Evseeva, O. Innovative approach to the development of mineral raw materials of the Arctic zone of the Russian Federation. *J. Min. Inst.* **2018**, *232*, 438–444.
12. Belyakova, L.; Bogatsky, B.; Bogdanov, E.; Dovzhikova, E.G.; Laskin, V.M. *Foundation of the Timan-Pechora Oil and Gas Basin*; Kirov Regional Printing House: Kirov, Russia, 2008; p. 288.
13. Cherepovitsyn, A.; Metkin, D.; Gladilin, A. An algorithm of management decision-making regarding the feasibility of investing in geological studies of forecasted hydrocarbon resources. *Resources* **2018**, *7*, 43. [CrossRef]
14. Kontorovich, A.E.; Burshtein, L.M.; Malyshev, N.A.; Safronov, P.I.; Guskov, S.A.; Ershov, S.V.; Kazanenkov, V.A.; Kim, N.S.; Kontorovich, V.A.; Kostyreva, E.A. Historical and geological modeling of naphthydogenesis processes. Historical and geological modeling of naphthydogenesis processes in Mesozoic-Cenozoic sedimentary basin of the Kara Sea. *Geol. Geofiz.* **2013**, *54*, 1179–1226.
15. Sabukevich, V.; Podoprigora, D.; Shagiakhmetov, A. Rationale for selection of an oil field optimal development system in the eastern part of the Pechora sea and its calculation. *Periódico Tchê Química* **2020**, *17*, 634–655. [CrossRef]

16. Demaison, G.J. The Generative Basin Concept—Petroleum Geochemistry and Basin Valuation. *AAPG Mem.* **1984**, *35*, 1–14.

17. Doust, H. Sedimentary Basin Evolution and Conventional and Unconventional Petroleum System Development. *Swiss Bull. Appl. Geol.* **2011**, *16*, 57–62. [CrossRef]

18. Grigorenko, Y.N.; Sobolev, V.S.; Kholodilov, V.A. Petroleum system and forecasting the largest oil and gas fields in the offshore Russian Arctic basins. *Geol. Soc. Lond. Mem.* **2011**, *35*, 433–442. [CrossRef]

19. Kontorovich, A.E. *Geochemical Methods of Quantitative Forecast of Oil-and-Gas Bearing Capacity*; Nedra: Moscow, USSR, 1976.

20. Espitalie, J.; Marquis, F.; Drouet, S. Critical Study of Kinetic Modelling Parameters. In *Basin Modelling: Advances and Applications: Norwegian Petroleum Society Special Publication*; Elsevier: Amsterdam, The Netherlands, 1993; pp. 233–242.

21. Jarvie, D.M. Components and Processes affecting producibility and commerciality of shale oil resource systems. In Proceedings of the HGS Applied Geoscience Conference, Houston, TX, USA, 6–9 April 2012; pp. 307–325.

22. Magoon, L.B.; Dow, W.G. The Petroleum System—From Source to trap. *AAPG* **1994**, *60*, 329–338.

23. Pepper, A.S.; Corvi, P.J. Simple kinetic models of petroleum formation. Part I: Oil and gas from kerogen. *Mar. Petr. Geol.* **1995**, *12*, 291–319. [CrossRef]

24. Peters, K.E.; Burnham, A.K.; Walters, C.C. Petroleum generation kinetics: Single versus multiple heating-ramp open-system pyrolysis. *AAPG Bull.* **2015**, *99*, 591–616. [CrossRef]

25. Neruchev, S.G.; Bazhenova, T.K.; Smirnov, S.V. *Assessment of Potential Hydrocarbon Resources Based on Modeling of Their Generation, Migration and Accumulation*; Nedra: St. Petersburg, Russia, 2006; 363p.

26. Neruchev, S.; Smirnov, S. Assessment of Potential Hydrocarbon Resources Based on Modeling of Their Generation Processes and Formation of Oil and Gas Fields. *Oil Gas Geol. Theory Pract.* **2007**, *2*, 1–34.

27. Bazhenova, T.K. Study of Ontogenesis of Hydrocarbon Systems as a Basis for Real Prediction of Oil and Gas Content of Sedimentary Basins. *Oil Gas Geol. Theory Pract.* **2007**, *2*, 1–17.

28. Vassoevich, N.B. Theory of sedimentary migration origin of oil. *J. Acad. Sci. USSR* **1967**, *11*, 135–156.

29. Vassoyevich, N.B.; Korchagina, Y.I.; Lopatin, N.V.; Chernyshev, V.V. The main phase of oil formation. *Vestn. MSU Geol.* **1969**, *6*, 3–27.

30. Bazhenova, T.K.; Shimansky, V.K.; Vasilyeva, V.F.; Shapiro, A.I. *Organic Geochemistry of the Timan-Pechora Basin*; VNIGRI: St. Petersburg, Russia, 2008; 164p.

31. Goncharov, I.V.; Samoylenko, V.V.; Oblasov, N.V.; Fadeeva, S.V.; Vekhlich, M.A.; Kashapov, R.S.; Trushkov, P.V.; Bahtina, E.S. Types and Catagenesis of Organic Matter of the Bazhenov Formation and Its Age Analogues. *Oil Ind.* **2016**, *10*, 20–25.

32. Kiryukhina, T.A.; Bol'shakova, M.A.; Stoupakova, A.; Korobova, N.; Pronina, N.; Sautkin, R.S.; Suslova, A.A.; Mal'tsev, V.V.; Slivko, I.E.; Luzhbina, M.C.; et al. Lithological and geochemical characteristics of domanik deposits of the Timan-Pechora basin. *Georesources* **2015**, *61*, 87–100. [CrossRef]

33. Kiryukhina, T.; Fadeeva, N.; Stupakova, A.; Poludetkina, E.; Sautkin, R. Domanik Deposits of the Timan-Pechora and Volga-Ural Basins. *Geol. Oil Gas* **2013**, *3*, 76–87.

34. Klimenko, S.S.; Anishchenko, L.A. Features of Naphthidogenesis in the Timan-Pechora Basin. *Izv. Komi NTS UrO RAN* **2010**, *2*, 61–69.

35. Palyanitsina, A.; Sukhikh, A. Peculiarities of assessing the reservoir properties of clayish reservoirs depending on the water of reservoir pressure maintenance system properties. *J. Appl. Eng. Sci.* **2020**, *18*, 10–14.

36. Putikov, O.; Kholmyanski, M.; Ivanov, G.; Senchina, N. Application of geoelectrochemical method for exploration of petroleum fields on the Arctic shelf. *Geochemistry* **2019**, *80*, 125498. [CrossRef]

37. Sannikova, I.; Kiryukhina, T.; Bolshakova, M. *Generation Potential of the Domanik Deposits of the Southern Part of the Timan-Pechora Oil and Gas Basin Geology in the Developing World: A Collection of Scientific Papers (Based on the Materials of the VIII Scientific and Practical Conference of Students, asp. and Young Scientists from the International Participation)*; State National Research University Perm: Perm, Russia, 2015; pp. 498–502.

38. Basov, V.A.; Vasilenko, L.V.; Viskunova, K.G. Evolution of sedimentation environments of the Barents-North Kara paleobasin in the Phanerozoic. *Oil Gas Geol. Theory Pract.* **2009**, *4*, 1–3.

39. Bogoslovskii, S.A. Modeling the Processes of Oil and Gas Formation in Korotaikhinsk Depression Using Temis 3D. In Proceedings of the Materials of the 2nd International Scientific and Practical Conference of Young Scientists and Specialists, St. Petersburg, Russia, 9–12 April 2011; pp. 142–146.

40. Egorov, A.; Prischepa, O.; Nefedov, Y.; Kontorovich, A.; Vinokurov, I. Deep structure, tectonics and petroleum potential of the western sector of the Russian arctic. *J. Mar. Sci. Eng.* **2021**, *9*, 258. [CrossRef]

41. Egorov, A.; Vinokurov, I.; Telegin, A. Scientific, and methodical approaches to increase prospecting efficiency of the Russian Arctic shelf state geological mapping. *J. Min. Inst.* **2018**, *233*, 447–458. [CrossRef]

42. Garetskii, R.G. Tectonics and geodynamics of the northeastern and southwestern margins of the East European craton and adjacent areas of young plates. In Proceedings of the International Scientific and Technical Conference "Oil and Gas of the Arctic", Moscow, Russia, 27–30 September 2007; pp. 42–49.

43. Kulpin, L.G.; Savchenko, V.P.; Obmorosheva, L.B. Shallow bays of the arctic seas: Features of the development of hydrocarbon resources. *Oil Gas J. Russ.* **2011**, *5*, 22–27.

44. Parmuzina, L.V. *Upper Devonian Structure of Timan-Pechora Province (Structure, Formation Conditions, Patterns of Reservoir Location and Oil and Gas Content)*; Nedra: St. Petersburg, Russia, 2007; 151p.

45. Repin, Y.S.; Fedorova, A.A.; Bystrova, V.V.; Kulikova, N.K.; Polubotko, I.V. Mesozoic of Barents sea sedimentary basin. In *Stratigraphy and Its Role in the Development of the Oil and Gas Complex of Russia*; Dmitrieva, T.V., Kirichkova, A.I., Eds.; VNIGRI: St. Petersburg, Russia, 2007; pp. 112–116.
46. Sannikova, I.; Bolshakova, M.; Stupakova, A. Modeling of the scale of generation of hydrocarbon fluids by the domanik oil source strata of the Timan-Pechora basin using various kinetic spectra of organic matter destruction. *Georesources* **2017**, *1*, 65–79. [CrossRef]
47. Astakhov, S. Refinement of the vitrinite maturation model in the dislocated areas. *Geol. Nefti I Gaza* **2014**, *3*, 64–74.
48. Bazhenova, T. Oil and gas source formations of ancient platforms of Russia and oil and gas potential. *Oil Gas Geol. Theory Pract.* **2016**, *11*, 1–29.
49. Averyanova, O.; Morariu, D. Variability of estimates of the hydrocarbon potential of oil and gas systems. *Pet. Geol.—Theor. Appl. Stud.* **2016**, *11*, 3.
50. Vasiltsov, V.; Vasiltsova, V. Strategig planning of Arctic shelf development using fractal theory tools. *J. Min. Inst.* **2018**, *234*, 663–672. [CrossRef]

Article

Digital Technologies in Arctic Oil and Gas Resources Extraction: Global Trends and Russian Experience

Ekaterina Samylovskaya [1,*], Alexey Makhovikov [2], Alexander Lutonin [2], Dmitry Medvedev [3] and Regina-Elizaveta Kudryavtseva [4]

[1] Department of History, Faculty of Fundamental and Humanitarian Disciplines, Saint Petersburg Mining University, 199106 Saint Petersburg, Russia
[2] Department of Computer Science and Computer Technology, Faculty of Fundamental and Humanitarian Disciplines, Saint Petersburg Mining University, 199106 Saint Petersburg, Russia; makhovikov_ab@pers.spmi.ru (A.M.); lutonin_as@pers.spmi.ru (A.L.)
[3] Department of Legal Security of the Fuel and Energy Complex, Faculty of Integrated Safety of the Fuel and Energy Complex, Gubkin Russian State University of Oil and Gas (Research Institute), 197349 Moscow, Russia; medvedev.d@gubkin.ru
[4] Department for Publication Activities, Saint Petersburg Mining University, 199106 Saint Petersburg, Russia; aethel@yandex.ru
*   Correspondence: katerina-samylovskaya88@yandex.ru; Tel.: +7-9213673791

Abstract: The paper is devoted to the analysis of the current and the forecast of the prospective state of introducing digital technologies into the oil and gas mining industry of the Russian Arctic. The authors of the paper analyzed the global trends that define the process of digital technologies' introduction into the oil and gas mining industry. They also reviewed the Russian companies' experience in this sphere. The main trends and prospects for the development of oil and gas resources extraction in the Russian Arctic in the digitalization sphere were identified. Together with this, the researchers considered prospects for digital technologies' introduction into the oil and gas industry, observing their competition with RES. As a result, the authors have come to the following conclusions: (1) in Russian companies, digitalization is being more actively introduced into the processes of general enterprise management. (2) The main purpose of Russian oil and gas sector digitalization is to increase the efficiency of business process management, while the key constraining factors of digitalization are the lack of qualified personnel, lack of material and technical base and cyber-security threats aggravation. (3) The prospects of introducing a new package of sanctions can become both an incentive for a qualitative leap in Russian software development/implementation and an obstacle to the development of Arctic projects due to their rise in price. (4) The COVID-19 pandemic factor is one of the incentives for the widespread introduction of production and various business processes automation in the oil and gas industry, as well as the development of digital communications. (5) The leader in the digital technology development industry among Russian oil and gas companies is "Gazprom Neft" PJSC, followed by "NK Rosneft" PJSC. (6) "Gazprom" PJSC continues to lag behind in the sphere of digitalization; however, qualitative changes here should be expected in 2022. (7) The "sensitivity parameters" influencing the industry digitalization process in the Arctic region are the high dependence on foreign technological solutions and software, characteristics of the entire Russian oil and gas industry, and the problem of ensuring cybersecurity in Arctic oil and gas projects and power outages. (8) For the Arctic regions, the use of RES as the main source of electricity is the most optimal and promising solution; however, hydrocarbon energy will still dominate the market in the foreseeable future.

Keywords: oil and gas; Arctic; digital technologies; digitalization

Citation: Samylovskaya, E.; Makhovikov, A.; Lutonin, A.; Medvedev, D.; Kudryavtseva, R.-E. Digital Technologies in Arctic Oil and Gas Resources Extraction: Global Trends and Russian Experience. *Resources* **2022**, *11*, 29. https://doi.org/10.3390/resources11030029

Academic Editor: Antonio A. R. Ioris

Received: 1 February 2022
Accepted: 9 March 2022
Published: 11 March 2022

Publisher's Note: MDPI stays neutral with regard to jurisdictional claims in published maps and institutional affiliations.

1. Introduction

The exhaustion of readily available hydrocarbons leads to the need to change the principles, strategy and tactics of the development and operation of deposits with hard-to-recover reserves. First, the issue is the optimization of the processes taking place at the enterprise and the introduction of new effective technologies that reduce the cost of production.

Under these conditions, there is a natural increase in the pace at which oil and gas companies introduce intelligent technologies for hydrocarbons production management. These companies ensure the optimal redistribution of resources and effective planning of production in the short and long term. Taking into account the peculiarities of Arctic hydrocarbon deposits (remoteness from existing infrastructure and consumption centers, harsh climatic conditions, etc.), such technologies become essential, since they aim to reduce downtime and drilling costs, reducing production losses, as well as increasing the oil recovery coefficient [1]. Digitalization is designed to reduce the costs associated with hydrocarbons' exploration and production, as well as with the development and production of special equipment. The received geological, technical, statistical and other data are available in a single cloud, where they are stored, processed and analyzed. This allows track production indicators to be tracked in real time and, among other things, prevents breakdowns and accidents in real time.

The Arctic is a region, which has a strong potential for the global oil and gas industry development. At the same time, there is a number of factors limiting the extractive industry development and other economic activities in the region, especially in the Russian sector: remoteness of territories, difficult climatic conditions, a fragile ecosystem, lack of trained specialists, poorly developed infrastructure, etc. These limitations make the introduction and widespread development of digital technologies a particularly promising concept. In this regard, the purpose of this study is to analyze the current and forecast the prospective state of introducing digital technologies into the oil and gas mining industry in the Russian Arctic. For this, it seems necessary to analyze the global trends of digital technologies' introduction into the oil and gas mining industry, consider the experience of Russian companies in this area, identify the main trends and prospects of oil and gas resources' development in the Russian Arctic from a digitalization point of view, and consider the prospect of digital technologies' introduction into the oil and gas industry under RES competition.

The structure of this paper is as follows: a description of research methods and approaches (including an expert survey of characteristics); analysis of the expert survey results; review of global trends in oil and gas resources production digitalization; analysis of the Russian oil and gas production digitalization experience; analysis of the specifics of hydrocarbon production in the Arctic (including "sensitivity parameters" that affect the process of digitalization in Russian Arctic oil and gas projects); analysis of prospects for the introduction of digital technologies in the Arctic oil and gas industry in competition with RES; research results; and an appendix (description of the expert survey).

2. Research Methodology

This research is based on the analysis of the modern, foreign and Russian scientific literature, as well as open sources (official websites of companies, statistical and cartographic materials).

The analysis of the scientific literature and strategic documents belonging to the largest energy companies (Shell, Chevron, BP, Petoro, Halliburton, Schlumberger, Gazpromneft, Rosneft, Lukoil, etc.), as well as the experience of implementing specific projects on digital solutions introduction ("Captain", "Cognitive Geologist", "Digital Core", etc.) allowed us to identify the main trends in oil and gas resources extraction digitalization, in order to highlight the main developments and formulate the vector of digitalization development in Russian companies, the specifics of hydrocarbon production in the Arctic, and Russia's experience in the use of digital technologies, in particular, in practical projects.

The analysis of the scientific literature and open statistical and cartographic materials made it possible to conduct a comparative analysis of Industry 4.0 technologies usage, in practice (Smart Field, Smart Wells, Digital Field and other concepts) in the oil and gas sector and renewable energy sources (RES) and to formulate prospects for digital technologies' introduction in oil and gas sector competing with RES.

As part of the study, the method of system and structural analysis was applied in terms of the study of digital technologies' use at various stages of the production, operation and maintenance of deposits.

In order to identify the qualitative characteristics of digitalization development in the Russian extractive oil and gas industry, an expert survey was conducted (Appendix A), which interviewed the leading Russian oil and gas companies' employees and heads of educational and scientific organizations involved in the training of Arctic personnel.

The expert survey was sent to 48 respondents from 17 institutions at the end of December 2021. The survey results were received by the authors in early January 2022. In total, 36 experts actually took part in the survey; unfortunately, 12 respondents refused to participate.

The survey was attended by heads of departments, shift supervisors, geologists and engineers of Russian companies specializing in oil and gas production, as well as heads of departments and researchers of scientific institutes (55.6%) dealing with the problems of resources, who trained the oil and gas industry specialists (30.5%). The authors purposefully gave a prediction (in a numerical ratio) to experts from oil and gas companies due to the need to understand the real picture emerging in this area (Table 1).

Table 1. The expert group description [1].

Type of Activity the Organization Does	Name of the Organization	The Number of Respondents	The Number of Those Who Refused
Companies specializing in oil and gas production (heads of departments, shift supervisors, geologists and engineers)	"Gazprom Neft" PJSC	3	1
	"Gazprom Neft Shelf" LLC	2	2
	"Gazprom" PJSC	3	1
	"NK" "Rosneft" PJSC	4	0
	"LUKOIL" PJSC	2	2
	"Tatneft" PJSC	3	1
	"NOVATEK" PJSC	3	1
Scientific institutes (heads of departments, researchers)	Kola Scientific Center at Russian Academy of Sciences	2	0
	Arctic and Antarctic Research Institute	2	0
	Institute of Oil and Gas at Russian Academy of Sciences	1	1
Universities (deans and heads of specialized departments)	Saint Petersburg Mining University	2	-
	Gubkin Russian State University of Oil and Gas	2	-
	Northern (Arctic) Federal University	2	-
	Northeastern Federal University named after M.K. Ammosov	2	-
	Ukhta State Technical University	1	1
	Ufa State Petroleum Technical University	2	-
	Kazan Federal University	2	-
	Kola Scientific Center at Russian Academy of Sciences	3	1
	Arctic and Antarctic Research Institute	2	2
	Institute of Oil and Gas at Russian Academy of Sciences	3	1

[1] Compiled by the authors.

3. Results of the Expert Survey

The survey showed that experts identify the following digital technologies as the most effective in the extraction of oil and gas resources: digital twins—seism-geological, hydrodynamic and integrated modeling (44.4%); Big Data (25%); and artificial intelligence (14%). Other technologies were indicated by 16.6% of respondents, among which they noted robotics, automated control systems, integrated development environments for modular cross-platform Eclipse applications, Petrel software platform, and digital communications.

At the same time, experts mainly noted digital twins (38.9%), robotics (30.3%), digital communications (14%), automated control systems (8.4%) and other technologies (8.4%) as among the most effective technologies in oil and gas projects' implementation in the Arctic. Among other technologies, in addition to Eclipse and Petrel, they emphasized additive technologies. Accordingly, the survey revealed specific features of digitalization in relation to the implementation of oil and gas projects in the Arctic.

The majority of experts noted that the leader in digital technologies' introduction in Russia is "Gazprom Neft" PJSC (52.8%). The level of digitalization of "NK 'Rosneft'" PJSC (22.2%) and "Gazprom" PJSC (19.4%) companies was estimated to be approximately equal. The experts also singled out "NOVATEK" PJSC Company (5.6%).

According to the respondents, Russian oil and gas companies most successfully and effectively implemented the following digital projects and technologies: "Digital Core" (33.3%), "Cognitive Geologist" (27.7%), "Cyber Hydraulic Fracturing" (22.2%) and "Digital Field" (22.2%). The respondents also noted the following digital projects and programs: tNavigator, Eclipse and Petrel, "CAPTAIN" system, and "Gazprom Neft" PJSC. Attention is drawn to the fact that the vast majority of projects were developed and implemented by "Gazprom Neft" PJSC, which correlates with the data given above.

In turn, assessing digital technologies and projects that are most effectively implemented by Russia today in the extraction of oil and gas resources in the Arctic, the respondents identified specialized Arctic projects: "CAPTAIN", a digital logistics management system (27.8%), and block chain technologies on the sea-ice-resistant stationary platform, "Prirazlomnaya" (hereinafter "Prirazlomnaya" SISP) (19.4%). "Cognitive geologist" (19.4%), "Digital core" (16.7%) and "Digital twin of the Vostochno-Messoyakhskoye field" (16.7%) were also noted. All the programs and projects were also developed and implemented by "Gazprom Neft" PJSC (the digital twin of the Vostochno-Messoyakhskoye field was created together with "NK Rosneft" PJSC). These results suggest that today's experts rate the company as a leader in digital projects' development and implementation in the Arctic (Table 2).

Table 2. Qualitative characteristics of the Russian extractive oil and gas industry development [1].

Qualitative Characteristics	Respondents' Opinions
Digital technologies are the most effective for the extraction of oil and gas resources in the entire industry	Digital twins—44.4%: companies' representatives—25% scientific institutes' representatives—5.5% universities' representatives—13.9% Big Data—25%: companies' representatives—13.9% scientific institutes' representatives—2.8% universities' representatives—8.3% Artificial intelligence—14% companies' representatives—5.6% scientific institutes' representatives—2.8% universities' representatives—5.6% Others—16.6% companies' representatives—11% scientific institutes' representatives—2.8% universities' representatives—2.8%

Table 2. *Cont.*

Qualitative Characteristics	Respondents' Opinions
Digital technologies are the most effective for the implementation of oil and gas projects in the Arctic	Digital twins—38.9% companies' representatives—22.2% scientific institutes' representatives—5.6% universities' representatives—11.1% Robotization—30.3% companies' representatives—16.4% scientific institutes' representatives—5.6% universities' representatives—8.3% Digital communication—14% companies' representatives—5.6% scientific institutes' representatives—2.8% universities' representatives—5.6% Automated control systems—8.4% companies' representatives—5.6% scientific institutes' representatives—0% universities' representatives—2.8% Others—8.4% companies' representatives—5.6% scientific institutes' representatives—0% universities' representatives—2.8%
A Russian oil and gas company that is a leader in digital technologies' introduction	Gazprom Neft—52.8% companies' representatives—30.6% scientific institutes' representatives—8.3% universities' representatives—13.9% Rosneft—22.2% companies' representatives—11.1% scientific institutes' representatives—2.8% universities' representatives—8.3% Gazprom—19.4% companies' representatives—11% scientific institutes' representatives—2.8% universities' representatives—5.6% NOVATEK—5.6% companies' representatives—2.8% scientific institutes' representatives—0% universities' representatives—2.8%
Digital technologies/projects/ programs that are successfully and effectively implemented by Russian oil and gas companies	Digital Core—33.3% companies' representatives—19.4% scientific institutes' representatives—2.8% universities' representatives—11.1% Cognitive Geologist—27.7% companies' representatives—16.6% scientific institutes' representatives—2.8% universities' representatives—8.3% Cyber Hydraulic Fracturing—22.2% companies' representatives—13.8% scientific institutes' representatives—2.8% universities' representatives—5.6% Digital Deposit—22.2% companies' representatives—13.8% scientific institutes' representatives—2.8% universities' representatives—5.6% Others—16.8% companies' representatives—5.6% scientific institutes' representatives—5.6% universities' representatives—5.6%

Table 2. *Cont.*

Qualitative Characteristics	Respondents' Opinions
Digital technologies/projects that are most effectively implemented today with the extraction of oil and gas resources in the Arctic	"CAPTAIN" system—27.8% companies' representatives—13.9% scientific institutes' representatives—5.6% universities' representatives—8.3% Block chain technology "Prirazlomnaya" SISP—19.4% companies' representatives—8.3% scientific institutes' representatives—2.8% universities' representatives—8.3% Cognitive Geologist—19.4% companies' representatives—8.3% scientific institutes' representatives—2.8% universities' representatives—8.3% Digital Core—16.7% companies' representatives—11.1% scientific institutes' representatives—2.8% universities' representatives—2.8% Digital twin of the Vostochno-Messoyakhskoye field—16.7% companies' representatives—11.1% scientific institutes' representatives—0% universities' representatives—5.6%

[1] Compiled by the authors.

According to the respondents, the main factors that influenced the intensification of digital technologies' introduction by Russian oil and gas companies are the need to improve the efficiency of business process management and the need to reduce companies' expenses, which naturally increase companies' competitiveness in the market. At the same time, Russian state policy in the digitalization field, the problem of import substitution and sanctions, as well as the COVID-19 pandemic are causing companies to revert to their second and even third plans. In turn, the factors hindering digital technologies' introduction in this area are mainly the lack of qualified personnel (41.7%) and the lack of an appropriate material and technical base (36.1%). Problems with ensuring cybersecurity (22.2%) can be considered a secondary factor (Table 3).

Table 3. Factors that influenced the intensification of digital technologies' introduction by Russian oil and gas companies in 2019–2021 [1].

Factors	Respondents' Opinions
Three factors that influenced the intensification of digital technologies' introduction in 2019–2021	Improving the efficiency of business process management—36.1% companies' representatives—19.4% scientific institutes' representatives—5.6% universities' representatives—11.1% The need to reduce companies' expenses—30.5% companies' representatives—16.6% scientific institutes' representatives—5.6% universities' representatives—8.3% Import substitution and sanctions—13.9% companies' representatives—8.3% scientific institutes' representatives—2.8% universities' representatives—2.8% "Digital Economy of the Russian Federation" national program adopted in 2017—13.9% companies' representatives—8.3% scientific institutes' representatives—0% universities' representatives—5.6%

Table 3. *Cont.*

Factors	Respondents' Opinions
	COVID-19 pandemic—5.6% companies' representatives—2.8% scientific institutes' representatives—0% universities' representatives—2.8%
The main factor influencing the intensification of digital technologies' introduction in 2019–2021	Improving the business process management efficiency—58.3% companies' representatives—38.8% scientific institutes' representatives—5.6% universities' representatives—13.9% The need to reduce companies' expenses—33.3% companies' representatives—16.6% scientific institutes' representatives—5.6% universities' representatives—11.1% "Digital Economy of the Russian Federation" national program adopted in 2017—8.4% companies' representatives—0% scientific institutes' representatives—2.8% universities' representatives—5.6% Import substitution and sanctions—0% companies' representatives—0% scientific institutes' representatives—0% universities' representatives—0% COVID-19 pandemic—0% companies' representatives—0% scientific institutes' representatives—0% universities' representatives—0%
Factors hindering the digital technologies' introduction in the extraction of oil and gas resources	Lack of qualified personnel—41.7% companies' representatives—22.2% scientific institutes' representatives—5.6% universities' representatives—13.9% Lack of an appropriate material and technical base—36.1% companies' representatives—22.2% scientific institutes' representatives—5.6% universities' representatives—8.3% Cyber security issues—22.2% companies' representatives—11.1% scientific institutes' representatives—2.8% universities' representatives—8.3% Low growth in companies' efficiency compared to expectations—0% companies' representatives—0% scientific institutes' representatives—0% universities' representatives—0% The high cost of technology—0% companies' representatives—0% scientific institutes' representatives—0% universities' representatives—0% Sanctions—0% companies' representatives—0% scientific institutes' representatives—0% universities' representatives—0%

[1] Compiled by the authors.

In order to identify the prospects for digital technologies' introduction in the oil and gas industry within its competition with RES, the respondents were asked questions about the prospects for the use of RES. The survey showed that, according to the respondents, renewable energy use is mainly advisable due to its economic benefits (52.7%). At the same time, RES is recognized to be the most optimal solution for providing an energy supply to hard-to-reach Arctic regions (66.6%). It is noteworthy that skepticism towards renewable energy is mainly typical of oil and gas companies' representatives (Table 4).

Table 4. The ratio of renewable energy and hydrocarbon sources' use (RES) [1].

Question	Respondents' Opinions
In which cases is the use of renewable energy more appropriate in comparison with the use of hydrocarbons?	If it is economically unprofitable to extract hydrocarbons—52.7% companies' representatives—27.8% scientific institutes' representatives—8.3% universities' representatives—16.6% None—25% companies' representatives—25% scientific institutes' representatives—0% universities' representatives—0% Ecological issues—16.7% companies' representatives—2.8% scientific institutes' representatives—2.8% universities' representatives—11.1% In the absence of hydrocarbon ones (remote stations)—5.6% companies' representatives—0% scientific institutes' representatives—2.8% universities' representatives—2.8%
Is RES the most optimal solution for providing an energy supply to hard-to-reach Arctic regions?	Yes—66.6% companies' representatives—38.8% scientific institutes' representatives—11.1% universities' representatives—16.7% No—19.5% companies' representatives—13.9% scientific institutes' representatives—0% universities' representatives—5.6% No idea—13.9% companies' representatives—2.8% scientific institutes' representatives—2.8% universities' representatives—8.3%
Will RES be able to completely replace hydrocarbon energy in the future?	Yes—0% companies' representatives—0% scientific institutes' representatives—0% universities' representatives—0% No—97.2% companies' representatives—55.6% scientific institutes' representatives—13.9% universities' representatives—27.7% No idea—2.8% companies' representatives—0% scientific institutes' representatives—0% universities' representatives—2.8%

[1] Compiled by the authors.

4. Transition to Digital and Smart Deposits: General Analysis of the Current State, the Leading Companies' Experience

Information technologies in the oil and gas industry were first introduced at the beginning of 1980, and in the early 2000s they were combined into a set of programs for reservoir modeling, the calculation of optimal logistics, analysis of financial and economic indicators, visualization of current processes, and others [2]. Digital fishing technologies evolved "from simple to complex", that is, from the processes of the primary collection, aggregation and analysis of fishing data to the introduction of complex analytical systems that solved current and future tasks in real time and were integrated into a single network of circulation and the analysis of information [2].

In 2006, Shell oil and gas company presented the concept of "smart field" technology on the Brunei shelf for the first time. Currently, the accelerated processing of heterogeneous data and the use of intelligent technologies are key factors of accelerating the search for optimal solutions in the sphere of oil and gas fields development and operation. Modern analytical systems provide an automation of data collection, storage and processing, physical processes description, forecast of expected hydrocarbon production and visualization of key parameters for the preparation and subsequent implementation of solutions at all levels of management [2].

The offshore oil platform has about 90 thousand sensors that generate up to 15 petabytes of data throughout the entire development period, that is, at least for 15–20 years. Digitalization makes it possible to monitor this array of oilfield information, process it and display it in an accessible form.

To date, almost all major oil and gas companies are implementing concepts for the software and hardware complex's introduction in production processes (Table 5).

Table 5. Digital technologies used by large oil and gas companies (prepared by the authors; the information was taken from the companies' official websites) [1].

Company	Technology
Shell	Smart Field
Chevron	i-field
BP	Field of the future
Petoro	Smart Operations
Halliburton	Real Time Operation
Schlumberger	Smart Wells
Gazpromneft, Rosneft	Digital field
Lukoil	Life-Field

[1] Compiled by the authors.

The analysis of the scientific literature allows us to identify several approaches to understanding the "digital field" concept essence:

- A system of interrelated technologies and business processes that ensure an increase in the efficiency of all elements of oil and gas assets' production and management;
- A software package that includes a set of applications that allow for modeling and managing processes in the field;
- A way to generate additional value of an oil and gas asset by improving the cycle of data collection, processing, modeling, decision making and their execution;
- A system of operational management of an oil and gas facility that aims to optimize production and reduce financial losses through a predictive analysis of problems and rapid responses based on data obtained online [2–4].

However, a focus on one of the segments or on a separate technology does not allow us to describe the essence of oil and gas facilities' digitalization. It is worth noting that various digital technologies are used in different segments of oil and gas production (upstream, midstream, downstream), so digitalization is a much more complex process than it may seem [5]. For example, in upstream processes, for the most part, this term rather means a complex of digital technologies introduced into operational processes, and the main directions of upstream digitalization include Big Data, the industrial Internet, robotics and artificial intelligence [6].

In other words, the concept of a "digital field" aims to integrate all of the stages (geology, development, drilling and completion of wells, oil production, construction, economics, ecology, risk analysis) into a single system and improve the efficiency of their implementation by optimizing processes online.

Innovative technical solutions for the digital modernization of oil and gas facilities include:

- Digital twin—cyber-physical oil and gas production system;
- The construction of super-heavy permanently operating systems for monitoring the seismological situation and combined active–passive systems of seismological monitoring;
- Building systems for the integrated control of development processes in various physical fields;
- The integration of borehole probes into self-organizing sensor networks;
- The construction of high-precision positioning systems and identification of deep processes;
- The construction of control and measurement systems for monitoring oil and gas fields and wells and others [6].

The digital twin acts as the basic digital technology of industrial modernization [7,8]. There are three important processes in this technology:

1. The transformation of physical data into digital, which includes the following processes of collecting information on a real object: verification and rejection of incorrect data and the formation of an archive of events that significantly affect the physical process.
2. The creation of a numerical model of a physical process, creation of specialized calculation modules of the digital twin, optimal solutions base formation;
3. The ranking of action options by decision-making levels using intelligent technologies [7–9].

Ideally, the digital model in the field has algorithms for obtaining and processing data from remote field development monitoring systems, and allows us to automate control processes and predict and plan the work of each of the field system components with the elimination of labor-intensive manual processes [2]. The purpose of constructing such models is to increase the efficiency of not only each individual system, but also the entire asset as a whole, taking into account the mutual influence of the systems [2].

Digitalization makes it possible to significantly reduce operating costs and increase the share of recoverable hydrocarbons. According to experts, the introduction of digital technologies makes it possible to increase the oil recovery rate by up to 50% compared to the global average of 30% [8,10]. At the same time, digitalization is only the initial stage of the industry digital transformation, which is characterized by qualitative changes in its structure and management models.

At the moment, the number of wells where digital technologies have been partially or fully implemented is about 20 thousand, and this indicator is expected to double in the next 5 years. According to experts, foreign leading oil and gas companies are moving to full digital control coverage of production wells. For comparison, "Shell" corporation is already managing its entire well fund in real time, and "BP" company is increasing the same management by 60% [11].

Thus, there is a rapid pace of the oil and gas sector digitalization, as the intellectualization and introduction of unpopulated technologies allows the profitability of oil production to be maintained at an acceptable level by increasing the efficiency of field operation and optimizing labor costs. This issue is especially relevant for deposits in the Arctic and the Russian Far East. In this regard, a fuel and energy complex acts as the "locomotive" of the country's economy digitalization, especially in its northern region.

The main digital technologies used in oil and gas enterprises are Big Data, neurotechnology and artificial intelligence, distributed registry systems, the industrial Internet of Things, robotics and sensor components, as well as virtual and augmented reality technologies [12,13].

5. Experience of Russian Oil and Gas Production Digitalization

Digital technologies first came to Russia in 2000. The use of "smart wells" technology in the Salym group of fields can be considered one of the first successful projects in the sphere of digital technologies in oil and gas industry production. "Salym Petroleum Development" company started the implementation of this project in 2006 based on Shell technologies [14,15].

The year of 2017 was an important period for digital technologies' development. This was the year when, taking into account the priorities outlined in "Digital Economy of the Russian Federation" national program framework, the Ministry of Energy created a departmental project "Digital Energy" (lasting until 2024), which increased the key fuel and energy sector organizations share up to 40%, using digital technologies and various platform solutions. The project provides for the oil and gas complex digitalization [16]. As a result, digital technologies are becoming drivers of oil and gas industry development. Since 2017, the largest Russian oil corporations have implemented several projects on the introduction of digital technologies.

To date, the leaders in digital technologies' implementation are the "Gazprom Neft" PJSC and "NK 'Rosneft'" PJSC companies, which is confirmed by the survey conducted by the authors. The digitalization of various technological and business processes (self-learning programs, neural networks, artificial intelligence, digital twins and Big Data) will play a key role in these companies' innovative development programs until 2025 [17,18].

Experts [19] identify the following areas of the digital transformation strategy of "Gazprom Neft" PJSC: cognitive exploration; cognitive engineering; drilling management corporate center; production management center; and digital twins [19]. The main digital strategies of "Rosneft-2022" are:

- The launch of a corporate data processing center with an industrial Internet platform and an integrated digital twin of fields;
- The testing of technology for monitoring production facilities using drones and machine vision;
- The use of artificial intelligence in field development;
- Tests of the ice rig monitoring system for offshore drilling;
- The implementation of predictive analytics systems and dynamic equipment status indicators [19].

One of the most popular technological trends is the creation of programs for the modeling of oil and gas production processes (for example, mathematical [20,21] and digital [22] modeling in the sphere of drilling, hydrocarbon systems modeling (basin modeling) [23])—the creation of "digital twins".

To date, it is the developments, digital technologies and projects of "Gazprom Neft" that are recognized by experts as the most successful in the sphere of Russian oil and gas production digitalization (this is also demonstrated by the results of an expert survey conducted by the authors of this paper). Back in 2007, the company established a Scientific and Technical Center ("Gazpromneft STC" LLC), which deals with the following:

- The creation of regional models of oil and gas basins; construction of structural–tectonic and geological models of deposits; and modeling in the sphere of exploration drilling;
- The creation of digital models of fields (for example, in Priobskoye field by "Gazprom Neft");
- Modeling the properties of reservoir fluids for the selection of optimal oil production technologies (they developed a self-learning program "Digital Core", which predicts the properties of rocks in new fields);
- Conceptual geological modeling;
- Cognitive programs ("Cognitive geologist" project) [24].

In 2014, the company began developing the "Digital Deposit" program. In 2017, the company established an Oil and Gas Production Management Center, on the basis of which a self-learning "digital twin" of the field was created [25]. The company is successfully implementing the "Digital Core" project (a digital laboratory for core material research, which allows data on the characteristics of the formation to be obtained and the selection of optimal solutions for production) [26]. In 2019, the company presented a hydraulic fracturing simulator "Cyber Hydraulic fracturing" (which simulates the creation of hydraulic fracturing and the search for the best algorithms for geological operations) [27], and in 2020, it presented an improved simulator "Cyber Hydraulic Fracturing 2.0", which allowed operation options to be developed for extreme conditions.

In the wake of "Gazprom Neft", "Rosneft Oil Company" PJSC is moving in this direction. Back in 2005, the company began the formation of a corporate scientific and technical complex.

In 2019, the company launched the "Digital Field" project on the basis of the Ilishevsky field, "Bashneft" ANC PJSC (a subsidiary of "NK 'Rosneft'" PJSC), covering all the main processes of oil production and logistics. In the same year, the company managed to test a drone monitoring system that could automatically detect the presence of people and equipment in protected areas and inform about oil spills. Another daughter company of "Rosneft", "Rospan International" JSC tested the technology of unmanned aerial vehicles for remote monitoring the liquidated and mothballed "Hermes" wells' fund state [28]. In 2021, Rosneft began using drones to control greenhouse gas emissions in oil and gas treatment and transportation facilities [29].

In 2019, "Rosneft" created a prototype and put into operation a modern system for repair crews and services management—the "Digital Crews TCRS" platform, which enabled the operational management of crews and services, processing of data from control systems, detecting anomalies, notifying about incidents and risks during well repairs, etc. [28].

Moreover, in the same year, the company developed the concept of a digital gas field. As a result, the technology of the operational integrated optimization of gas production at the Wheatgrass gas condensate field was successfully tested using machine-learning methods. A prototype of the integrated model of the "Odin" deposit was created, an information technology platform implementing a digital twin of the Wheatgrass fishery was tested, and a prototype with a three-dimensional visualization of "Mitra" fishery was created [28].

In 2021, "Rosneft" company demonstrated its new developments: models and algorithms of geomechanics, hydrodynamics, petrophysics, and geology, which formed the basis of the software products "RN-GRID" (a hydraulic fracturing simulator), "RN-SIGMA" (risk management of drilling wells), "RN-KIM" (a hydrodynamic simulator of hydrocarbon deposits), etc. [30].

In addition, "LUKOIL" PJSC also achieved certain results in the sphere of digitalization. In 2018, the "LUKOIL" Board of Directors approved the functional program "Information Strategy of the 'LUKOIL' Group", which became a part of "LUKOIL" Group's Strategic Development Program for 2018–2027. "LUKOIL-Technologies" LLC created an intelligent platform of the Unified Information Space (UIS), and in 2021, they began to improve their infrastructure (UIS) [31]. The company is working on the creation of integrated operations centers of a single digital platforms for data analysis and management. It is modeling drilling conditions and processes, as well as the automation of various management, planning and control processes ("Integrated Management Systems"). New integrated models of fields under development are also being created (for example, in 2020, they developed the Imilorskoye field and fields named after V. Vinogradov and V. Greifer) [32].

"Tatneft" PJSC has achieved some success by developing a program to create a digital platform for managing large geological, geophysical and field data—a project (jointly with "ChemTech" company) on the depth of oil refining at "TANECO" oil refining complex [25]. In 2019, the company began to create a unified data collection system (USDC). In 2020, "Tatneft", supported by Huawei, successfully implemented the Huawei OceanStor Dorado V6 data storage system [33].

Experts also highlight the achievements in the digitalization sphere by "NOVATEK" PJSC (according to the conducted survey, the company ranks third in terms of digital technology development after "Gazprom Neft" PJSC and "NK 'Rosneft'" PJSC). In 2010, the company established "NOVATEK Scientific and Technical Center" LLC (hereinafter, NOVATEK STC), which is engaged in the development and implementation of digital technologies, among its other projects. Back in 2017, the company commissioned the "Digital Field Production Management System" to manage production and ensure a prompt response to technological incidents at "Yamal SPG" OJSC [34]. Starting in 2019, the comprehensive digitalization of drilling and down-hole operations, also with the use of artificial intelligence technologies, was carried out on an ongoing basis. In 2020, the process of seismic exploration digitalization was launched on an ongoing basis. The company also conducted research in the sphere of "unattended/unmanned technologies" introduction—a system for remote monitoring and conducting the technological process of extracting hydrocarbons from wells [35].

In 2021, NOVATEK STC, together with Umbrella IT company, developed a voice assistant, the "Nova" application, which in a few seconds provides company employees with access to data on the processes regarding wells (including offline mode) [36].

According to experts [37], "NK 'Rosneft'" PJSC and "LUKOIL" PJSC occupied the leading positions in the implementation of digital technologies in the field of gas production in 2020, while "Gazprom" PJSC occupied only the third place. Today, it is obvious that, by the end of 2021, "NOVATEK" PJSC was also among the leaders. At the same time, "Gazprom" definitely ranks first in terms of investments in digitalization. However, in comparison to other countries around the world, the Russian Federation lags far behind foreign competitors in terms of business processes for the digitalization of natural gas production and LNG production in terms of the number of projects and their quality (about 30% of the projects are implemented with significant delays) [37]. In turn, only "Gazprom" PJSC has a product capable of developing according to the laws of Industry 4.0; at the end of 2020, among the platforms of industrial digitalization, the most successful Russian solution was EvOil from "Gazprom" PJSC [37].

In 2020, "Gazprom" put information management systems (IMS) into permanent operation. Since 2020, the company has also been developing a Unified Digital Platform (UDP), the main purpose of which is to increase the efficiency of business processes in "Gazprom"'s investment activities through the use of digital technologies [38], as well as the "Company's Digital Twin" project.

In December 2021, the "Gazprom" Group's digital transformation strategy for 2022–2026 was approved. Its main purpose is to create a Unified Data Model of "Gazprom" Group companies that will be integrated with the National Data Management System. As a result, it plans to create digital platforms, which would become the basis for the digital ecosystems of the gas, oil and electric power business. Each platform is a group of specialized IT solutions and services united by a single regulatory reference information. At the same time, the company stated that the implementation of the strategy will be provided mainly by domestic technological solutions [39].

In the process of transition to digital technologies in the oil and gas industry, the most important directions are in the sphere of exploration and production drilling of wells and their subsequent maintenance and support. However, unfortunately, the transition to digital technologies in these areas in Russia is slow. Nevertheless, as shown above, there is certain progress in this development direction demonstrated by Russian companies.

According to experts (including the survey conducted by the authors of this paper), one of the key factors hindering the introduction of digital technologies in the extraction of oil and gas resources in Russia is the lack of qualified personnel. Therefore, leading Russian companies cooperate with leading universities and institutes in their training of specialists in the digital technologies sphere. Thus, in 2019, with the support of "Gazprom Neft" PJSC, the Department of Mathematics and Computer Science was established at St. Petersburg State University. The company employs its best graduates. It also cooperates with Moscow Institute of Physics and Technology and St. Petersburg Polytechnic University in the sphere of digital technologies. "Rosneft" cooperates with Gubkin Russian State University of Oil and Gas (Research Center), Lomonosov Moscow State University, Moscow Institute of Physics and Technology, "Higher School of Economics" Research University, etc. [40]. Therefore, it is not surprising that today these companies are Russian leaders in the sphere of oil and gas resources extraction digitalization. However, it should be noted that all oil and gas companies are trying to maintain contacts and implement scientific and technical projects with leading Russian universities.

In addition, the factors hindering the process of the widespread use of digital technologies are the lack of appropriate material and technical bases, as well as the problem of ensuring cyber security [25]. It can be noted that, with sector digitalization, the issues of neutralizing emerging threats are being actualized. Such threats can be present in the form of both information flows distortion (changes in the real values of critical parameters, false protection triggers, false fault definitions) and inconsistencies of the digital model with the original one, etc. [41].

The main reason for the existing problems in the sphere of information security, as well as in the sphere of digitalization in general, is the lack of highly qualified personnel, which is especially obvious in companies operating in hard-to-reach regions, in particular, in the Arctic. This leads to the fact that companies are forced to outsource the task of ensuring digital services operation and provide third-party organizations with remote access to their digital systems, which, in some cases, can lead to serious problems in the sphere of information security.

Despite the fact that the expert survey (conducted from the end of December 2021 to the beginning of January 2022) showed the secondary role of import substitution and sanctions problems, as well as the role of the COVID-19 pandemic in the process of intensifying the digital technologies' introduction in Russian oil and gas producing companies (Table 3), it is not possible to reduce the impacts of these factors or to completely exclude them.

It is noteworthy that the COVID-19 pandemic can also be considered a stimulating factor for the accelerated transition to digital solutions in the extractive sector. Restrictions imposed by most countries, including Russia, have led to the search for opportunities to introduce "unattended/unmanned" technologies, including digital communications development. According to EY surveys, more than 70% of multinational companies' managers plan to increase investments in digital technologies, and more than 35% are already actively investing in business process automation [42]. Moreover, the reports and company development strategies studied by the authors also mention COVID-19 as one of the accelerated digitalization factors [32,40].

Speaking about the problem of import substitution, we note that until recently, companies mainly used foreign practices and software in the sphere of security, which imposed additional difficulties. However, an analysis of the companies' annual reports and development strategies has shown that the problem of import substitution, including software and technological solutions, comes to the fore under the conditions of a possible tightening of sanctions. The new developments by "Gazprom Neft", "Rosneft", "Gazprom" and "LUKOIL", described above, indicate the beginning of a qualitative shift in this direction.

One of the decisive factors here is the growth in international tension since the end of 2021 and the prospect of a new package of sanctions being introduced against Russian oil and gas companies in connection with the Russian–Ukrainian conflict that broke out at the end of February 2022. Despite the fact that Western countries, in particular the United States, refused to impose sanctions against the Russian oil sector, it is clear that software and equipment could fall under this sanctions regime, which could cause significant damage to the development of the industry, since digital technologies are introduced mainly when implementing joint projects with foreign companies. For example, 'Gazprom Neft', the leader in the digitalization of the Russian oil and gas industry, is implementing its flagship projects together with foreign companies (IBM Services, Brasil and PwC).

Of course, in recent years, there has been progress in the production of Russian microprocessors and operating systems. Only recently were Russian operating systems based on the Linux operating system (for example, Astra Linux, Rosa Linux, etc.) actively introduced in Russia [43,44]. Russia also produced and is beginning to widely introduce its own microprocessors, such as Elbrus and Baikal [45].

However, in today's realities, the prospect of the development of these processors becomes a big question: both microprocessors are manufactured by TSMC (Taiwan Semiconductor Manufacturing Company, Taiwan, China), but the company stopped its deliveries to Russia and its suppliers due to the imposition of sanctions against Russia following the Russian–Ukrainian conflict [46].

6. Specifics of Hydrocarbon Production in the Arctic

Huge mineral resources are concentrated in the Arctic. Currently, 10% of the world's oil and one-quarter of natural gas are produced in this region. The Russian Arctic zone now produces about 70% of gas and up to 20% of oil from the total production of the country [47]. At the same time, the Russian part of the Arctic has the largest projected oil reserves (Figure 1) (according to experts, there are 7.3 billion tons of oil reserves in the Arctic zone of the Russian Federation [48]). In other words, the Arctic region will be of key importance for the Russian fuel and energy complex in the medium term.

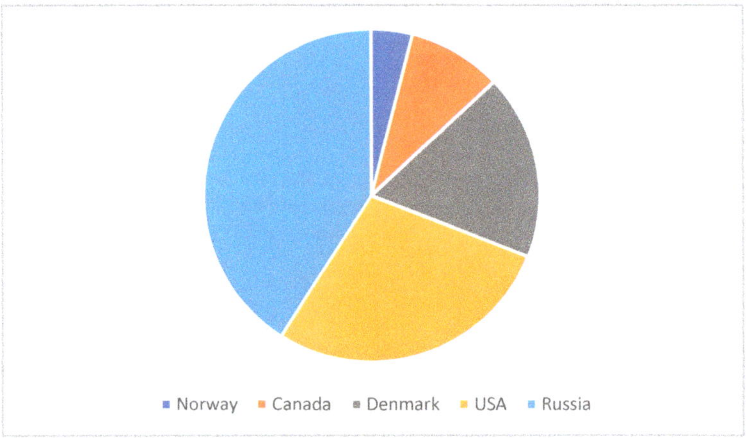

Figure 1. Projected oil reserves in the Arctic by country [49].

The peculiarities of hydrocarbon production in the Arctic include:

- Extreme natural and climatic conditions (polar nights, strong winds combined with low air temperatures, frequent magnetic anomalies in the atmosphere, etc.);
- Ice formations of various natures (icebergs, ice cover of various cohesion);
- The significant (up to 600 km) remoteness of supply bases and the practical lack of infrastructure to ensure the development of offshore fields;
- High sensitivity of the Arctic ecosystem to man-made impacts [50].

The above-mentioned natural and climatic features of the Arctic entail adaptation and optimization of technologies and production processes. In fact, the Arctic region provides companies with unique opportunities to develop not only new technologies but also innovative strategies for the development of economic activities—the formation of clusters, which increases the efficiency of capital-intensive and technically complex projects, accelerates the development of high-tech technologies ("Messoyakha", "Yamal LNG", "New Port" Arctic projects) [51].

As mentioned earlier, the integrated application of digital technologies and effective management algorithms serve as the basis for optimizing operating costs and provides the possibility of remote management of production facilities, which is extremely important for facilities in the Arctic.

As the first experience of the development of offshore fields located in the waters of freezing seas has shown, the direct transfer of technologies and equipment for the development and operations of offshore oil and gas fields in other areas of the world's oceans to the Arctic and subarctic regions is not always the best solution. An example of such a transfer is the Shtokman Gas Condensate field (GCF) development project, which was eventually frozen and has been postponed until 2028 [52].

One of the factors that influenced the freezing of the Shtokman Gas Processing Plant was the high dependence on foreign technologies, which negatively affected the prospects of the project after the 2014 Ukrainian crisis. According to experts [53], the technological dependence of the Russian oil and gas sector on the Arctic ranges from 80 to 95%. This "technological gap" leads to an additional increase in the cost of Arctic offshore projects. In this regard, nowadays, the problem of import substitution in the sphere of technical and software solutions for Arctic projects is quite acute.

It is important to note that the deposits digitalization in the Arctic is associated with a comprehensive solution of tasks: not only oil production, but also the development of all related infrastructure.

In this regard, special attention is paid to the development of a digital logistics management system. It is the digital projects in this area that are recognized by experts (in correlation with the results of the survey conducted by the authors) as the most effective in the extraction of oil and gas resources by Russian companies in the Arctic.

"Gazpromneft-Snabzheniye" LLC (a subsidiary of "Gazprom Neft" PJSC) has successfully implemented a project on block-chain use at the "Prirazlomnaya" sea-ice-resistant stationary platform [19]. So, in order to work with data on the movement of goods, radio frequency (RFID) tags and GPS sensors were installed on them at the plant in Veliky Novgorod city. This made it possible to process data on all the routes, the transportation speed and the location of goods. Due to the use of block-chain technology, all data on the cargo movement were collected in a single information space, but not on a centralized server, while each user connected to the block-chain network had a copy of the records and could confirm the addition of new blocks. All participants in the system had access to the data while the possibility of errors, substitution or correction of records was eliminated using block-chain; the logistics became as transparent as possible for all participants in the process [54].

At the moment, the most successful example of the digital logistics management system development in the Russian Arctic is the "CAPTAIN" system (a complex for automatic planning of interactive Arctic oil transportation), developed and implemented by "Gazprom Neft". As the company officially reported, the economic efficiency from the introduction of "CAPTAIN", the digital management system of the Arctic logistics, amounted to RUB 0.9 billion in 2019–2020 [53]. This is a result of optimizing the tanker operations costs by choosing the best routes, saving fuel, reducing the cost of icebreaking wiring, as well as of reducing downtime and maximizing oil exports.

The "CAPTAIN" system consolidates data from all the objects of the Arctic logistics chain and analyzes about 15 thousand input parameters online, including telemetry from support vessels and tanker fleets, cargo tanks of floating oil storage facilities and terminals, information about weather conditions, etc. Every 15 min, the system recalculates the movement schedule of tankers and oil shipments from terminals, choosing the best solution (Figure 2) [55].

The conditions for the implementation of large-scale Arctic oil and gas projects are associated with solving non-trivial tasks and finding ways to reduce costs while maintaining environmental friendliness and safety at a high level. As noted earlier, the transition to digital deposits aims to increase production, minimizing expenses, labor costs, and the likelihood of human error [56].

The introduction of digital technologies reduces costs, but requires significant investments in technology and changes in the approach to managing production processes. The Arctic is a unique innovative "testing ground" for the Russian oil and gas industry.

In the sphere of hard-to-recover oil reserves extraction in the Arctic, according to experts, the use of the "Cognitive Geologist" platform, the flagship project by "Gazprom Neft" (developed jointly with IBM Services Brasil), is extremely promising. The IT platform is a self-learning program that allowed the company to obtain additional oil at Vyngapurovskoye field (Yamalo-Nenets Autonomous Okrug). In June 2021, at the 24th St. Petersburg Economic Forum, "Gazprom Neft" and PwC consulting company representation in Russia signed a memorandum of cooperation in the development of intelligent digital solutions for geology and oil production. The key place is given to the "Cognitive Geologist" [57].

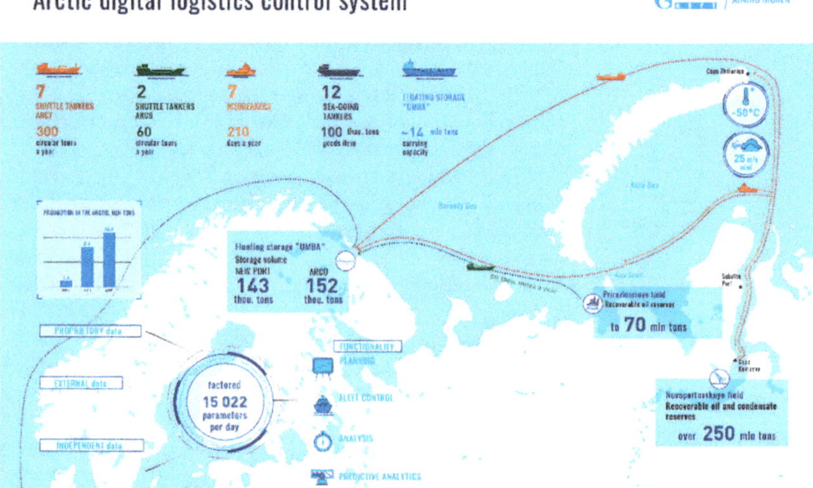

Figure 2. "CAPTAIN", a digital logistics management system [54].

Experts also highlight the "Digital Core" program (a self-learning program presented in the form of a single database of core samples and grinds that predicts the properties of rocks in new fields). It has been successfully implemented in the study of Achimov deposits of hard-to-recover oil at Yamburskoye oil and gas field (creating a digital model of Achimov core) [58].

In 2021, "Messoyakhaneftegaz" JSC (a joint venture of "Gazprom Neft" and "NK 'Rosneft'") created a digital twin of the Vostochno-Messoyakhskoye field and, in partnership with "Schlumberger", simulated a full drilling cycle using the DrillPlan platform, which allowed for automated calculations and optimized well planning. As a result, the terms of well design for the hard-to-recover Arctic continental reserves development were halved [59,60].

Nevertheless, despite the implementation of a number of successful digital oil and gas projects in the Arctic, we need to note the critical parameters that influenced the industry digitalization process in the region.

"Sensitivity parameters" influencing the digitalization process in Arctic oil and gas projects.

High dependence on foreign technological solutions and software, characteristic of the entire Russian oil and gas industry (discussed above).

6.1. The Problem of Ensuring Cybersecurity in Arctic Oil and Gas Projects

Ensuring information security in the Arctic, in comparison with the central territories of Russia, encounters a number of problems, primarily caused by the inaccessibility of the Arctic territory. For example, there is a problem with ensuring security during data transmission. The fact is that laying secure wired communication lines is often impossible, and wireless data transmission is obviously the most vulnerable to attacks, in particular, to a man-in-the-middle attack. The introduction of the full encryption of all transmitted data along the entire transmission path, and reliable and non-compromised algorithms such as GOST 28147-89 [61] are the only solution to this problem.

Another problem is related to the provision of information security specialists. The fact is that IT specialists, unlike other specialists in the extractive industry, receive a decent salary in the central regions of Russia and do not seek to move to work in the Arctic at all. Accordingly, companies are forced to employ these specialists remotely and outsource this work, which leads to additional security problems, since the work towards setting up equipment protection is carried out by unverified people, often in insufficient numbers.

6.2. Power Outages

In Arctic conditions, power outages can occur [62] as a result of extremely low temperatures. This, in turn, will lead to the failure of all digital systems.

The solution to this problem may be the use of smart power distribution systems, which will allow a reconfiguring of network topology depending on the possible occurrence of a failure on the line and the even distribution of the load on power lines. To date, there is a project [63] of a smart grid system designed for the Arctic region, which, in addition to the above-mentioned requirements, makes it possible to increase the reliability of the power supply system by connecting a backup battery in the event of a failure. However, the use of this technology is possible only if there are prediction systems based on digital twins that would allow the battery to be connected shortly before the upcoming failure.

Special attention should be paid to secondary power supply sources intended to supply end users, which include actuators responsible for the operation of individual system nodes. The main requirements for the design of secondary power supply sources are the use of a modular architecture with parallel channels to implement a redundancy system in case of failure, as well as the development of unified mechanisms for the interaction of all segments with a common centralized management system [64].

7. Prospects for the Digital Technologies' Introduction in the Oil and Gas Industry within Its Competition with RES

In the energy sector, despite the rapid growth of RES shares, the oil/gas sector remains the dominant supplier of electricity [65] and will remain such for at least the next 20–30 years [66]. At the same time, in 2020, the share of renewable energy in global energy consumption increased by 9.7% (which is its largest ever increase), while gas and oil consumption fell by 9.3% and 2.3%, respectively [65]. The reason for this is the high environmental friendliness of obtaining energy from renewable sources. It is a good alternative to extracting energy from hydrocarbons, producing up to three quarters of all carbon dioxide emissions into the atmosphere [67]. RES is also most appropriate technology for use in conditions where centralized management is difficult, for example, in the Arctic zone [68].

Thus, one of the main aspects of Industry 4.0 is the direction towards the rational use of natural resources, which implies an energy-efficient production process with the achievement of its maximum environmental friendliness [69].

The main technologies of Industry 4.0 are robotics, modeling (digital twins), system integration, industrial Internet of Things, cyber security, cloud computing, 3D printing, and augmented reality [70,71].

Considering the introduction of Industry 4.0 technologies in the oil and gas sector, which competes with RES, a comparative analysis of the above-mentioned use of technologies at each stage of production was carried out.

7.1. Application of Industry 4.0 Technologies in the Oil and Gas Sector

The oil/gas industry is a complex system that includes a large number of interconnected processes that must control the extraction of raw materials, its processing, productivity and equipment wear, the efficiency of personnel, and be clearly coordinated with each other. The optimization and strict control of each of the above-mentioned items helps to reduce the cost of equipment operation, increase overall productivity, as well as reducing the risks of accidents [72].

Thus, the following problems stand in the way of creating a centralized automatic control system [73]:

- A variety of protocols, as well as a variety of data types. As the number of devices and sensors grows, the number of protocols for data collection increases, which urges the need to create new interfaces for organizing device networks and integrating them with existing data ecosystems. In addition, there is a need for a centralized data management system, which should be able to integrate disparate data types to create their single representations.

- Increasing the number of devices and sensors. Hubs, aggregators, gateways and other network equipment are needed to manage the lifecycle of new devices and sensors. At the same time, the amount of data created in the course of work and their dynamic nature may exceed the capabilities of systems used for operational decision support. Sensors, and the data produced by them, must be ordered, combined, matched and transformed.

At the same time, traditional control systems such as PLC, DCS and SCADA can be inherently complex and expensive to upgrade or expand. The solution to this is to use Internet of Things technologies, Big Data, as well as cloud computing, through the use of which it is possible to create a unified system of control, management, as well as support in making operational decisions. For a more detailed analysis of digital technologies' introduction in oil and gas production automation, we shall consider the process of raw materials extraction in the oil and gas industry. The whole process can be divided into three main parts (Figure 3):

1. Search, exploration and development (upstream).
2. Storage and transportation (midstream).
3. Processing, marketing and sales (downstream).

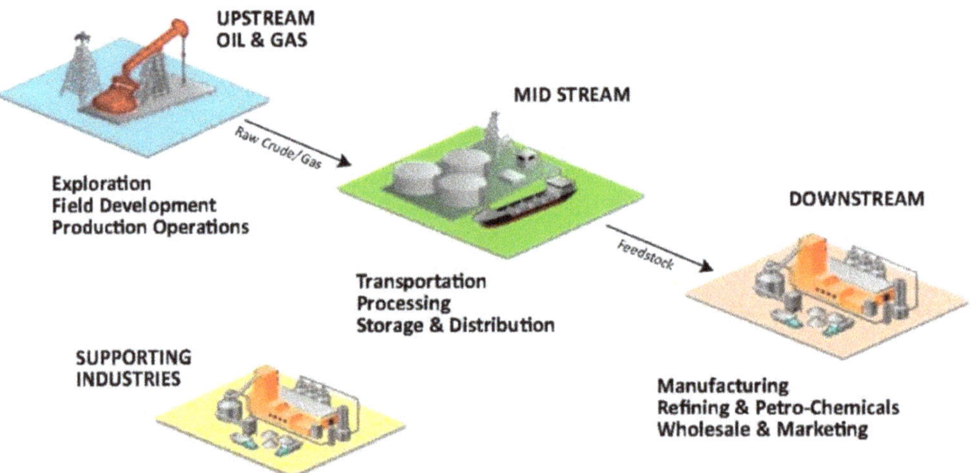

Figure 3. The production process in the oil and gas industry system [74].

Currently, the process of introducing new technologies for each of the sectors is not regulated due to the complexity of technological processes, as well as their close correlations with each other.

The conceptual model of transition to the concept of Industry 4.0 should focus on the following areas:

- The increase in the oil recovery coefficient;
- The reduction in operating and capital costs;
- The reduction in pollutant emissions;
- The increase in high-tech jobs;
- Improving the energy efficiency of production [75].

Based on the data given in Table 6, it can be concluded that the main focus of the research on Industry 4.0 introduction in the oil and gas sector is on the following technologies: industrial Internet of Things; digital twins; Big Data; and augmented reality.

Table 6. The use of Industry 4.0 technologies in the oil and gas sector [1].

Oil and Gas Production Steps	Technology	Examples
Search, exploration and development	Big Data; Industrial Internet of Things; Digital twins; Augmented reality;	1, 2 and 3D geological maps [76]; Forecasting possible technological threats [77]; Seismic data processing [78]; Improving drilling efficiency [79,80]; Analysis of oil layers [81]; Smart cleaning of petroleum products [82]; Application of digital twins for drilling [83]; The use of augmented and virtual reality technologies for drilling wells [84] and pumping equipment maintenance [85].
Storage and transportation	Big Data Industrial Internet of Things Digital twin Augmented reality	The use of Big Data to optimize the movement of ships [86]; The use of digital technologies for the pipeline system [87]; Use of robots for ship inspection [88,89]; Autonomous pipe leak detection system [90].
Processing, marketing and sales	Industrial Internet of Things; Digital twins; Big Data	Introduction of digital technologies into oil refining [82].

[1] Compiled by the authors.

7.2. Application of Industry 4.0 Technologies for Renewable Energy

According to experts [91], in the context of Industry 4.0 technologies' use in renewable energy, the main emphasis should be placed on:

- Improving the stability of the energy system state;
- Providing additional flexibility for renewable energy systems;
- Improving energy efficiency;
- Reducing energy consumption.

Similar to the oil and gas sector, the process of obtaining energy from renewable sources can be divided into:

- Generation;
- Buffering;
- Transmission;
- Consumption.

Industry 4.0 technologies, presented in Table 7, contribute to the achievement of impressive indicators for each of the designated points.

Thus, the main focus in the renewable energy sector is on the following technologies: industrial Internet of Things, Big Data, digital twins.

Table 7. Use of Industry 4.0 technologies for RES [1].

Renewable Energy Production Steps	Technology	Examples
Generation	Industrial Internet of Things; Big Data; Cloud computing; Digital twin;	Increasing the stability of renewable energy generation by improving weather forecasting [92]; Modernization of batteries [93,94]
Buffering	Industrial Internet of Things; Big Data	Energy storage by means of hydrogen [95,96] and thermal [97,98] energy accumulators.
Transmission	Industrial Internet of Things; Big Data	Implementation of inter-network connections [99]; Equalization of consumption and generation over long distances using main power lines [100–102]
Consumption	Industrial Internet of Things	Optimization of the distributed electricity system operation depending on the consumer's needs [103,104]; The use of storage devices on the consumer side in order to compensate for the additional load on the network [105,106];

[1] Compiled by the authors.

7.3. Competitiveness of Renewable Energy in Russia

Despite the fact that the bulk of Russian energy consumption is accounted for by hydrocarbon energy sources (the total share of energy consumption from renewable sources does not exceed 0.15%, according to Rosstat data for 2019 [107]), the share of renewable energy is growing rapidly every year (Figure 4). By 2030, the share of renewable energy consumption in Russia will increase by 5% [108]. This is facilitated by the problem of climate change associated with the extraction of energy from fossil energy sources, and, as a result, increased public interest in the use of "green" energy sources [109].

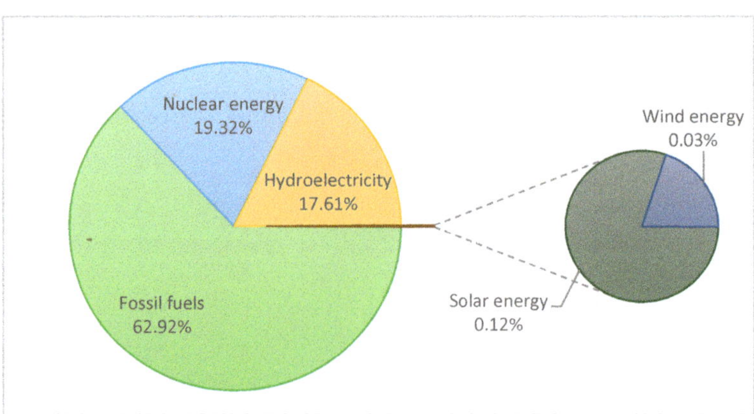

Figure 4. Electricity consumption in Russia [107].

According to the maps of wind and solar energy presented in [110], the greatest intensity of solar and wind energy falls on regions remote from the central part of Russia (Figures 5 and 6). It is also worth noting that, nowadays, the cost of energy extracted from petroleum products is significantly lower than analogues from renewable energy sources. However, the estimated cost of hydrocarbon fuel, in the case of its delivery to remote parts of the Russian Federation, may significantly exceed the cost of energy supplied by RES [111].

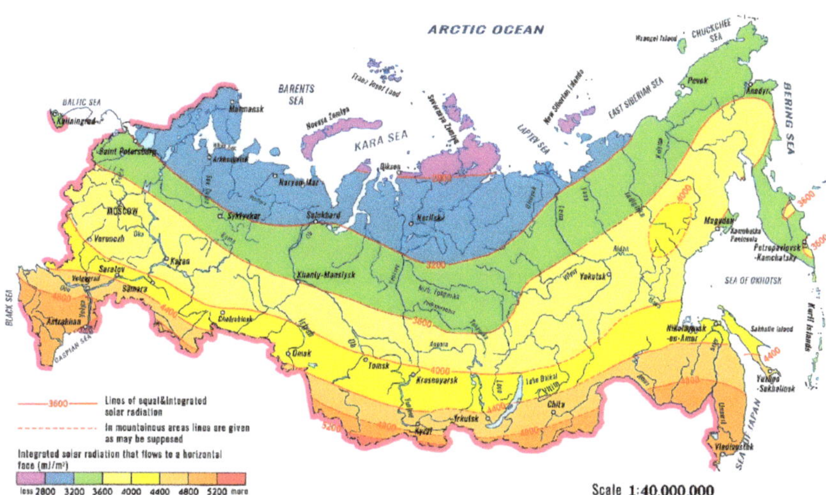

Figure 5. Average annual solar radiation intensity map in Russia [110].

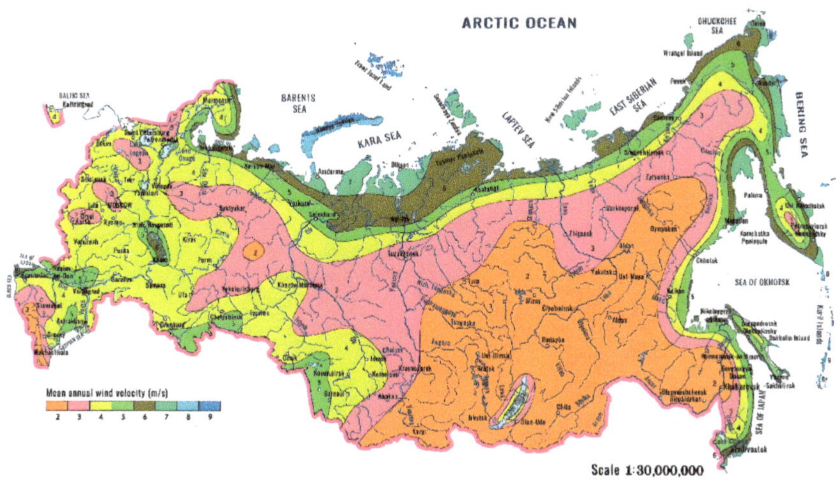

Figure 6. Average annual wind intensity map in Russia [110].

Therefore, for the regions located on the periphery of the Russian Federation, especially for the Far North region of Russia, which is the most promising region for the development of rich mineral deposits [112–114], the use of renewable energy as the main source of electricity is the most optimal and promising solution.

Based on the above, it can be concluded that the issue of introducing digital technologies into the oil and gas sector is now most difficult, since, despite the increasing demand for renewable energy, the use of which is limited to regions that are remote from the central region of Russia, hydrocarbon energy remains dominant in the market. At the same time, the digitalization of the oil and gas sector is a more time-consuming task due to the presence of more complex technological processes, as well as the large working personnel's involvement in the production processes. In accordance with this, the use of technologies such as virtual and augmented reality are most relevant for the oil and gas sector, as these technologies contribute to reducing the influence of human factors, as well as increasing the final speed and efficiency of performing the required operations. At the same time, virtual reality, the Internet of Things and artificial intelligence are currently the least-used technologies, compared with Big Data and robotics, due to the complexity of developing and implementing the final product for an introduction into the production process [25]. Therefore, research that focuses on the above-mentioned introduction of technologies into the oil and gas industry is the most relevant.

8. Conclusions

The paper is a review and analysis of the main approaches to the introduction of digital technologies in the modern Russian oil and gas industry under Arctic conditions as well as the problems that arise in this case. The study is based on information obtained from open sources, as well as opinions and assessments provided by experts working in the Arctic and those who are becoming training specialists in this region.

The paper brings together information about the digital technologies currently used in Russian oil and gas production in the Arctic, as well as a survey of experts who are directly involved in their implementation. Based on the information provided, the significance of these technologies for improving oil exploration and production is demonstrated, as well as the main problems that prevent their large-scale application in the Arctic region.

The need to reduce the costs of operating deposits in the Far North region, due to natural, climatic and other factors, stimulates the development of the remote management capabilities of production facilities and the use of unmanned technologies. The technologies that are most actively implemented in oil and gas enterprises include working with Big Data, neurotechnologies, distributed registry systems, industrial Internet of Things, robotics and sensor components, and, in the future, with the use of virtual and augmented reality technologies, which are currently used mainly for training events.

The main purpose of the digitalization of the Russian oil and gas sector is to increase the efficiency of business process management. The key constraining factors of digitalization are the lack of qualified personnel, lack of material and technical bases and increase in cyber security threats. The latter factor often requires a transformation of the entire enterprise security system. The main reason for these existing problems in the sphere of information security is, once again, the shortage of highly qualified personnel, which is especially clear in companies that operate in hard-to-reach regions, particularly in oil and gas facilities in the Arctic region.

Russian experts (especially representatives of oil and gas companies) underestimate the sanctions regimes against their country. Nevertheless, the prospects of a new package of sanctions being introduced against Russia (due to the Russia–Ukraine conflict), could, on the one hand, become an incentive for a qualitative leap in the development and implementation of Russian software, including in Arctic projects. On the other hand, what is more likely in the short and medium term is that these sanctions will become obstacles to the development of Arctic projects' due to rising costs and could even become a reason to freeze a number of them.

Another underestimated factor of the COVID-19 pandemic is one of the incentives for the widespread introduction of production automation and various business processes in the oil and gas industry, as well as the development of digital communications, which is most relevant to remote Arctic territories.

To date, the leader in the digital technology development industry among Russian oil and gas companies is "Gazprom Neft" PJSC (the company has held the leading positions for the past 4 years), with "NK "Rosneft" PJSC in second place. These companies invest in the development and implementation of domestic software solutions and platforms.

According to the results of an expert survey, the most successful Russian digital projects in the Arctic are the projects by the "Gazprom Neft" PJSC: "CAPTAIN" logistics management system, the use of block-chain technologies in "Prirazlomnaya" SISP, "Cognitive Geologist" IT platform, "Digital Core" self-learning program and the digital twin of the Vostochno-Messoyakhskoye field (together with "NK 'Rosneft'" PJSC).

"Gazprom" PJSC continues to lag behind in the sphere of digitalization; however, given the "Gazprom" group's digital transformation strategy adopted at the end of 2021, we should expect a "breakthrough" in this direction in 2022.

The "sensitivity parameters" influencing the industry digitalization process in the Arctic region are:

- A high dependence on foreign technological software solutions, characteristic of the entire Russian oil and gas industry;
- The problem of ensuring cybersecurity in Arctic oil and gas projects (ensuring security during data transmission and providing companies with information security specialists);
- Power outages.

For the Arctic regions, the use of renewable energy as the main source of electricity is the most optimal and promising solution. However, despite the increasing demand for renewable energy, hydrocarbon energy will dominate the market in the near future. At the same time, it is in the oil and gas sector that the economic effect of digitalization is expected to be higher, and some technologies, such as virtual reality, the Internet of Things and artificial intelligence, are more applicable to oil and gas enterprises.

The authors of this study plan to focus on identifying and studying in more depth the specifics of the digitalization of Russian oil and gas projects in the Arctic in the context of international sanctions in the near future.

The expert survey presented in the paper is an integral part (the first stage) of a larger study to identify the qualitative characteristics of digitalization development in the Russian extractive oil and gas industry in the Arctic and the prospects for its development. As the next stage of the study, the authors plan to conduct a large-scale survey (at least 500 respondents and an expanded list of questions) among employees of Russian oil and gas companies and representatives of scientific and educational institutions, as well as foreign specialists, which will significantly expand our understanding of the development of digitalization in the Russian oil and gas industry as a whole, particularly in Arctic oil and gas projects, as well as deepening our understanding of the ongoing digitalization processes under the current sanctions regime.

Author Contributions: Conceptualization, E.S. and D.M.; methodology, E.S.; validation, A.M.; formal analysis, E.S. and A.L.; resources, R.-E.K.; writing—original draft preparation, E.S., D.M. and A.L.; writing—review and editing, A.M.; visualization, R.-E.K.; project administration, E.S. All authors have read and agreed to the published version of the manuscript.

Funding: This research received no external funding.

Institutional Review Board Statement: The study did not require ethical approval.

Informed Consent Statement: Informed consent was obtained from all subjects involved in the study.

Data Availability Statement: The data presented in this study are available on request from the corresponding author.

Conflicts of Interest: The authors declare no conflict of interest.

Appendix A

Expert Survey

1. Specify your company/work place/institution
2. Specify your position
3. Specify the digital technologies that, in your opinion, are most effective in the extraction of oil and gas resources in the entire industry.

    a. Big Data
    b. Digital twins
    c. Artificial intelligence
    d. Robotization
    e. Automated control systems
    f. Digital communication
    g. Specify your own option

4. Specify the digital technologies that, in your opinion, are most effective when implementing oil and gas projects in the Arctics.

    a. Big Data
    b. Digital twins
    c. Artificial intelligence
    d. Robotization
    e. Automated control systems
    f. Digital communication
    g. Specify your own option

5. Specify the Russian oil and gas company that, in your opinion, is the leader in digital technologies' introduction

    a. Gazprom Neft
    b. Gazprom
    c. Rosneft
    d. NOVATEK
    e. LUKOIL
    f. TATNEFT
    g. Specify your own option

6. Specify three digital technologies/projects/programs that, in your opinion, are successfully and effectively implemented by Russian oil and gas companies.

Open question (without provided options)

7. Specify three digital technologies/projects that, in your opinion, are most effectively implemented today in the extraction of oil and gas resources in the Arctic.

Open question (without provided options)

8. Specify 3 factors that, in your opinion, influenced the intensification of the introduction of digital technologies in 2019–2021.

    a. The need to reduce the costs for companies
    b. Improving the efficiency of business process management
    c. Adoption of the national program "Digital Economy of the Russian Federation" 2017
    d. Import substitution and sanctions
    e. COVID-19 pandemic
    f. Specify your own option

9. Specify the main factor for the digital technologies' introduction intensification in 2019–2021.

    a. The need to reduce the costs of companies
    b. Improving the efficiency of business process management
    c. The adoption of "Digital Economy of the Russian Federation" 2017 national program
    d. Import substitution and sanctions
    e. The COVID-19 pandemic

10. In what cases, in your opinion, is the use of renewable electricity sources more appropriate in comparison with the use of hydrocarbons?

    a. If it is economically unprofitable to extract hydrocarbons
    b. In the absence of hydrocarbon resources (remote stations)
    c. If it is necessary to ensure a favorable environmental situation (Ecological issues)

d.   No way

11.   Is renewable energy, in your opinion, the most optimal solution for providing energy supply to hard-to-reach Arctic regions?

a.   Yes
b.   No
c.   I find it difficult to answer

12.   Will RES, in your opinion, be able to completely replace hydrocarbon energy in the future?

a.   Yes
b.   No
c.   I find it difficult to answer

References

1. Eryomin, N.A.; Stolyarov, V.E.; Sardanashvili, O.N.; Chernikov, A.D. Intelligent drilling in digital field development. *Autom. Telemech. Commun. Oil Ind.* **2020**, *5*, 26–36. [CrossRef]
2. Vlasov, A.I.; Mozhchil, A.F. Technology overview: From digital to intelligent oilfield PRONEFT. *Prof. Oil* **2018**, *3*, 68–74.
3. Tcharo, H.; Vorobyev, A.E.; Vorobyev, K.A. Digitalization of the oil industry: Basic approaches and rationale for "intelligent" technologies. *Eurasian Sci. J.* **2018**, *2*, 1–17. Available online: https://esj.today/PDF/88NZVN218.pdf (accessed on 30 January 2022).
4. Mazakov, E.B. Knowledge representation and processing in information automated systems of intellectual deposits. *J. Min. Inst.* **2014**, *208*, 256–262.
5. Makhovikov, A.B.; Katuntsov, E.V.; Kosarev, O.V.; Tsvetkov, P.S. Digital Transformation in Oil and Gas Extraction. In *Innovation-Based Development of the Mineral Resources Sector: Challenges and Prospects—11th Conference of the Russian-German Raw Materials, Potsdam, Germany, 7–8 November 2018*; CRC Press: Boca Raton, FL, USA, 2018; pp. 531–538.
6. Eryomin, N.A.; Dmitrievsky, A.N. Digital development of Russian Arctic zone: Status and best practices. *Reg. Energy Energy Conserv.* **2018**, *3*, 60–61.
7. Pashali, A.A.; Kolonskikh, A.V.; Khalfin, R.S.; Silnov, D.V.; Topolnikov, A.S.; Latypov, B.M.; Urazakov, K.R.; Katermin, A.V.; Palaguta, A.A.; Enikeev, R.M. A digital twin of well as a tool of digitalization of bringing the well on to stable production in bashneft pjsoc. *Neftyanoe Khozyaystvo Oil Ind.* **2021**, *3*, 80–84. [CrossRef]
8. Clemens, T.; Viechtbauer-Gruber, M. Impact of digitalization on the way of working and skills development in hydrocarbon production forecasting and project decision analysis. *SPE Reserv. Eval. Eng.* **2020**, *23*, 1358–1372. [CrossRef]
9. Shishkin, A.N.; Timashev, E.O.; Solovykh, V.I.; Volkov, M.G.; Kolonskikh, A.V. Bashneft digital transformation: From concept design to implementation. *Oil Ind.* **2019**, *3*, 7–12. [CrossRef]
10. Litvinenko, V.S.; Tsvetkov, P.S.; Dvoynikov, M.V.; Buslaev, G.V. Barriers to implementation of hydrogen initiatives in the context of global energy sustainable development. *J. Min. Inst.* **2020**, *244*, 428–438. [CrossRef]
11. Linnik, Y.N.; Kiryukhin, M.A. Digital technologies in the oil and gas complex. *State Manag. Univ. Bull.* **2019**, *7*, 37–40.
12. Eryomin, N.A.; Dmitrievsky, A.N.; Tikhomirov, L.I. Present and future of intellectual deposits. *Oil Gas Innov.* **2015**, *12*, 44–49.
13. Dmitrievskiy, A.N.; Eryomin, N.A.; Stolyarov, V.E. Digital transformation of gas production. *IOP Conf. Ser. Mater. Sci. Eng.* **2019**, *700*, 012052. [CrossRef]
14. Timchuk, D.D. "Smart wells" technology application at Salym group of deposits illustrated with the example of Western Salym. *Sci. Forum. Sib.* **2017**, *3*, 11.
15. Kondeikina, K.V.; Tsoi, I.V. "Smart wells" technology application illustrated with the example of Western Salym. *Acad. J. Sib.* **2017**, *13*, 9.
16. The Results of Russian Ministry of Energy Work and the Main Results of the Fuel and Energy Complex Functioning in 2020. Tasks for 2021 and the Medium Term. Available online: https://minenergo.gov.ru/node/20515 (accessed on 30 January 2022).
17. Passport of the Innovative Development Program by "Gazprom Neft" PJSC until 2025. Saint Petersburg: "Gazprom Neft" PJSC. 2020. Available online: https://www.gazprom.ru/f/posts/97/653302/prir-passport-2018-2025.pdf (accessed on 30 January 2022).
18. Passport of the Innovative Development Program by "NK 'Rosneft'" PJSC. Moscow: "NK 'Rosneft'" PJSC. 2016. Available online: https://www.rosneft.ru/upload/site1/document_file/FU6HdSZ3da.pdf (accessed on 30 January 2022).
19. Zaichenko, I.M.; Fadeev, A.M.; Kostyuchenko, A.I. Building trends in the development of the fuel and energy complex enterprises in the Russian Federation in the context of digital business transformation. *South Russ. State Polytech. Univ. (NPI) Bull.* **2021**, *3*, 162–181.
20. Dvoynikov, M.V.; Kunshin, A.A.; Blinov, P.A.; Morozov, V.A. Development of Mathematical Model for Controlling Drilling Parameters with Screw Downhole Motor. *Int. J. Eng. IJE Trans. A Basics* **2020**, *7*, 1423–1430.
21. Litvinenko, V.S.; Dvoynikov, M.V. Methodology for determining the parameters of drilling mode for directional straight sections of well using screw downhole motors. *J. Min. Inst.* **2020**, *241*, 105–112. [CrossRef]
22. Belozerov, I.P.; Gubaidullin, V. On the concept of technology for determining filtration-capacitance properties of terrigenous reservoirs on a digital core model. *J. Min. Inst.* **2020**, *244*, 402–407. [CrossRef]

23. Prishchepa, O.M.; Borovikov, I.S.; Grokhotov, E.I. Oil and gas potential of the little-studied part of the north-west of the Timan-Pechora oil and gas province according to the results of basin modeling. *J. Min. Inst.* **2021**, *247*, 66–81. [CrossRef]

24. Official Website of "Gazprom Neft" Scientific and Technical Center. Available online: https://ntc.gazprom-neft.ru/ (accessed on 30 January 2022).

25. Razmanova, S.V.; Andrukhova, O.V. Oilfield service companies in the framework of economy digitalization: Assessment of innovative development prospects. *J. Min. Inst.* **2020**, *244*, 482–492. [CrossRef]

26. Idrisova, S.A.; Tugarova, M.A.; Stremichev, E.V.; Belozerov, B.V. Digital core. integration of carbonate rocks thin section studies with results of routine core tests. *PRONEFT'. Professional'no o Nefti* **2018**, *2*, 36–41. [CrossRef]

27. Erofeev, A.A.; Nikitin, R.N.; Mitrushkin, D.A.; Golovin, S.V.; Baykin, A.N.; Osiptsov, A.A.; Paderin, G.V.; Shel, E.V. BCYBER FRAC—Software platform for modeling, optimization and monitoring of hydraulic fracturing operations. *Neftyanoye Khozyastvo* **2019**, *12*, 64–68. Available online: https://www.oil-industry.net/Journal/archive_detail.php?art=235137 (accessed on 30 January 2022). [CrossRef]

28. "NK 'Rosneft's'" Annual Report for Year 2019. Available online: https://www.rosneft.ru/upload/site1/document_file/a_report_2019.pdf (accessed on 30 January 2022).

29. Rosneft Expands the Geography of the Use of Unmanned Aerial Vehicles to Control the Level of Greenhouse Gases. Available online: https://www.rosneft.ru/press/news/item/207499 (accessed on 30 January 2022).

30. RN Digital. Software by "NK 'ROSNEFT'" PJSC in the Sphere of Field Development. Available online: https://rn.digital/ (accessed on 30 January 2022).

31. Lukoil-Technologies Expand the Possibilities of a Unified Information Space. LUKOIL-Technologies. Available online: https://technologies.lukoil.ru/ru/News/News?rid=539160 (accessed on 30 January 2022).

32. "LUKOIL" PJSC's Annual Report. Available online: https://lukoil.ru/FileSystem/9/551394.pdf (accessed on 30 January 2022).

33. "Tatneft's" Official Website. Available online: https://tatneft.ru/press-tsentr/press-relizi/more/8020/?lang=ru (accessed on 30 January 2022).

34. "NOVATEK's" Annual Report 2017. Transformation into a Global Gas Company. Available online: https://docplayer.com/80367186-Transformaciya-v-globalnuyu-gazovuyu-kompaniyu-godovoy-otchet.html (accessed on 30 January 2022).

35. "NOVATEK's" Sustainability Report for Year 2020. Available online: https://www.novatek.ru/common/upload/doc/NOVATEK_SR_2020_RUS.pdf (accessed on 30 January 2022).

36. Digitalization of the Industry: Why Oil and Gas Companies Are Introducing Voice Assistants? Available online: https://neftegaz.ru/science/tsifrovizatsiya/711911-tsifrovizatsiya-otrasli-zachem-neftegazovye-kompanii-vnedryayut-golosovykh-pomoshchnikov/ (accessed on 30 January 2022).

37. Titkov, I.A. Digital gap in the sphere of technologies on extraction and production of liquefied natural gas: A strategic factor in weakening the economic security of the country. *Econ. Soc. Mod. Models Dev.* **2020**, *10*, 309–329.

38. Gazprom Digital Project Services. A Single Digital Platform. Available online: https://gazpromcps.ru/?page_id=29#section-briefcase (accessed on 30 January 2022).

39. The Management Board Approved Gazprom Group's Digital Transformation Strategy for 2022–2026. Available online: https://www.gazprom.ru/press/news/2021/december/article545124/ (accessed on 30 January 2022).

40. "NK 'Rosneft's'" Annual Report for Year 2020. Available online: https://www.rosneft.ru/upload/site1/document_file/a_report_2020.pdf (accessed on 30 January 2022).

41. Pravikov, D.I. Actual approaches to ensuring the safety of industrial automation systems. *Methods Tech. Means Ensuring Inf. Secur.* **2020**, *29*, 15–17.

42. Steve Krouskos Our Capital Confidence Barometer Survey Forms Part of a Wider Range of Insights on the COVID-19 Crisis. Available online: https://www.ey.com/en_ru/ccb/how-do-you-find-clarity-in-the-midst-of-covid-19-crisis (accessed on 30 January 2022).

43. Made in Russia: Overview of 20 Russian Operating Systems. Available online: https://3dnews.ru/958857/made-in-russia-obzor20-rossiyskih-operatsionnih-sistem (accessed on 26 February 2022).

44. Is There Life on the Russian OS Market? Overview of Popular Russian OS. Available online: https://habr.com/ru/company/digdes/blog/442906/ (accessed on 26 February 2022).

45. Comparison of Baikal-M and Elbrus-8SV Processors. Available online: https://habr.com/ru/company/icl_services/blog/558564/ (accessed on 26 February 2022).

46. CNA: Semiconductor Manufacturer TSMC Has Stopped Deliveries to Russia. Available online: https://ria.ru/20220227/tayvan-1775376385.html (accessed on 30 January 2022).

47. Fedorov, V.P.; Zhuravel, V.P.; Grinyaev, S.N.; Medvedev, D.A. The Northern Sea Route: Problems and prospects of development of transport route in the Arctic. *IOP Conf. Ser. Earth Environ. Sci.* **2020**, *434*, 012007. [CrossRef]

48. Savard, C.; Nikulina, A.Y.; Mécemmène, C.; Mokhova, E. The Electrification of Ships Using the Northern Sea Route: An Approach. *J. Open Innov. Technol. Mark. Complex* **2020**, *6*, 13. [CrossRef]

49. Lindholt, L.; Glomsrød, S. The Role of the Arctic in Future Global Petroleum Supply. Discussion Papers. 2011, p. 645. Available online: https://www.ssb.no/a/publikasjoner/pdf/DP/dp645.pdf (accessed on 30 January 2022).

50. Gafurov, A.; Skotarenko, O.; Nikitin, Y.; Plotnikov, V. Digital transformation prospects for the offshore project supply chain in the Russian Arctic. *IOP Conf. Ser. Earth Environ. Sci.* **2020**, *539*, 012163. [CrossRef]

51. Cherepovitsyn, A.E.; Lipina, S.A.; Evseeva, O.O. Innovative Approach to the Development of Mineral Raw Materials of the Arctic Zone of the Russian Federation. *J. Min. Inst.* **2018**, *232*, 438–444. [CrossRef]

52. Sochneva, I.O. Evolution of technical solutions for the Shtokman Gas Condensate Field and a modern view of its arrangement. *Neftegaz Bus. Mag.* **2021**, *4*, 74–83.

53. "CAPTAIN", a Digital Logistics Management System. "Gazpromneft"'s Official Website. Available online: https://www.gazprom-neft.ru/press-center/news/gazprom-neft-vnedrila-pervuyu-v-mire-tsifrovuyu-sistemu-upravleniya-logistikoy-v-arktike/ (accessed on 30 January 2022).

54. Shurupov, N. Block-chain in logistics: "Gazprom Neft's" experience. *Sib. Oil* **2018**, *150*. Available online: https://www.gazprom-neft.ru/press-center/sibneft-online/archive/2018-april/1533012/ (accessed on 30 January 2022).

55. "Gazpromneft" Official Website. Available online: https://digital.gazprom-neft.ru/about-project?id=capitan (accessed on 30 January 2022).

56. Cherepovitsyn, A.E.; Tsvetkov, P.S.; Evseeva, O.O. Critical analysis of methodological approaches to assessing the sustainability of Arctic oil and gas projects. *J. Min. Inst.* **2021**, *249*, 463–479. [CrossRef]

57. "Gazprom Neft" and PwC in Russia Will Jointly Develop Technologies for Oil Production. "Gazprom Neft"'s Official Website. Available online: https://www.gazprom-neft.ru/press-center/news/gazprom_neft_i_pwc_v_rossii_budut_vmeste_razvivat_tekhnologii_dlya_neftedobychi/ (accessed on 30 January 2022).

58. Core and Reservoir Fluids Research. "Gazprom Neft" Scientific and Technical Center. Available online: https://ntc.gazprom-neft.ru/business/exploration/core-analysis/ (accessed on 30 January 2022).

59. "Messoyakhaneftegaz" Specialists Have Created Russia's First Digital Well Construction Project in Partnership with "Schlumberger". Schlumberger. Available online: https://www.slb.ru/press-center/press-releases/spetsialisty-messoyakhaneftegaza-sozdali-pervyy-v-rossii-tsifrovoy-proekt-stroitelstva-skvazhiny-v-p/ (accessed on 30 January 2022).

60. Stolyarova, M.Y.; Valshin, O.K. Digital transformation of drilling planning: A new level of team interaction and the possibility of optimizing wells in a cloud solution DrillPlan. *Oil Gas Innov.* **2021**, *8*, 44–48.

61. GOST 28147-89 Information Processing Systems. Cryptographic Protection. Cryptographic Conversion Algorithm. Available online: https://docs.cntd.ru/document/1200007350 (accessed on 30 January 2022).

62. Zmieva, K.A. Problems of energy supply in the Arctic regions. *Russ. Arct.* **2020**, *1*, 5–14. [CrossRef]

63. Eikeland, O.F.; Maria Bianchi, F.; Holmstrand, I.S.; Bakkejord, S.; Santos, S.; Chiesa, M. Uncovering Contributing Factors to Interruptions in the Power Grid: An Arctic Case. *Energies* **2022**, *15*, 305. [CrossRef]

64. Shpenst, V.; Orel, E. Methods of ensuring the operational stability of dc-dc power supply in arctic conditions. *Power Eng. Res. Equip. Technol.* **2021**, *23*, 166–179. [CrossRef]

65. Looney, B. BP statistical review of world energy. In *Statistical Review of World Energy*, 70th ed.; Whitehouse Associates: London, UK, 2021; p. 72.

66. Blockchain Technology in the Oil and Gas Industry: A Review of Applications, Opportunities, Challenges, and Risks. *IEEE J.* **2019**, *7*, 41426–41444. Available online: https://ieeexplore.ieee.org/document/8675726 (accessed on 30 January 2022).

67. Stocker, T. *Climate Change 2013: The Physical Science Basis: Working Group I Contribution to the Fifth Assessment Report of the Intergovernmental Panel on Climate Change*; Cambridge University Press: Cambridge, UK, 2014; pp. 465–570.

68. Kirsanova, N.; Lenkovets, O.; Nikulina, A. The Role and Future Outlook for Renewable Energy in the Arctic Zone of Russian Federation. *Eur. Res. Stud. J.* **2018**, *XXI*, 356–368.

69. Anna, B. Classification of the European Union member states according to the relative level of sustainable development. *Qual. Quant.* **2016**, *50*, 2591–2605. [CrossRef]

70. Lu, H.; Guo, L.; Azimi, M.; Huang, K. Oil and Gas 4.0 era: A systematic review and outlook. *Comput. Ind.* **2019**, *111*, 68–90. [CrossRef]

71. Erboz, G. How to define industry 4.0: Main pillars of industry 4.0. In Proceedings of the Managerial trends in the development of enterprises in globalization era: 7th International Conference on Management (ICoM 2017), Nitra, Slovakia, 1–2 June 2017; 2017; pp. 761–767. Available online: https://spu.fem.uniag.sk/fem/ICoM_2017/files/international_scientific_conference_icom_2017.pdf (accessed on 30 January 2022).

72. The Internet of Things: Mapping the Value Beyond the Hype. Available online: https://www.mckinsey.com (accessed on 30 January 2022).

73. Challenges, Opportunities and Strategies for Integrating IIoT Sensors with Industrial Data Ecosystems. Available online: https://resources.osisoft.com/white-papers/challenges,-opportunities-and-strategies-for-integrating-iiot-sensors-with-industrial-data-ecosystems/ (accessed on 30 January 2022).

74. Udie, J.; Bhattacharyya, S.; Ozawa-Meida, L. A Conceptual Framework for Vulnerability Assessment of Climate Change Impact on Critical Oil and Gas Infrastructure in the Niger Delta. *Climate* **2018**, *6*, 11. [CrossRef]

75. Alexandrova, T.V.; Prudsky, V.G. On the conceptual model of oil and gas business transformation in the transitional conditions to the Industry 4.0. *Sci. Pap. Univ. Pardubice. Ser. D Fac. Econ. Adm.* **2019**, *46*. Available online: https://www.semanticscholar.org/paper/On-the-conseptual-model-of-oil-and-gas-business-in-Alexandrova-Prudsky/6ad138bb8af0bc6260bcff7d831bbad8fdd15d98 (accessed on 30 January 2022).

76. Olneva, T.; Kuzmin, D.; Rasskazova, S.; Timirgalin, A. Big data approach for geological study of the big region West Siberia. In Proceedings of the SPE Annual Technical Conference and Exhibition 2018, Dallas, TX, USA, 24–26 September 2018. [CrossRef]

77. Cadei, L.; Montini, M.; Landi, F.; Porcelli, F.; Michetti, V.; Origgi, M.; Tonegutti, M.; Duranton, S. Big data advanced anlytics to forecast operational upsets in upstream production system. In Proceedings of the Abu Dhabi International Petroleum Exhibition & Conference 2018, Abu Dhabi, United Arab Emirates, 12–15 November 2018. [CrossRef]

78. Alfaleh, A.; Wang, Y.; Yan, B.; Killough, J.; Song, H.; Wei, C. Topological data analysis to solve big data problem in reservoir engineering: Application to inverted 4D seismic data. In Proceedings of the SPE Annual Technical Conference and Exhibition 2015, Houston, TX, USA, 28–30 September 2015. [CrossRef]

79. Duffy, W.; Rigg, J.; Maidla, E. Efficiency improvement in the bakken realized through drilling data processing automation and the recognition and standardization of best safe practices. In Proceedings of the SPE/IADC Drilling Conference and Exhibition 2017, Hague, The Netherlands, 14–16 March 2017. [CrossRef]

80. Hutchinson, M.; Thornton, B.; Theys, P.; Bolt, H. Optimizing drilling by simulation and automation with big data. In Proceedings of the SPE Annual Technical Conference and Exhibition 2018, Dallas, TX, USA, 24–26 September 2018. [CrossRef]

81. Lin, A. Principles of big data algorithms and application for unconventional oil and gas resources. In Proceedings of the SPE Large Scale Computing and Big Data Challenges in Reservoir Simulation Conference and Exhibition 2014, Istanbul, Turkey, 15–17 September 2014. [CrossRef]

82. Yuan, Z.; Qin, W.; Zhao, J. Smart manufacturing for the oil refining and petrochemical industry. *Engineering* **2017**, *3*, 179–182. [CrossRef]

83. Mayani, M.G.; Svendsen, M.; Oedegaard, S. Drilling digital twin success stories the last 10 years. In Proceedings of the SPE Norway One Day Seminar 2018, Bergen, Norway, 18 April 2018. [CrossRef]

84. Clarke, S.; Kapila, K.; Stephen, M. AR and VR Applications Improve Engineering Collaboration, Personnel Optimization, and Equipment Accuracy for Separation Solutions. In Proceedings of the SPE Offshore Europe Conference and Exhibition 2019, Aberdeen, UK, 3–6 September 2019. [CrossRef]

85. Koteleva, N.; Buslaev, G.; Valnev, V.; Kunshin, A. Augmented reality system and maintenance of oil pumps. *Int. J. Eng.* **2020**, *33*, 1620–1628.

86. Anagnostopoulos, A. Big data techniques for ship performance study. In Proceedings of the 28th International Ocean and Polar Engineering Conference 2018, Sapporo, Japan, 10–15 June 2018.

87. MohamadiBaghmolaei, M.; Mahmoudy, M.; Jafari, D.; MohamadiBaghmolaei, R.; Tabkhi, F. Assessing and optimization of pipeline system performance using intelligent systems. *J. Nat. Gas Sci. Eng.* **2014**, *18*, 64–76. [CrossRef]

88. Menegaldo, L.L.; Santos, M.; Ferreira, G.A.N.; Siqueira, R.G.; Moscato, L. SIRUS: A mobile robot for Floating Production Storage and Offloading (FPSO) ship hull inspection. In Proceedings of the 2008 10th IEEE International Workshop on Advanced Motion Control, Trento, Italy, 26–28 March 2008; pp. 27–32. [CrossRef]

89. Menegaldo, L.L.; Ferreira, G.A.N.; Santos, M.F.; Guerato, R.S. Development and Navigation of a Mobile Robot for Floating Production Storage and Offloading Ship Hull Inspection. *IEEE Trans. Ind. Electron.* **2009**, *56*, 3717–3722. [CrossRef]

90. Belsky, A.A.; Dobush, V.S.; Ivanchenko, D.I.; Gluhanich, D.Y. Electrotechnical complex for autonomous power supply of oil leakage detection systems and stop valves drive control systems for pipelines in arctic region. *J. Phys. Conf. Ser.* **2021**, *1753*, 012062. [CrossRef]

91. Scharl, S.; Praktiknjo, A. The Role of a Digital Industry 4.0 in a Renewable Energy System. *Int. J. Energy Res.* **2019**, *43*, 3891–3904. [CrossRef]

92. Polhamus, M. Joint Venture Has IBM Tackling Renewable Energy's Grid Effects. VTDigger 2017. Available online: https://vtdigger.org/2017/02/28/joint-venture-ibm-tackling-renewable-energys-grid-effects/ (accessed on 30 January 2022).

93. Wang, G.; Konstantinou, G.; Townsend, C.D.; Pou, J.; Vazquez, S.; Demetriades, G.D.; Agelidis, V.G. A Review of Power Electronics for Grid Connection of Utility-Scale Battery Energy Storage Systems. *IEEE Trans. Sustain. Energy* **2016**, *7*, 1778–1790. [CrossRef]

94. Miller, N.; Manz, D.; Roedel, J.; Marken, P.; Kronbeck, E. Utility scale Battery Energy Storage Systems. In Proceedings of the IEEE PES General Meeting, Minneapolis, MN, USA, 25–29 July 2010; pp. 1–7. [CrossRef]

95. Glenk, G.; Reichelstein, S. Economics of converting renewable power to hydrogen. *Nat. Energy* **2019**, *4*, 216–222. [CrossRef]

96. Carmo, M.; Stolten, D. Energy storage using hydrogen produced from excess renewable electricity: Power to hydrogen. In *Science and Engineering of Hydrogen-Based Energy Technologies*; Elsevier: Amsterdam, The Netherlands, 2019; pp. 165–199.

97. Bloess, A.; Schill, W.-P.; Zerrahn, A. Power-to-heat for renewable energy integration: A review of technologies, modeling approaches, and flexibility potentials. *Appl. Energy* **2018**, *212*, 1611–1626. [CrossRef]

98. Kirkerud, J.G.; Bolkesjø, T.F.; Trømborg, E. Power-to-heat as a flexibility measure for integration of renewable energy. *Energy* **2017**, *128*, 776–784. [CrossRef]

99. O'Leary, D.T.; Charpentier, J.-P.; Minogue, D. *Promoting Regional Power Trade: The Southern African Power Pool*; World Bank: Washington, DC, USA, 1998.

100. Saborío-Romano, O.; Bidadfar, A.; Sakamuri, J.N.; Göksu, Ö.; Cutululis, N.A. Novel Energisation Method for Offshore Wind Farms Connected to HVdc via Diode Rectifiers. In Proceedings of the IECON 2019-45th Annual Conference of the IEEE Industrial Electronics Society 2019, Lisbon, Portugal, 14–17 October 2019; pp. 4837–4841. [CrossRef]

101. Hedayati, M.; Jovcic, D. Scaled 500A, 900V, Hardware Model Demonstrator of Mechanical DC CB with Current Injection. In Proceedings of the 2018 IEEE PES Innovative Smart Grid Technologies Conference Europe (ISGT-Europe) 2018, Sarajevo, Bosnia and Herzegovina, 21–25 October 2018; pp. 1–5. [CrossRef]

102. Shukla, R.; Chakrabarti, R.; Narasimhan, S.R.; Soonee, S.K. Indian Power System Operation Utilizing Multiple HVDCs and WAMS. In *Power System Grid Operation Using Synchrophasor Technology, Springer International Publishing*; Springer: Cham, Switzerland, 2019; pp. 403–432. [CrossRef]
103. Wang, J.; Chen, C.; Lu, X. *Guidelines for Implementing Advanced Distribution Management Systems-Requirements for DMS Integration with DERMS and Microgrids*; Argonne National Lab.(ANL): Argonne, IL, USA, 2015; pp. 1–76.
104. Ilic, M.D.; Jaddivada, R.; Korpas, M. Interactive protocols for distributed energy resource management systems (DERMS). *IET Gener. Transm. Distrib.* **2020**, *14*, 2065–2081. [CrossRef]
105. Kim, Y.-J.; Del-Rosario-Calaf, G.; Norford, L.K. Analysis and experimental implementation of grid frequency regulation using behind-the-meter batteries compensating for fast load demand variations. *IEEE Trans. Power Syst.* **2016**, *32*, 484–498. [CrossRef]
106. Wu, D.; Kintner-Meyer, M.; Yang, T.; Balducci, P. Economic analysis and optimal sizing for behind-the-meter battery storage. In Proceedings of the 2016 IEEE Power and Energy Society General Meeting, Boston, MA, USA, 17–21 July 2016; pp. 1–5. [CrossRef]
107. Electricity Consumption in Russia 2019. Rosinfostat. Available online: https://rosinfostat.ru/potreblenie-elektroenergii/ (accessed on 30 January 2022).
108. Dolf, G.; Deger, S. REMAP 2030 Renewable Energy Prospects for the Russian Federation. IRENA. Available online: https://www.irena.org/-/media/Files/IRENA/Agency/Publication/2017/Apr/IRENA_REmap_Russia_paper_2017.pdf (accessed on 30 January 2022).
109. Public Perceptions on Climate Change and Energy in Europe and Russia: Evidence from Round 8 of the European Social Survey | Resul Umit. Available online: https://resulumit.com/publications/perceptions-climate-energy/ (accessed on 30 January 2022).
110. Avenarius, I.G. National Atlas of Russia. Available online: https://elibrary.ru/item.asp?id=23554805 (accessed on 30 January 2022).
111. Lanshina, T.A.; John, A.; Potashnikov, V.Y.; Barinova, V.A. The slow expansion of renewable energy in Russia: Competitiveness and regulation issues. *Energy Policy* **2018**, *120*, 600–609. [CrossRef]
112. Ilinova, A.; Dmitrieva, D. Sustainable Development of the Arctic zone of the Russian Federation: Ecological Aspect. *Biosci. Biotechnol. Res. Asia* **2016**, *13*, 2101–2106. [CrossRef]
113. Ilinova, A.; Chanysheva, A. Algorithm for assessing the prospects of offshore oil and gas projects in the Arctic. *Energy Rep.* **2020**, *6*, 504–509. [CrossRef]
114. Gold–Sulphide Deposits of the Russian Arctic Zone: Mineralogical Features and Prospects of Ore Benefication—ScienceDirect. Available online: https://www.sciencedirect.com/science/article/pii/S0009281918301375 (accessed on 30 January 2022).

MDPI
St. Alban-Anlage 66
4052 Basel
Switzerland
Tel. +41 61 683 77 34
Fax +41 61 302 89 18
www.mdpi.com

Resources Editorial Office
E-mail: resources@mdpi.com
www.mdpi.com/journal/resources

www.ingramcontent.com/pod-product-compliance
Lightning Source LLC
LaVergne TN
LVHW070117170726
843515LV00007B/1707